# International Conference on Energy, Chemical and Materials Science (ECMS 2019)



IOP Conference Series: Materials Science and Engineering
Volume 738

Malacca, Malaysia
6 - 8 December 2019

ISBN: 978-1-7138-2654-5
ISSN: 1757-8981

**Printed from e-media with permission by:**

Curran Associates, Inc.
57 Morehouse Lane
Red Hook, NY 12571

**Some format issues inherent in the e-media version may also appear in this print version.**



Institute of Physics
Dirac House, Temple Back
Bristol BS1 6BE UK

Phone: 44 1 17 929 7481
Fax: 44 1 17 920 0979

techtracking@iop.org

**Additional copies of this publication are available from:**

Curran Associates, Inc.
57 Morehouse Lane
Red Hook, NY 12571 USA
Phone: 845-758-0400
Fax: 845-758-2633
Email: curran@proceedings.com
Web: www.proceedings.com

# TABLE OF CONTENTS

IOP Conf. Series: Materials Science and Engineering **738** (2020) 011001 doi:10.1088/1757-899X/738/1/011001

# Preface

2019 International Conference on Energy, Chemical and Materials Science (ECMS 2019) was successfully held on Dec 6-8, 2019 in Malacca, Malaysia. This conference gained great supports from Universiti Malaysia Pahang, Universiti Tun Hussein Onn Malaysia, Universiti Kuala Lumpur and National Defence University of Malaysia. We were honored to have A. Prof. Mohd Arif Agam from University Tun Hussein Onn Malaysia as our Conference Chair and over forty experts at home and abroad as our committee members. We were pleased to invite Prof. Rajan Jose form Universiti Malaysia Pahang, who presented us a marvelous academic report with topic of Future of Materials Research: Renewable Materials & Smart Discovery Platforms, and Prof. Mailadil Sebastian from National Institute for Interdisciplinary Science & Technology, India, who also gave a report on Materials for 5 th generation (5G) wireless communication and internet of things (IOT).

The aim of this conference is to provide a platform for researchers, engineers, lecturers and practitioners in this filed all over the world to present their research achievements and development trend in the fields of energy, chemical and materials. It also provides chances for lecturers to exchange new thoughts, face-to-face communicate, and enhance business and research correlation for future collaboration with global partners. All papers submitted in the conference have been gone through rigorous review, and a volume of them with high quality has been selected to publish. We sincerely hope the achievements of this conference can make a contribution to the advance research in this scientific field.

At last, we would like to express our gratitude to all staff of the conference for their diligent work, committee members for reviewing and selecting the submitted papers, all guests from all over the world for attending the conference, and all staff from IOP. We sincerely hope 2019 International Conference on Energy, Chemical and Materials Science (ECMS 2019) will become an excellent forum and give a reference for all lecturers in this filed.

The Organizing Committee of ECMS 2019

IOP Conf. Series: Materials Science and Engineering **738** (2020) 011001 doi:10.1088/1757-899X/738/1/011001

# Committee member

## Honorary Chairman

A.Prof.Mohd Arif Agam, University Tun Hussein Onn Malaysia, Malaysia

## Committee

Pro.Shahid Hussain, Jiangsu University, China
Prof. Mohammad Mansoob Khan, Universiti Brunei Darussalam, Brunei
Prof. Rajan Jose, Universiti Malaysia Pahang, Malaysia
Prof. Kim Choon NG, King Abdullah University of Science & technology, Saudi Arabia
Prof. Mamoun M. Bader, Alfaisal University, Kingdom of Saudi Arabia
A.Prof. Muhammad Kashif, Tianjin University, China
Prof. Sandip Mahajan, Dr. B.A. M. University, India
Prof. Kasim Uthman, The Federal University of Technology, Akure
Prof. Uda Bin Hashim, University Malaysia Perlis, Malaysia
Prof. Md Saidin Wahab, Universiti Tun Hussein Onn Malaysia, Malaysia
A. Prof. Pratama Jujur Wibawa, Diponegoro University, India
A. Prof. Lokman Hakim Ismail, Universiti Tun Hussein Onn Malaysia, Malaysia
A. Prof. Hairul Anuar Tajuddin, University of Malaya, Malaysia
A. Prof. Nafarizal Bin Nayan, Universiti Tun Hussein Onn Malaysia, Malaysia
Prof. RAMLI BIN NAZIR, Universiti Teknologi Malaysia, Malaysia
A. Prof. Dr Razali Hassan, Universiti Tun Hussein Onn Malaysia, Malaysia
A. Prof. Mohammed Alias Yusof, Universiti Pertahanan Nasional, Malaysia
A. Prof. Yongkang Xing, Chongqing University, China
Assis.Prof. Uzair Majeed, NED University of Engineering and Technology, Pakistan
Assis.Prof. Mohammed Abdo Hashem, Universiti Malaysia Pahang, Malaysia
Assis.Prof. Shamzani Affendy Mohd.Din, International Islamic University, Malaysia
Dr. Tammar H. Ali, Al-Muthanna University, Iraq
Dr. Syed Sajjad Haider Rizvi, Universiti Teknologi Malaysia, Malaysia
Dr. Kelihol Tase, Nagaland University, India
Assis.Prof. Qadir Bux alias Imran Latif, University of Nizwa (UON), OMAN
Trustee Desmond Asamoah, The Maame Efua Kids Foundation, Ghana
Dr. MUHAMMAD WAKIL SHAHZAD, King Abdullah University of Science & technology, Saudi Arabia
Dr. Deni Shidqi KHAERUDINI, National Institute of Advanced Industrial Science and Technology, Japan
Dr. M.S. ROSLAN, Universiti Tun Hussein Onn Malaysia, Malaysia
Dr. Anil V. Ghule, Shivaji University, India
Dr. Khairul Anuar, University of Kuala Lumpur, Malaysia
Dr. Nur Farhana Diyana Binti Mohd Yunos, University Malaysia Perlis, Malaysia
Dr. Hendrik Oktendy Lintang, Universiti Teknologi Malaysia, Malaysia
Dr. Biswajit Paik, R&D Manager, Physonit Inc., Japan
Dr. Abdul Kareem, Universiti Teknologi Malaysia, Malaysia
Assoc. Prof. Ching Yern Chee, University of Malaya, Malaysia

## Campus committees

Mr.Abdullah Hasan Jabbar, Universiti Tun Hussein Onn Malaysia, Malaysia
Mr.Maytham Qabel Hamzah, Universiti Tun Hussein Onn Malaysia, Malaysia
Researcher Romel Hidayat, Sejong University, Korea
Dr. MOHAMMED, Isah Kimpa, The Federal University of Technology, Akure
Ms.Siti Ashikin Hassan, Universiti Teknologi Malaysia, Malaysia

IOP Conf. Series: Materials Science and Engineering **738** (2020) 011002 doi:10.1088/1757-899X/738/1/011002

# Peer review statement

All papers published in this volume of *IOP Conference Series: Materials Science and Engineering* have been peer reviewed through processes administered by the proceedings Editors. Reviews were conducted by expert referees to the professional and scientific standards expected of a proceedings journal published by IOP Publishing.

IOP Conf. Series: Materials Science and Engineering **738** (2020) 012001 doi:10.1088/1757-899X/738/1/012001

# Synthesis and Photocatalytic Activity of Superfine $Cu_2O$ Sphere with Smooth Surface on Degradation of Methyl Orang

**Yurong Wu[1], Longshan Xu[1, *], Yanling Hu[1], Yanying Jiang[2]**

[1]School of Materials Science and Engineering, Xiamen University of Technology, Xiamen, Fujian, China

[2]School of Information and Communication, Guilin University of Electronic Technology, Guilin, Guangxi, China

[*]Corresponding author's e-mail address: winxls1@126.com

**Abstract**. Superfine cuprous oxide ($Cu_2O$) spheres with smooth surface have been successfully synthesized by using gelatin as template. The possible mechanism for the formation of superfine $Cu_2O$ sphere is revealed. The results of X-ray diffraction (XRD), scanning electron microscopy (SEM) and Transmission electron microscope (TEM) indicated that the as-obtained powder was cubic structure, pure, uniform, monodispersed $Cu_2O$ spherical polycrystalline with smooth surface. What's more, the photocatalytic activity of the superfine $Cu_2O$ spherical polycrystalline is investigated under visible light irradiation. It is worth noting that the superfine $Cu_2O$ microspheres have a highly selective adsorption for anionic dyes, such as methyl orange, and the mechanisms of absorption were mainly attributed to electrostatic adsorption.

## 1. Introduction

Cuprous oxide, an important p-type semiconductor with the narrow direct band gap of 2.0-2.2 eV, has attracted great interest of researches, because it has many potential applications in solar energy conversion [1], photocatalytic degradation of organic pollutants [2], marine antifouling coating [3], lithium ion batteries [4], carbon dioxide [5], gas sensors for $NO_2$[6], and so on.

Compared with some conventional photocatalyst, such as ZnO and $TiO_2$, the electrons of cuprous oxide at band gap are more easily excited under visible light irradiation. The suitable position of its energy levels which make it a more suitable visible photocatalyst. Meanwhile, cuprous oxide is non-toxic, has low preparation cost, so it is an ideal semiconductor for water decomposition under visible light and degradation of dye contaminants. However, it is disadvantageous that the easy recombination of electron and hole results in low catalytic activity of $Cu_2O$. In order to control the rate of recombination, the semiconductor materials possessing different morphology were prepared and found to be able to suppress the electron-hole recombination, resulting in the better catalytic performance. As known, morphology is an important factor in determining the physical and chemical properties of particles. Therefore morphology controlled synthesis has become a focus. The different morphologies of cuprous oxide were studied intensively by changing the reaction conditions and processes in the experiment. Up to now, cuprous oxide has been prepared by several different methods, such as electrodeposition, thermal relaxation, sonar-chemical methods, γ-irradiation, vacuum evaporation, and the liquid-phase reduction of a metal salt [7-9]. Moreover, these various morphologies of $Cu_2O$, such as octahedrons [10-11], nano-cube, nanocages [12-15], nanowire [16-17], solid and hollow spheres

[18-19], double tower-like tip nanostructures, and multi pods [20] have been synthesized by the method mentioned above. Recently, it has been reported in the previous literature that inorganic hollow spheres are prepared using microorganisms or gelatins as templates. Zhou et al [21] reported that ZnO hollow spheres by employing lactobacillus as template. Yang et al [18] fabricated hollow spheres $Cu_2O$ by a reaction between CuO suspension and $NH_2OH$ in the presence of gelatin. In this paper, we reported the preparation approach of $Cu_2O$ possessing ultrafine, uniform, monodispersed and spherical polycrystalline using gelatin as template and their characterizations by using X-ray diffraction (XRD), scanning electron microscopy (SEM) and Transmission electron microscope (TEM). Moreover, the photocatalytic activity of synthesized superfine spherical $Cu_2O$ was studied.

## 2. Experimental details

### *2.1. Chemicals*
$CuSO_4 \cdot 5H_2O$ (≥99.7%, AR), NaOH (≥99.7%, AR), Polyethylene glycol (PEG, ≥99.7%, CP), and glucose were purchased from XiLong Scientific Co. LTD. Gelatin was supplied from Shanghai Jinsui Biotechnology Co., Ltd.

### *2.2. Synthesized of superfine $Cu_2O$ sphere*
The $Cu_2O$ spherical polycrystalline were prepared by the following procedure. First, solutions of 1g gelatin in 100ml deionized water were prepared, and then sonicated for 30 min, labeled as A solution. A set of solutions were prepared by adding 0.2mol/l Copper sulfate pentahydrate ($Cu\ (SO_4)_2 \cdot 5H_2O$) to 100ml deionized water, in which a certain amount of polyethylene glycol (PEG) as surfactant, labeled as B solution, Second, B solutions were added to A solutions drop by drop, and then heated at 50°C in a water bath and stirred continuously. Third, 100ml 0.1mol/l aqueous glucose solution was added to the above mixture drop by drop. And then 100ml 3mol/l NaOH solution was added to the mixture drop by drop. And the mixture was kept at 50°C for another 30 min. Magnetic stirring was continuously applied throughout the entire process of reduction and particle growth. The $Cu_2O$ particles were separated from the solution by centrifugation at 2000rpm for 5 min. They were then suspended again in hot water and the centrifugation was repeated twice so as to remove the surfactant. After the $Cu_2O$ particles were filtered and washed with hot deionized another two times, the $Cu_2O$ ultrafine microspheres by using gelatin as templated were obtained. For comparison, the $Cu_2O$ was synthesized with the same parameters without using gelatin as templated.

### *2.3. Characterization*
The structure of the prepared powders was characterized by X-ray diffraction (XRD) with Cu Kα radiation (λ=1.5418 Å) (Philips, PANalytical X'pert). The morphology of powders was examined using a field mission scanning electron microscope (SEM) (Hitachi-S4800, Japan) and transmission electron microscopy (TEM) (FEI Talos, F200S, American).

### *2.4. Photocatalytic activities*
Adsorption performance of the synthesized $Cu_2O$ microsphere was assessed by organic dye, such as rhodamine B, methylene blue and methyl orange. 100mg of catalyst was dispersed in 100mL of a 10mg/L simulated pollutants solutions in a quartz tube under the irradiation of visible light (PLS-SXE300C, 300W, λ≥420nm) with ultrasonication during reaction. Before irradiation, the mixture was magnetically stirred for 30 min in the dark to establish adsorption-desorption equilibrium between the dye and the catalyst. During illumination, at given time intervals (30min), 12ml solution was sampled for photocatalytic examination by absorption band maximum. The curves of absorbance vs wavelength for organic dye solutions were recorded using a UV-vis spectrophotometer (UV-9000A, Metash).

## 3. Results and Discussion

Figure 1 shows typical XRD patterns of the synthesized-$Cu_2O$. There are five peaks with 2θ values of 29.58, 36.48, 42.39, 52.61, 61.52, 73.71 and 77.59, corresponding to (110), (111), (200), (211), (220), (311) and (222) crystal planes of $Cu_2O$ with cubic structure (standard JCPDS file no. 99-0041). The lattice constant calculated from this pattern was 4.2623Å, which is in agreement with the standard value 4.269 Å. And the XRD pattern of the as-synthesized powders in Figure 1 shows the sharp peak shape, indicating that the powders obtained by the current one-pot synthetic approach were well-crystallized. No characteristic peaks of CuO and Cu are observed in XRD patterns, indicating high-purity $Cu_2O$ crystalline is successfully synthesized by this method. The pure crystal phase is attributed to the existence of glucose in the process of preparation. Because glucose is a weak reducing agent, only in a strongly basic solution can the $Cu^{2+}$ be reduced to $Cu^{+}$.

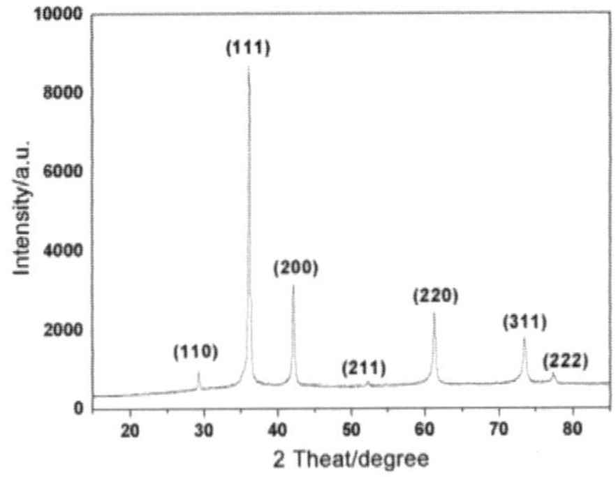


Figure 1. Powder x-ray diffraction pattern of the $Cu_2O$ sphere.

Figure 2 (a)-(b) show the SEM images of as-synthetic $Cu_2O$ particles under different magnifications. Figure 2 (c) is the TEM images of as-synthetic $Cu_2O$. It can be clearly seen that the powders have uniform, monodispersed and spherical particle with the crystal size of 200-500 nm, because of PEG as surfactant. When the micrograph of microsphere is further enlarged to high magnification (Figure 2 (b)), it is found the surface of microsphere is very smooth, this is different from other literature [13, 19], which the surface of the reported $Cu_2O$ particles is very rough, with small flakes and large particles of irregular morphology, or composed of abundant nanoparticles. In addition, its selected area electron diffraction (SAED) pattern is shown in Figure 2(d), the as-synthetic $Cu_2O$ is polycrystalline rather than amorphous, because the XRD of as-prepared $Cu_2O$ shows crystal material (as shown in Figure 1).

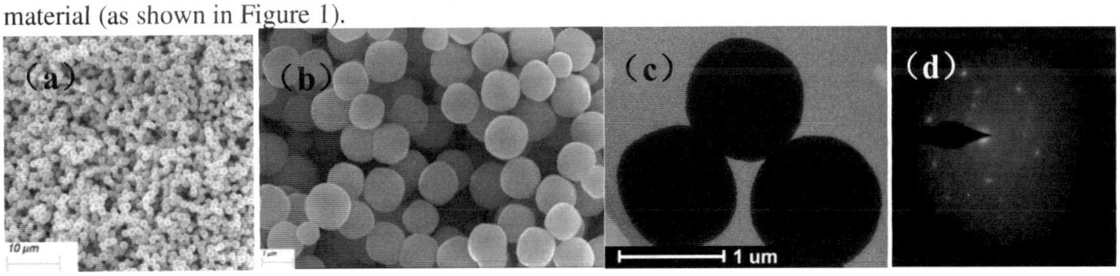


Figure 2. (a)-(b) SEM image of as-synthesized $Cu_2O$ microsphere using gelatin as templated, (c) TEM image of $Cu_2O$ microsphere using gelatin as templated, (d) Selected area electron diffraction (SAED) pattern of the $Cu_2O$.

For comparison, $Cu_2O$ was synthesized without using gelatin as templated, the SEM image of $Cu_2O$ was shown in Figure 3. From the Figure 3, it can be seen that $Cu_2O$ particles exhibit octahedral-like morphology, and aggregation. According to the Figure 2 and Figure 3, the two above mentioned samples indicates that gelatin play important roles in the formation of $Cu_2O$ sphere. The mechanism of formation of $Cu_2O$ sphere was analyzed as follows. When the temperature is relatively high (50°C), the gelatin sol can exist in the form of droplets in the solution. These droplets should serve as templates for the development of spherical structures. The gelatin contain large amounts of carboxyl and ammonia, the positive $Cu^{2+}$ is easily absorbed on the surface of the gelatin droplets, due to electrostatic interaction. The reduction of $Cu^{2+}$ by glucose makes $Cu^{+}$ to nucleate at the appropriate position and isotropic grow into polycrystalline structure. The mechanism of formation of $Cu_2O$ spherical polycrystalline was shown in Figure 4.

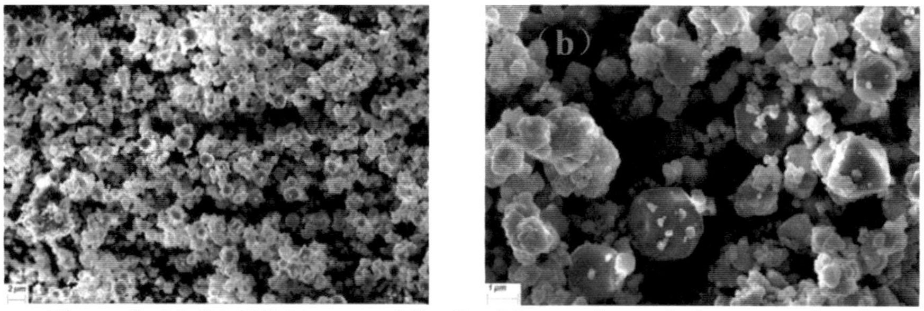


Figure 3. (a)-(b) SEM image of $Cu_2O$ without using gelatin as templated.

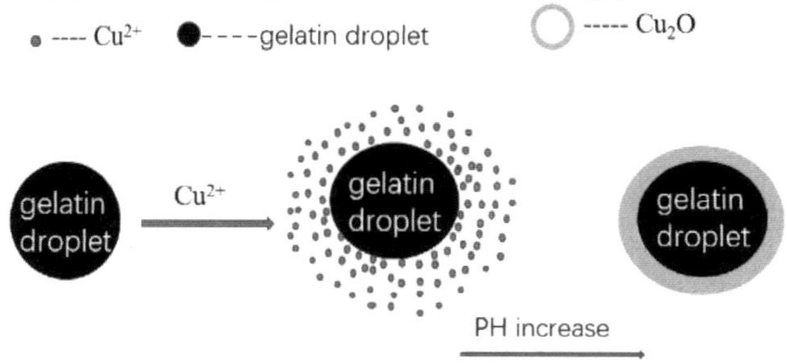


Figure 4. Schematic illustration of the formation of $Cu_2O$ spherical polycrystalline.

Cuprous oxide is an important P-type semiconductor with a band gap of 2.0-2.2eV and has important applications in photocatalytic degradation of organic pollutants, lithium ion batteries, gas sensors for $NO_2$ [2,4,6,13]. Especially, cuprous oxide is a promising material with potential applications in catalytic degradation of organic pollutants under visible light [2,13]. Recently, Xu et al [22] reported that the adsorption and photodegradation of methyl orange with octahedral $Cu_2O$ was more that of cubic $Cu_2O$. In addition, Huan et al [13] demonstrated that nanostructure $Cu_2O$ shows a highly selective adsorption for anionic dyes. Herein, this phenomenon with selective adsorption for anionic dyes for as-synthesized $Cu_2O$ polycrystalline was also found. Figure 5 (a)-(c) show the absorption spectra of different organic dye solutions after photodegradation for various durations over prepared $Cu_2O$ polycrystalline microsphere. Figure 5 (d) gives the degradation ration of different organic dye at 120min. The degradation ratio is calculated using the relationship: $1-A/A_0$, where $A_0$ is the initial absorbance of the organic dye, and A is the residual absorbance after an irradiation time. Under visible light, as shown in Figure 5 (a)-(c), the absorption peak of rhodamine B (RB), methylene blue (MB) and methyl orange (MO) located at 553nm, 664nm and 460nm, respectively. the absorption peak of RB located at 553nm almost unchanged, and the absorption peak of MB located at 664nm weaken slightly. However, the absorption peak of MO located at 460nm weaken gradually as the illumination time increase. From recorded absorption spectra, we have calculated the residual absorbance of organic dye as a function of irradiation time, the results are presented in Figure 5 (d). It is obvious from Figure 5(d) that the photodegradation of RB is very low (~ 0.15%) after 120min of visible light irradiation. This indicated the degradation of RB dye is not so easy under visible light in the presence of the as-synthesized $Cu_2O$ photocatalyst. The photodegradation rate of MB dye is also low (~ 12%) after 120min of visible light irradiation. This exhibits that the as-synthesized $Cu_2O$ microsphere has very low photocatalytic activity towards MB dye. However, the photodegradation rate of MO dye in presence of as-synthesized $Cu_2O$ microsphere is calculated to be ~84.89% during the same time duration of visible light irradiation. We presume that the smooth surface of the prepared

$Cu_2O$ polycrystalline microsphere can increase the multiple reflections to improve the visible light utilization, and smooth spherical structure has good dispersion and fluidity, which will also help to improve the photocatalytic effect. All the results certificated that the prepared $Cu_2O$ microspheres show a highly selective adsorption for anionic dyes, such as methyl orange. Similar phenomena were also observed in previous studies, and the mechanisms of adsorption mainly attribute to electrostatic adsorption [14]. It surmises that as-synthesized superfine $Cu_2O$ spherical polycrystalline may have potential application in degradation of anionic organic dye by visible light driven photocatalysis.

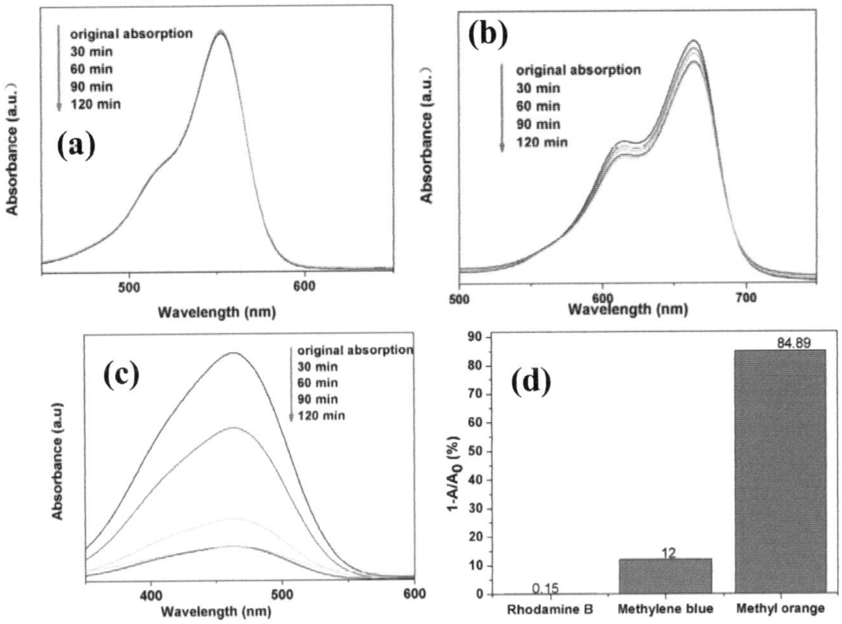


Figure 5. Variation in the light absorption of the different organic dye solution in the presence of the as-synthesized $Cu_2O$ photocatalysts using gelatin as template under visible light irradiation for different irradiation times: (a) RB, (b) MB, and (c)MO; (d) the degradation ratio of with/without gelatin as template for different organic dyes at 120 min.

## 4. Conclusions

We have successfully fabricated the superfine phase-pure $Cu_2O$ by using gelatin as template. As-synthesized $Cu_2O$ is cubic structure, pure phase, uniform, monodispersed and spherical polycrystalline with smooth surface. The diameter of sphere is about 200-500nm. Generated gelatin droplets, serving as template, play a key role in the formation of microspheres. As-prepared superfine $Cu_2O$ sphere has been verified to have a highly selective adsorption for anionic dyes, such as methyl orange, and the mechanisms of absorption were mainly awe to electrostatic adsorption. This suggests that superfine $Cu_2O$ spherical polycrystalline are promising and interesting materials with potential applications in adsorption and photocatalytic degradation of anionic organic dye.

## Acknowledgements

This research was financially supported by Nature Science Foundation of China (No.51201063), Fujian Provincial Natural Science Foundation of China (No. 2017J01720) and Xiamen University of Technology High-level Talent Project (No. YKJ14001R).

## References

[1] Briskman R N. 1992 Sol Energy Mater Sol Cells, **273** 361
[2] Jongh P E, Vanmaekelbergh D and Kelly J J. 1999 Chem Commun, **12** 1069

[3] Elsaeed A M, Abd El-Fattah M, Azzam A M and Dardir M M. 2016 Inter J Biological Macromolecules, **89** 190
[4] Fu L J, Gao J, Zhang T, Cao Q, Yang L C, Wu Y P, Holze R and Wu H Q. 2007 J Power Sources, **174** 1197
[5] Samarasekara P, Yapa N U S, Kumara N T R N and Perera M V K. 2007 $CO_2$ Bulletin Mater Sci, **30** 113
[6] Deng S Z, Verawati T, Hai M F, Tan H R, Sayle D C, Olivo M, Mhaisalkar S, Wei J and Sow C H. 2012 Journal of the American Chemical Society, **134** 4905
[7] Lu C, Qi L, Yang J, Wang X, Zhang D, Xie J and Ma J. 2005 Adv Mater, **17** 2562
[8] Yin M M, Wu C K, Lou Y, Burda C, Koberstein J T and Zhu Y. 2005 J.Am. Chem.So., **127** 9506
[9] Siegfried M J and Choi K S. 2006 J American Chemical Society, **128** 10356
[10] Wei X F, Pan J Q, Mei J, Chen Y Y, Cui C and Li C R. 2018 Photonics and Nanostructures Fundamentals and Applications, **30** 20
[11] He P, Shen X H and Gao H C. 2006 J. colloid and interface science, **284** 510
[12] Kuo C H and Huang M H. 2008 J.American Chemical Society, **130** 12815
[13] Huang L, Peng F, Yu H and Wang H J. 2008 Materials Research Bulletin, **43** 3047
[14] Yang Z, Chiang C K and Chang H T. 2008 Nanotechnology, **19** 025604
[15] Hong X, Wang G Z, Wei Z, Shen X S and Wang Y. 2009 J phys chem C, **113** 14172
[16] Das K and De S K. 2009 Journal of Luminescence, **129** 1015
[17] Sui Y M, Zhang Y Y, Fu W Y, Yang H B, Zhao Q, Sun P, Ma D, Yuan M X, Li Y X and Zou G T. 2009 J.Crystal Growth, **311** 2285
[18] Yang M M and Zhu J J. 2003 J. Crystal growth, **256** 14
[19] Zhang H W, Zhang X, Li H Y, Z K. Qiu, Fan S and Ji M Y. 2007 Cryst Growth Des, **7** 820
[20] Zhang L and Wang H. 2011 ACS Nano, **5** 3257
[21] Zhou H, Fan T and Zhang D. 2007 Microporous and mesoporous materials, **100** 322
[22] Xu H L, Wang W Z and Zhu W. 2006 The Journal of Physical Chemistry B, **110** 13829

IOP Conf. Series: Materials Science and Engineering **738** (2020) 012002 doi:10.1088/1757-899X/738/1/012002

# Research on Detection of Gunshot Residues on Textile by Infrared Spectroscopic Imaging

**Zhang Miaomiao[1], Zhang Jian [1] and Bi Dayu [2*]**

[1]Department of Forensic Science and Criminal Intelligence, Hubei University of Police, Wuhan, Hubei, 430034, China

[2]School of Management, South-Center University For Nationalities, Wuhan, Hubei, 430074, China

[*]Corresponding author's e-mail: xiaomiao69@163.com

**Abstract.** The different chemical components of gunshot residues and the corresponding carrier medium give rise to different spectroscopic features. By utilising this property, infrared spectroscopic imaging is employed to detect the gunshot residues on the surfaces of different textiles. Compared to traditional imaging methods, this technique produces higher-contrast images. Factors such as the composition of gunpowder, material burned, surface property of the carrier medium, range of excitation light, and starting and ending wavelengths are analysed. This research offers a novel idea for the detection of gunshot residues.

## 1. Introduction

It is well-established that non-destructive imaging is the most effective technique for the inspection of various forms of physical evidences [1].Infrared non-destructive testing is a technique that has been newly developed. It is a non-contact testing method, which is able to inspect a large area quickly; moreover, real-time testing is also possible with this technique [2]. Gunshot residue detection is a research topic that has just emerged in the field of infrared spectroscopic imaging. In this paper, experiments are carried out to explore this technique. As demonstrated in the results, infrared spectroscopic imaging has potential to be defined as a novel method for detecting gunshot residues on textiles.

## 2. Experimental principle

When an object is irradiated by an infrared wave with continuously varying frequency, its molecules absorb radiation of certain frequencies. Certain vibrational and rotational modes are triggered, causing changes in the dipole moment of the molecule. Molecular vibrational or rotational energy level transits from the ground state to the excited state. As a result of these transitions, less light is transmitted at the wavelengths of molecular absorption [3-5]. An infrared spectrum can be obtained by plotting the percentage transmission of infrared wave against wavenumber or wavelength. Distinct infrared spectra are obtained from different materials, which makes it possible to use infrared spectroscopic imaging for the qualitative analysis of materials.

The absorption capacity of gunshot residues in visible light area is on par with that of textiles but differs greatly in the infrared range. This difference forms the basis of infrared spectroscopic imaging.

## 3. Experimental conditions

### *3.1. Experimental equipment*
The equipment and specifications used in the experiment were: (1) Nuance multispectral imaging system (CRI, USA); (2) Varispec400 Liquid crystal infrared filter; (3) Samsung notebook, processing software for Nuance developed by CRI, and the processing software by Xingbo; (4) mini400 multi-band light source; (5) iodine tungsten lamp; and, (6) Jingguang Type-I multi-functional light source.

### *3.2. Modification of the infrared spectroscopic imaging system*
To use the spectroscopic imaging system, the liquid crystal infrared filter was placed in front of the objective lens, and the channel was set to "white". The spectral readings changed subsequently to 00-680 nm, 450-700 nm, 500-750 nm, 550-800 nm, 600-850 nm, 650-900 nm, and 700-950 nm, as shown in Figure 1.

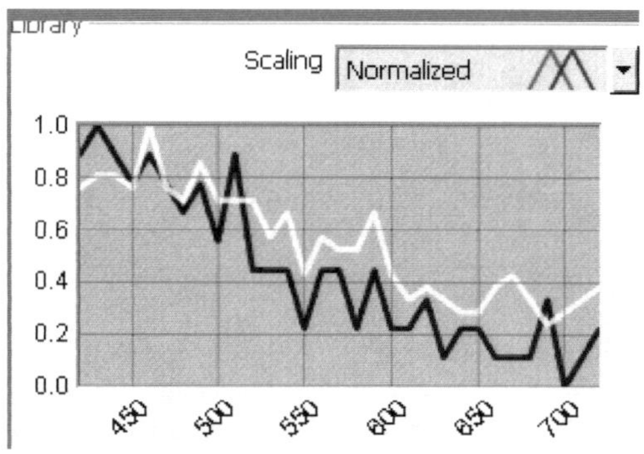


Figure 1. Spectral distribution of the imaging system in white channel

### *3.3. Test sample*

*3.3.1. Experimental materials.* Six types of textiles (polyester, pure cotton, wool, etc.) with different textures and patterns were used as the carrier media of gunshot residues. For each type of textile, 4 pieces of 20 cm×15 cm fabric were used.

*3.3.2. Test sample preparation.* The following set of equipment was used in the experiment: Type 64 7.62 mm pistol, conventional digital measuring tape, target plate, earmuffs, goggles, etc. For each group of tests, gunshot residues were produced with the Type 64 pistol on four pieces of the same fabric, at a distance of 5 cm, 30 cm, 60 cm and 150 cm respectively. The obtained four samples were considered as one group. This step was repeated for different textiles to prepare a total of 6 groups, namely 24 samples. Details of the sample preparation procedure can be found in Table 1.

Table 1. Fabric texture and shooting distance of gunshot residue samples.

| Sample number | Sample texture | Number of samples at each shooting distance | | | | Subtotal |
|---|---|---|---|---|---|---|
| | | 5 cm | 30 cm | 60 cm | 150 cm | |
| 1 | Pure cotton (red and black checks) | 1 | 1 | 1 | 1 | 4 |
| 2 | Satin (ivory) | 1 | 1 | 1 | 1 | 4 |
| 3 | Polyester (pale green with woven pattern) | 1 | 1 | 1 | 1 | 4 |
| 4 | Nylon (coffee with woven pattern | 1 | 1 | 1 | 1 | 4 |

| | | | | | | |
|---|---|---|---|---|---|---|
| 5 | Synthetic knits (black) | 1 | 1 | 1 | 1 | 4 |
| 6 | Woolen knits (dark red) | 1 | 1 | 1 | 1 | 4 |
| | Total | Total number of samples: 6 groups, 24 samples | | | | |

### *3.4. Experimental procedure*

*3.4.1. Image acquisition.* Image acquisition was carried out at room temperature. The prepared sample was placed on copy stand. A suitable light source was chosen for optimal lighting (white light from a 200 W iodine tungsten lamp was used in this experiment, at an angle of incidence of approximately 45 degrees). The objective was directed at the region of interest on the sample. The sample position, magnification and focus of the lens were adjusted until a clear image of the area to be inspected was formed on the computer screen.

*3.4.2 Image scans.* Parameter values such as the scan range and the resolution were entered into the image scanning interface of the CRI software. The Vs Software was launched, and the corresponding parameters such as the scan range and scan time, were specified. The scan range of the Vs processing software was set at 650 nm-1100 nm, and the scan interval was 20 nm. Each scan cycle was 5000 ms. Scan range of the CRI processing software was set to 420-720 nm. Synchronization between the two programs was enforced. On the image scanning interface, IR scans were performed at both wavelength ranges on the area to be inspected. The exposure time was determined by the image recording software of the imaging system.

*3.4.3 Image analysis.* The software developed by Cambridge Research & Instrumentation, Inc. (CRI) that comes with the imaging system was used to perform de-mixing and adjustment on the information contained in the input image. The sample was scanned by IR radiation to generate spectroscopic image. This image was a pseudo-color image automatically formed by the spectrometer. Wavelength ranges can be selected to produce multiple combinations of the three primary colors, and multiple pseudo-color images, which can be displayed on the screen for analysis.

## 4. Results

### *4.1. Detection rate of gunshot residues*

Images of textiles with gunshot residues produced at different shooting distances are collected and imported into the data library of the software. These images are then transformed into their pseudo-color counterparts and analyzed for gunshot residue detection. The detection results are shown in Table 2 and are compared with the observations made with stereomicroscope.

Table 2. Detection results of gunshot residue samples formed at different distances

| **Sample number** | | **1** | **2** | **3** | **4** | **5** | **6** |
|---|---|---|---|---|---|---|---|
| Sample texture and original color | | Pure cotton (red and black checks) | Satin (ivory) | Polyester (pale green,shadow pattern) | Nylon (coffee, shadow pattern) | Synthetic knits (black) | Woolen knits (dark red) |
| 5cm | Stereomicro-scope | √ | √ | √ | - | - | - |
| | Spectroscopic imaging | √ | √ | √ | √ | √ | √ |
| 30cm | Stereomicro-scope | - | - | - | -- | -- | -- |
| | Spectroscopic imaging | √ | √ | √ | √ | √ | √ |

| | | | | | | | |
|---|---|---|---|---|---|---|---|
| 60cm | Stereomicro-scope | -- | -- | -- | -- | -- | -- |
| | Spectroscopic imaging | √ | √ | √ | √ | √ | √ |
| 150cm | Stereomicro-scope | -- | -- | -- | -- | -- | -- |
| | Spectroscopic imaging | √ | √ | √ | √ | √ | √ |

Note: "√" indicates detection of gunshot residues, "-" indicates detection of weak gunshot residues, and "--" indicates detection failure.

As shown in the experimental results of the above table, gunshot residue detection by both stereomicroscope and spectroscopic imaging is influenced by shooting distance. However, as noticed in the observation of textiles with different color and texture under stereomicroscope, textile color interferes to a considerable extent with the detection of gunshot residues. On some dark-colored textiles, it is difficult to discern gunshot residues produced even at a shooting distance of 5 cm. This, however, is not a problem with the pseudo-color images formed by infrared spectroscopic imaging, as unambiguous detection is always possible. Representative images are shown below for comparison.

### *4.2. Gunshot residue detection on different carrier media*

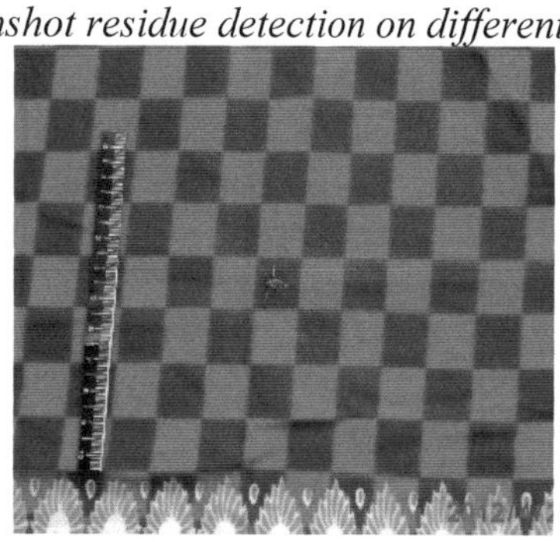

Figure 2. Gunshot residues left on printed pure cotton fabric from 30 cm away, as seen under natural light.

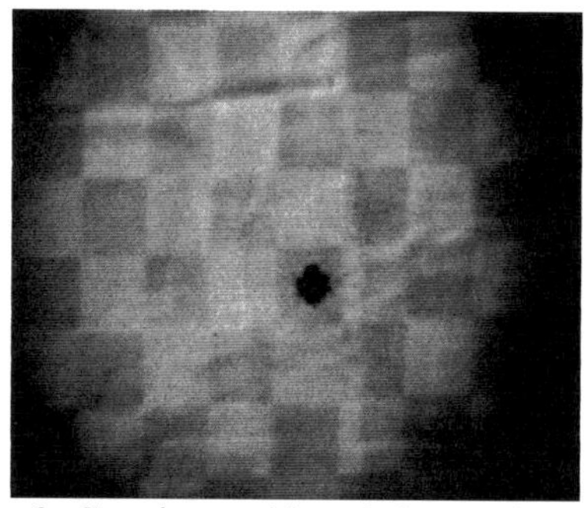

Figure 3. Gunshot residues left on printed pure cotton fabric from 30 cm away, as seen on the image formed by scanning the sample at 575 nm wavelength.

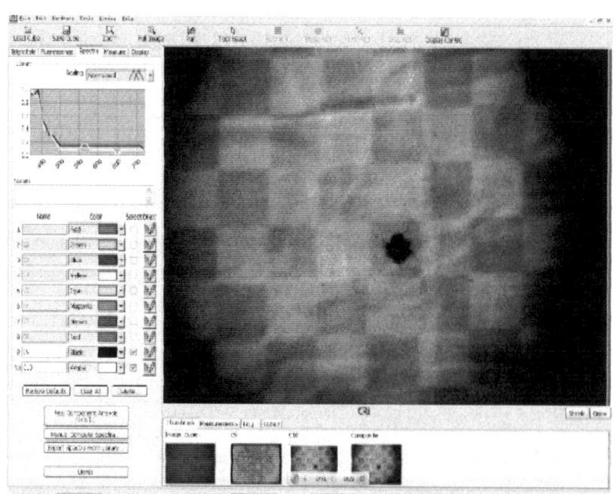

Figure 4. An analysis of the spectroscopic image in Figure 3 (dark speckles around the bullet hole are the gunshot residues).

*4.2.1. Comparison of gunshot residue detection on colored pure cotton fabric with prints.*
Owing to the long shooting distance and interference from printed pattern, gunshot residues left on pure cotton fabric from 30 cm away are no longer observable by naked eye and stereomicroscope under ambient lighting (Figure 2). With infrared imaging, the residues are clearly viewable after the sample is scanned at 575 nm wavelength and the image is processed. Even the unburned gunpowder particles can be seen (Figure 3). An analysis of the spectroscopic image is shown in Figure 4.

*4.2.2. Comparison of gunshot residue detection on pale green polyester fabric.*
Owing to the interference of fabric color and rubbing of the fabric, gunshot residues left on polyester fabric from 30 cm away are barely observable by naked eye and stereomicroscope (Figure 5). With infrared imaging, diffused dark masses of the residues are clearly viewable after the sample is scanned at 575 nm wavelength and the image is processed. Even the unburned gunpowder particles can be seen (Figure 6). An analysis of the spectroscopic image is shown in Figure 7.

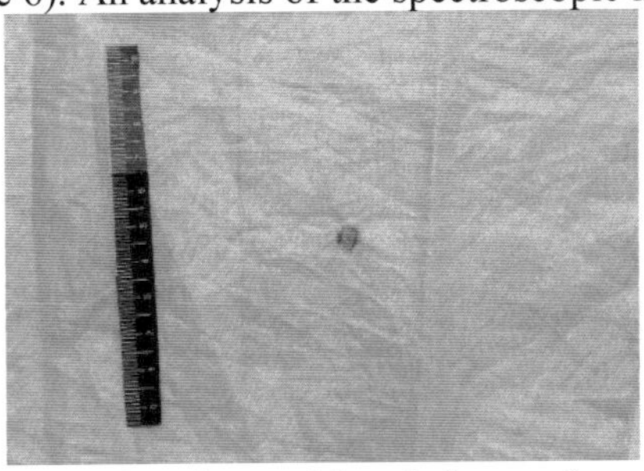

Figure 5. Gunshot residues left on pale green polyester fabric from 30 cm away, as seen under natural light

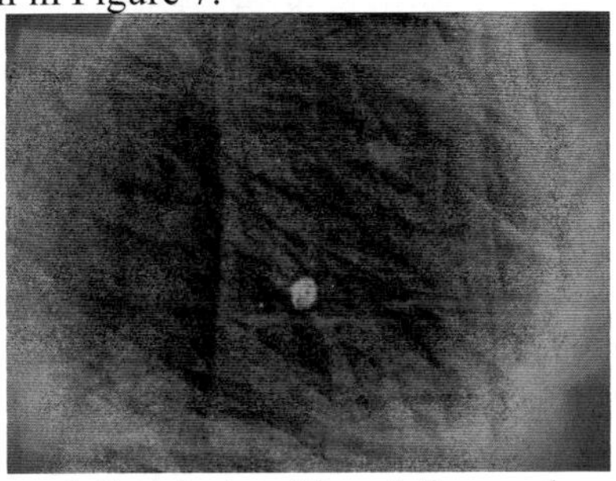

Figure 6 Gunshot residues left on pale green polyester fabric from 30 cm away, as seen on the image formed by scanning the sample at 575 nm wavelength.

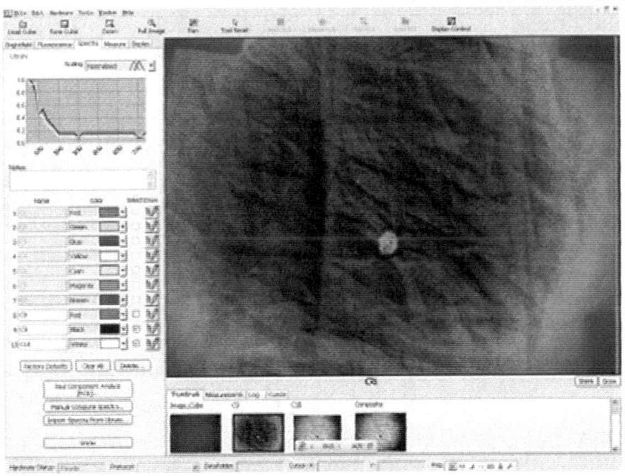

Figure 7. An analysis of the spectroscopic image in Figure 6 (dark speckles around the bullet hole are the gunshot residues).

*4.2.3. Precautions of the experiment.*

- The test samples collected should be kept in separate bags to avoid contamination;
- During the operation of CRI and Vs software, the scan range and scan time should be set such that the two programs run in a synchronized manner;
- For de-mixing of the input image using the software developed by CRI, several reference points should be selected among the gunshot residue pixels of the pseudo-color image and one in the background.

## 5. Conclusions

• Reports on the use of spectroscopic imaging function of infrared spectroscopy for the inspection of physical evidence are rarely seen in literature. The adoption of infrared spectroscopic imaging in forensic science, such as the detection of trace physical evidence including gunshot residues, has yet to be realized. This study is the first case in China that used infrared spectroscopic imaging for the detection of gunshot residues on textiles. Moreover, a special liquid crystal tunable wavelength filter covering the entire infrared region is also used. This study is expected to open up new directions for the research of both gunshot residue detection and infrared spectroscopic imaging.

• The proposed gunshot residue detection by infrared spectroscopic imaging is a novel method for the fast detection of gunshot residues. The NIR image generated also gives the identity and distribution of the chemical compounds found in the residues. The qualitative analysis of specific target components could also be realized through statistical analysis, enabling the qualitative analysis of gunshot residues by infrared spectroscopic imaging.

• Gunshot residues are almost always found on the scene when shooting takes place within a certain distance. However, little gunshot residue could be extracted due to shooting distances, extraction methods, and other factors. Direct observation of the residues is also difficult. Infrared spectroscopic imaging is a quick and intuitive method that allows for the easy and accurate detection of shooting, and the inspection of the state of gunshot residues. It provides a basis for analyzing the shooting distance, gun type, gun power, and muzzle devices.

• Common gunshot residue detection methods used at present include the paraffin method, neutron activation analysis, atomic absorption, and scanning electron microscopy/Energy-dispersive X-ray spectroscopy (SEM-EDX) [6]. These techniques all have their strengths but are also difficult to implement. Infrared spectroscopic imaging does not place high requirements on samples, and sample collection could be performed in a variety of ways. The detection is also simple and fast. Compared with other methods, it offers more comprehensive information, and can be used in the cross-verification with traditional methods, giving it unique advantages.

**Acknowledgments**

The authors would like to thank Hubei provincial key laboratory of forensic science for its great help in providing the lab.

This paper is supported by Key Project of Natural Science Foundation of Hubei Province (No. 2014CFA132).

## References

[1] Jin-Ling H,Jian Z.(2014)Study on adding words in infrared transmission spectral imaging examination.Journal of Hubei University of Police,27(4):162-164.

[2] Dong-Ting Q.(2010)Study and modeling of infrared nondestructive testing system.Harbin Institute of Technology.Chinese Journal of Light Scattering,17(2):180-186.

[3] Wu L,zhao-hui O,Shu-chao C,et al..(2005)Research development and application of raman scattering technology.

[4] Rebuffo-Scheer C A,Dietrich J,Wenning M,et al..(2008)Identification of five listeriaspecies based on infrared spectra (FTIR) using macrosamples is superior to a microsample approach. Analytical and Bioanalytical Chemistry, 390(6):1629-1635.

[5] Zhi-Guo W,Su-Qin S,Qun Z,et al..(2003)Examination of Dyestuffs in Dyed Fiber by NIR FT-Raman Spectroscopy.Chinese Journal of Light Scattering,14(4):212-215.

[6] Stich S,Bard D,Gros L,et al..(1998)Raman microscopic identification of gunshot residues. Journal of Raman Spectroscopy,29(9):787-790.

IOP Conf. Series: Materials Science and Engineering **738** (2020) 012003 doi:10.1088/1757-899X/738/1/012003

# Comparison between GaN and SiC for power switching transistor application

**Haoxiang Yu**

Material science and engineering, University of California, Davis, California, 95618, USA

Corresponding E-mail: angela@cas-harbour.org

**Abstract.** Semiconductor has been widely used in industry ever since its invention and has experienced three generations. However, there are still many problems existing in the third generation semiconductors industry such as making process and cost. This paper mainly discusses the differences between GaN and SiC in material properties and fabrication process. Material properties include band-gap, critical field strength, carrier mobility, and thermal conductivity. For fabrication process, GaN is normally used homoepitaxy approach and SiC is used heteroepitaxy approach. The details in these two approaches are given in the paper.

## 1. Introduction

Semiconductor has been widely used in industry ever since its invention and has experienced three generations. Each of these generations bring a great progress to our electronic devices. From silicon and germanium to semiconductors compounds such as GaN and SiC, electronic devices involved semiconductors are becoming smaller but more advanced. With the advent of third generation semiconductors which are composed of compounds whose band gap is larger than 2.3Ev, third generation semiconductors have emerged and show great advantages in high power and high switching frequency devices such as lasers and detectors. Compared with the first and the second-generation, the third-generation materials have outstanding features in higher breakdown electric field and thermal conductivity. Therefore, they are more suitable to work in high power device under high temperature and high radiation condition.

Although the introduction of third generation semiconductor allowed us to break through the limitation and obtained higher performance device, it's not as widely applicable as first and second generation semiconductors in industry. There are still many problems existing in the third generation semiconductors industry such as making process and cost. Many conflicts in making process reduced the stability of the device and increased cost. Thus it's impractical to extensive production with a great cost. However, in some specific areas, such as aerospace, third generation semiconductors play an important role. In our daily life, the combination of first and second generation semiconductors can complement shortcomings of each other and reduced cost. More properties and applications of third generation semiconductors are still developed.

In this paper, GaN and SiC, which are two representatives of third generation semiconductors, are mainly discussed in comparison between their material properties and structures. This paper serves to differentiate these two semiconductors' function and provides some problems exiting in MOSFET making in order to give readers better understanding of these two materials.

IOP Conf. Series: Materials Science and Engineering **738** (2020) 012003 doi:10.1088/1757-899X/738/1/012003

## 2. Comparison in Material properties

There are four important material properties related to semiconductor, including band-gap, critical field strength, carrier mobility, and thermal conductivity. These four properties determine the performance of the devices [1].

### *2.1. Band gap*

Band gap is the energy difference between valence band and conduction band. In general, it is the minimum energy that is required to excite an electron from valence band to conduction band. Only electrons in conduction band can conduct electricity. For semiconductors, their band gap is normally less than 5ev. In table 1, it shows that band gap of SiC is almost same as GaN's. Electrons in GaN and SiC require almost same energy to jump from valence band to conduction band. If band gap is the only consideration, both two materials should operate at high temperature and high voltage conditions [6].

### *2.2. Critical field*

In semiconductor, critical field is also called breakdown field. This parameter indicates the minimum voltage the material become electrically conductive. Breakdown field depends on band gap. In the field, electrons accelerate to a very high velocity and create an electron-hole pair. These created pairs can accelerate themselves and create more pairs. Therefore, this process would produce an avalanche of particles. The equation between band gap and velocity of electrons is given:

$$\frac{1}{2}mv^2 = E_g$$

m is the mass of electrons. v is velocity of electrons. Eg is the band gap energy. With similar band gap energy between GaN and SiC, their critical field is close. But SiC's is slightly higher. With same length of material, SiC has a higher breakdown voltage.

### *2.3. Electron mobility*

Electron mobility indicates how quickly the electron can move through the semiconductor. Electron mobility can influence the material in two aspects. First aspect is the conductivity of material. Conductivity is proportional to the product of electron mobility and carrier concentration.

$$\sigma = e(n\mu_e + p\mu_h)$$

σ is conductivity. e is electron charge. n and p are concentration of electrons and holes [5]. $\mu_e$ is $\mu_h$ are mobility of electrons and holes. In table 1, GaN's electron mobility is much higher than SiC's. This means the conductivity of GaN is much better than SiC. This characteristic help GaN device have much less energy lost. Second aspect is influencing frequency response. This influences include two ways. First, the carrier velocity is proportional to the mobility in low electric fields. Therefore, higher mobility corresponds to higher frequency because carriers take less time to go through the device. Second, under same condition higher mobility means higher current. Higher current can charge capacitance more quickly. Hence, it results in higher frequency response. GaN with high carrier mobility indicates that it is a better option for high frequencies device.

### *2.4. Thermal conductivity*

Thermal conductivity refers to the ability of material to conduct heat. In making device, it is an important factor because it defines the limitation of the material. In semiconductor devices, high thermal conductivity means the material needs less help from external environment to dissipate the heat. In most circuits, the resistance in each components generates a lot of heats. With good thermal conductivity, the material can work at higher temperature and higher power. In Table 1, SiC's thermal conductivity almost 4 times higher than GaN's. This gives SiC device a great advantage in the real application.

In summary, for SiC transistor and GaN transistor, the material properties we are interested are band-gap, critical field strength, carrier mobility, and thermal conductivity. High critical field of both SiC and GaN allows them to work at a higher voltages and lower leakage currents. High electron mobility allows them to work in high frequency of operation. Meanwhile, GaN and SiC have different advantages in these properties. GaN is more suitable for high frequency device due to its higher electron mobility, while SiC, has higher thermal conductivity thus having advantages in operating higher power.

Table 1. Material Properties of Si, SiC and GaN

**Table I. Material properties of Si, SiC and GaN[7,8]**

| Material property | Si | 4H-SiC | GaN |
|---|---|---|---|
| Bandgap (eV) | 1.12 | 3.26 | 3.4 |
| Critical field ($10^6$ V/cm) | 0.3 | 3.5 | 3.3 |
| Carrier mobility ($cm^2/V \times sec$) | 1500 | 650 | 990, 2000[a] |
| Electron saturation velocity($10^6$ cm/sec) | 10 | 20 | 25 |
| Thermal conductivity ($W/cm^2 \times K$) | 1.5 | 5 | 1.3 |

[a]In bulk GaN/2D electron gas region of GaN/AlGaN HEMT, respectively.

## 3. Comparison in SiC and GaN MOSFET structure

MOSFET structures for SiC and GaN are very different. As showed in figure 2, in power switching applications, transistor requires an epitaxial layer of either SiC or GaN, which is going to grow or deposit on a substrate that either the same material as the epitaxial layer (homoepitaxy) or a different one (heteroepitaxy).

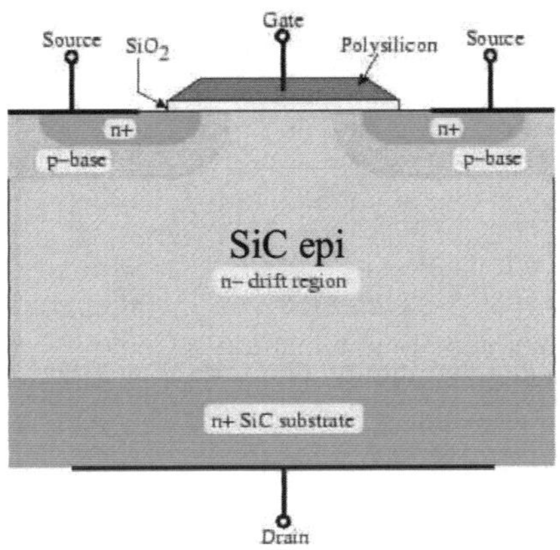


Figure 1. Vertical DMOS SiC MOSFET

For SiC, homoepitaxial approach is the best approach. The most commonly used epitaxial technique for growth of SiC is chemical vapor deposition[2]. It is a process in which gaseous compounds transfer to the substrate surface resulting information and growth of the desired material. The temperature for 6H-SiC layer growing on 6H-SiC substrate by CVD is range from 1500-1850°C. One way to lower the temperature and stabilize the material is to disorient substrate by a few degrees off {0001} in the <$11\bar{2}0$> direction. This growth is so-called step-controlled epitaxy and it can make the poly-type structure more stable. The result turned to be successful in conducing ability from top to bottom of the wafer.

For GaN, although homoepitaxy approach has advantages over the heteroepitaxy approach that has less defects on the material, today's technology can't produce high quality (low defect) GaN substrates with homoepitaxy approach. So most devices made of GaN are produced by heteroepitaxy approach, which is a GaN epi on SiC, Si, and sapphire substrate [3]. The most widely used method to produce heteroepitaxial GaN bulk is HVPE (hybrid vapor phase epitaxy). It can reach high growth rates (<300 um/h), control of impurities, high volume output, high purity substances and large wafer size. However, there are also many disadvantages in HVPE method[4]. Under high strain condition, GaN substrates are easily bowing and cracking at the interface of GaN and non-native substrate materials because of their different thermal expansion coefficients. However, there are some new researches that reduced some limitations caused by bowing and cracking in GaN on Si substrate and GaN on sapphire substrate. The GaN bulk creating with HVPE process has dislocation densities lower than $10^6\ cm^{-2}$.

## 4. Comparison in cost for making SiC device and GaN device

Cost is another important factor when designers made their choices. In general, GaN MOSFET requires more money than SiC MOSFET, because GaN MOSFET needs a buffer layer to accommodate differences in crystallography. An extra buffer layer not only increase the production cost but also creates plenty of defects between non-native substrates. Manufacturers also need to improve and spend more money on dealing with these defects. Thus building GaN devices, more money is required to overcome these problems in order to keep devices operate normally.So in general, GaN MOSFET cost 20% more than SiC MOSFET. Because SiC doesn't have problem with matching substrate and epitaxial layer, the power efficiency can increased about 20 percent.

## 5. Suggestion

From comparison in material properties, MOSFET structure and cost, it's obvious that SiC have advantages in thermal conductivity and cost and GaN has advantages in electron mobility. And these two semiconductors also have many aspects in common such as band gap and critical field. Therefore, I suggest choosing SiC when both materials can satisfy requirements or device are needed to operate at high temperature environment. With same amount of material, cost of SiC can save up to 20%. These savings can be used to improve other aspects of the device. But GaN is still very important when devices are required high frequency switching especially in computers. That's why scientists are still working on developing homoepitaxial approach for GaN. In GaN made devices, manufacturers need to pay more attention on heat radiation.

## 6. Conclusion

Both GaN and SiC are important semiconductors in the industry. The technology of producing SiC is relatively more mature than that of GaN. However, GaN has many advantages over SiC in various aspects. A brief comparison of GaN and SiC in material property and their fabrication method is the main focus of this paper. For material properties, there are differences in band-gap, critical field strength, carrier mobility, and thermal conductivity between GaN and SiC. For fabrication process, homoepitaxy approach is normally used for SiC while heteroepitaxy approach is widely used for GaN.

## Reference

[1] Flack, T. J., Pushpakaran, B. N., & Bayne, S. B. (2016). GaN Technology for Power Electronic Applications: A Review. Journal of Electronic Materials, 45(6): 2673–2682. doi: 10.1007/s11664-016-4435-3
[2] Park, Y. S. (1998). SiC materials and devices. San Diego: Acad. Press.
[3] Aida, H., Takeda, H., Aota, N., and Koyama, K. (2012). Reduction of Bowing in GaN-on-Sapphire and GaN-on-Silicon Substrates by Stress Implantation by Internally Focused Laser Processing. Japanese Journal of Applied Physics, 51(1R): 016504. doi: 10.7567/jjap.51.016504
[4] Aida, H., Koyama, K., Martin, Ikejiri, K., Aoyagi, T., Takeuchi, M., etc. (2013). Growth of Thick GaN Layers by Hydride Vapor Phase Epitaxy on Sapphire Substrate with Internally Focused Laser Processing. Applied Physics Express, 6(3): 035502. doi: 10.7567/apex.6.035502
[5] Ramsden, E. (2011). Hall-Effect Sensors: Theory and Application. Newnes.
[6] A.A Burk Kr., M.j O'Loughlin, R.R Siergiej, A.K Agarwal, S. Sriram, R.C Clarke, M.F MacMillan, V Balakrishna, C.D Brandt (1999) SiC and GaN wide bandgap semiconductor materials and devices, 43(8): 1459-1464.

IOP Conf. Series: Materials Science and Engineering **738** (2020) 012004 doi:10.1088/1757-899X/738/1/012004

# Comparison Between Competing Requirements of GaN and SiC Family of Power Switching Devices

**Yuanhang Zhang**

School of Electronic Information and Communication, Huazhong University of Science and Technology, Wuhan, Hubei, 430074, China

Corresponding E-mail: angela@cas-harbour.org

**Abstract.** GaN and SiC have been widely investigated for future power switching systems with high efficiencies. So far, prototypes of working transistors using these wide bandgap materials have demonstrated the superior performances suggesting the great potential. This article compares two materials over three main requirements in the power switching realm: the blocking voltage, the conduction loss (power efficiency) and the cost. Based on the comparison, a conclusion is drawn that GaN and SiC will be coexistent for each of them has own advantages and neither of them can replace the other.

## 1. Introduction

Since the 1950s wide band-gap semiconductors have been announced to be the next step, once silicon reaches its limits [1]. The advantages of power switching devices are low conduction and switching losses, high temperature capability, and high thermal conductivity. Nowadays, Gallium Nitride (GaN) and Silicon Carbide (SiC) are very promising wide band gap semiconductors, more suitable for power switching applications than silicon[8]. So far, working transistors with low on-state resistances and high breakdown voltages by these two materials have been demonstrated together with highly efficient switching systems with them[8]. Now the question becomes how these two materials and the various devices made out of them will prevail on the market[7]. To answer it, a comprehensive comparison between them is necessary. However, it is always some particular pivotal qualities that make a significant difference. Therefore, in the following part, three core factors, namely, the blocking voltage, the conduction loss (power efficiency) and the cost, will be analyzed and a comparison based on these aspects will be conducted. This article aims to find out whether GaN or SiC has superiority over the other one or neither of them can replace the other one in the future.

## 2. Comparison Between GaN and SiC Devices

### *2.1. Material*

There is a great disparity between Silicon carbide and gallium nitride, as SiC has usually held the heated competent position on devices for power and high temperature and never got an important status in the field of high frequency or optoelectronic. By contrast, optoelectronics is the major market for GaN[9]. However, SiC and GaN share similar material characteristics compared to the standard material silicon and the ultimate semiconductor diamond (table 1)[9]. In most parameters, GaN is slightly superior to SiC and it provides a three times higher Baliga's FoM for power devices[9]. However, GaN's thermal conductivity is lower than that of SiC, which means SiC device should have a better performance under high temperature conditions. However, due to the poor material quality, some of the theoretically superior properties of GaN in the table are not yet achieved in commercial devices or at least not simultaneously achieved in the same device[9].

Table 1. Material properties of SiC and GaN in comparison with those of silicon and diamond [3].

| Parameter | Silicon | 4H-SiC | GaN | Diamond |
|---|---|---|---|---|
| $W_g$ [eV] | 1.12 | 3.26 | 3.39 | 5.47 |
| $E_{crit}$ [MV/cm] | 0.23 | 2.2 | 3.3 | 5.6 |
| $\varepsilon_r$ | 11.8 | 9.7 | 9.0 | 5.7 |
| $\mu_n$ [cm²/V·s] | 1400 | 950 | 800/ 1700[2] | 1800 |
| BFoM[1] relative to Si | 1 | 500 | 1300/ 2700[2] | 9000 |
| $n_i$ [cm$^{-3}$] | $1\cdot10^{10}$ | $8\cdot10^{-9}$ | $2\cdot10^{-10}$ | $1\cdot10^{-20}$ |
| λ [W/cm·K] | 1.5 | 3.8 | 1.3/3[3] | 20 |

As table 2 is shown, the performance of different material is compared. Secondly, the salient differences between bulk and 2DEG is presented. Thirdly, there is a comparison between epitaxial layers and the bulk material.

The epitaxial growth of GaN layers is largely mastered the pivotal question with respect to the material is the substrate. Four substrates are mentioned, GaN itself, SiC, Sapphire and silicon. Of course, the ideal substrate would be GaN itself, bulk GaN, since GaN shows significant properties superior to other materials. Bulk GaN allows for homo-epitaxy without any mismatches between substrate and epitaxial layer while other substrates suffer different level of mismatch. Those mismatch will lead to degradation of GaN device's superiority. But the cost of bulk GaN is prohibitively high and still only available in small size wafer.

Table 2. Pros and cons of typical substrate materials for GaN-epitaxy (data relative to GaN properties) [4].

| Substrate | Bulk GaN | SiC | Sapphire | Silicon |
|---|---|---|---|---|
| Lattice mismatch | none | +3.5% | -16% | -17%, |
| Thermal mismatch | none | +33% | -25% | +116% |
| Electrical resistance | low | low[1] | ∞ | very low[1] |
| Thermal resistance | same | 0.3x[1] | 3x | 0.9x[1] |
| Available wafer size | 2" (3") | 4" (6") | up to 8" | any |
| Cost [€/cm²] | 100 | 10 | 1 | 0.1 |

### *2.2. Blocking Voltage*

The blocking voltage can be used across a switch in OFF mode, which is also the reverse-bias voltage that can be applied across a device[2]. The maximum accessible blocking voltage is in accordance with the breakdown voltage of a planar structure and in line with the lowest possible doping concentration of the semiconductor region to support the electric field because of the reverse bias[2]. Under the situation of SiC and GaN, this region is n- type doped and is called the drift region[2]. For a planar structure, the relationship between the breakdown voltage and the critical field that leads to breakdown can be obtained by solving the one-dimensional Poisson equation [5]:

$$V_B = \frac{\varepsilon_s E_{cr}^{\ 2}}{2qN_D}$$

In this equation, VB stands for the breakdown voltage and Ecr represents the critical or breakdown electric field, ND means the n- type doping level, εs is the semiconductor permittivity, and q is the value of electron charge[2]. From table 1, the critical field of SiC is about Ecr = 2.2MV/cm, nearly ten times bigger than it of silicon. The critical field is related to the strength of the atomic bonds, also further related to the energy of band gap of the material (1.12 eV in Si)[2]. The wider energy band gaps in SiC(3.26eV) and GaN(3.39eV) is conformed with stronger atomic bonds and critical field[2]. Consider the doping of ND= 5×1014 $cm^{-3}$ keeps same for each material and the breakdown voltage of silicon is about 600V, thus the breakdown voltage of SiC should be about 45000V. Nevertheless, the value of GaN increases to 94000V roughly, which is higher than SiC. According to these values of breakdown voltages, it is obvious that Si-based power electronics are more suitable for applications below 600 V, which can be labeled as power management in information technology and consumer electronics[2]. For those applications needing blocking voltages higher than 600 V can be sorted as switching applications, which is the core discussion content of the paper, can only be accessed by materials with wider energy band gaps than Si, especially SiC and GaN[2].

The breakdown voltage of GaN is one times higher than SiC, dose this mean GaN devices are more suitable than SiC devices when it comes to high-voltage applications? Well the answer is not really. If the GaN device is a pure GaN device without a substrate, it do have much higher breakdown voltage than a SiC one, but a pure GaN device is to hard to yield. Nearly all of the GaN devices in the market nowadays are Si-substrate, which is easier for volume production, and they have about 1000V breakdown voltage because of the mismatches between substrate. Before we find the way to eliminate those mismatches in Si-substrate GaN devices, or use pure GaN devices in a wide range, SiC devices are always better than GaN devices when it comes to breakdown voltage.

### *2.3. Conduction Loss (Power Efficiency)*

The requirement for high power efficiency, which should be possibly approach to 100%, is the main reason for the use of switching circuits for converting power. During the process of conversion, power losses can be categorized as static and dynamic[6]. The dynamic losses are due to charging and discharging of parasitic capacitance (C) during switching[2]. In the case of transistors, the parasitic capacitance can be considered as the equivalent input (Cin) and the equivalent output (Cout) capacitance[2]. The equivalent input capacitance is usually higher, and it may dominate the dynamic power loss in low-voltage applications[2]. However, the dynamic power loss due to the output capacitance will dominate and can be very large when these capacitance are charged to very high voltages in high-voltage applications[2]. So the dynamic losses is:

$$P_{dynamic} = fC_{out}V_B^{\ 2}$$

Where f is the switching frequency and VB is the breakdown voltage.

The static losses are due to non-zero voltages (V) when the switches conduct current ( I ) in ON mode (Pstatic= VI). In the case of both diodes and transistors, the main source of static power losses is the On resistance of the drift region, which is necessary for achieving high blocking voltages. There's always a tradeoff between ON resistance and breakdown voltage in a semiconductor device. If the device wants a higher breakdown voltage, it'll have to make its drift region bigger, which means the increase of the length and cross sectional area. As the length should not be smaller than the width, the increase of the length of drift area lead to the increase of ON resistance, is applied:

$$R_{SP} = \rho L_N = \frac{L_N}{q\mu_n N_D}$$

where ρ is the resistivity of the material, and the following relationship is obtained between the specific ON resistance and the blocking voltage.

$$R_{SP} = \frac{4V_B^{\ 2}}{\varepsilon_s \mu_n E_{cr}^{\ 3}}$$

Therefore, a higher blocking voltage requires a wider depletion layer, which results in a longer drift-region with larger resistance when the device is turned ON.

In equations, μn is the mobility of electrons in the drift region. Commercial GaN power transistors are based on a structure that is known as the high-electron-mobility transistor (HEMT). The name HEMT indicates that the mobility in the channel of the transistor is high, typically around 2000cm$^2$/Vs, which is almost 100 times higher than in the case of electron mobility in the channel of SiC MOSFETs. This is a huge advantage in terms of the ability to achieve low ON resistance values by devices with a much smaller area. The relationship between the specific ON resistance and the breakdown voltage, which determines the maximum blocking voltage by the drift region, is illustrated in figure 1. Based on the plot for SiC and GaN, if we apply a voltage of 1000V on both a SiC and a GaN device, the ON resistance of GaN device is about 0.1mΩ•cm$^2$, but the resistance of SiC device is up to over 1mΩ•cm$^2$, which is ten times higher than GaN device. So the static loss of SiC device is 10 times higher than GaN device, which leads to a lower power efficiency.

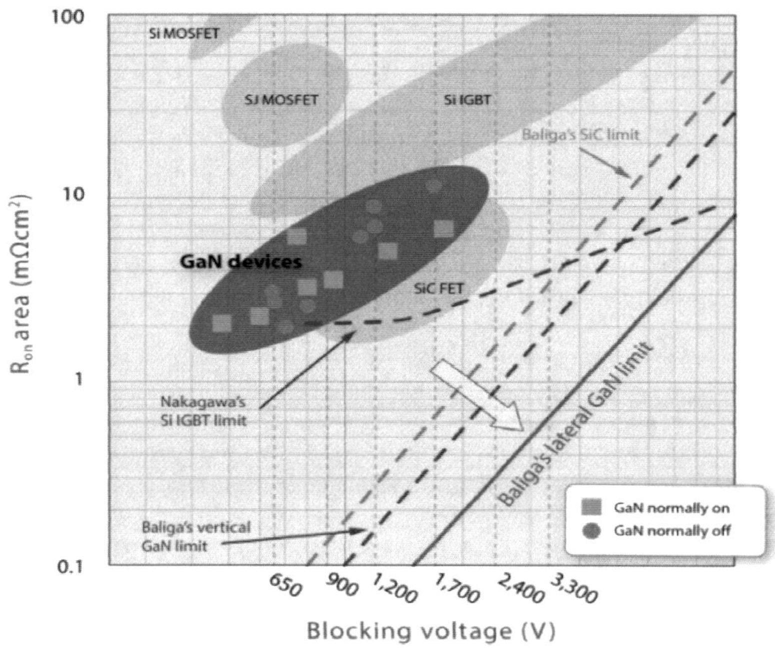


Figure 1. The relationship between blocking voltage and $R_{on}$ area of GaN devices [2].

### *2.4. Cost*

Every competition of products always mentions cost, since the expense of the products is a great concern for companies.

For SiC devices, the cost of main material SiC is about 10€/cm$^2$. As for GaN, pure GaN is about 100€/cm$^2$, 10 times more expensive than SiC. That's one reason the commercial GaN devices nowadays do not use bulk GaN, but Si-substrate GaN device instead. The content of GaN on a Si-substrate is way too low and since silicon is such an ideal material for a substrate with the cost of only 0.1€/cm$^2$, the price of Si-substrate GaN device offers an unrivalled cost advantage over other substrate and SiC device. Also due to the high mobility of GaN device, it become more smaller on size than a SiC one.

Besides, the cost reduction is not only due to the size of the device and the substrate, but also because of the inherently simpler fabrication of GaN-HEMTs compared with the SiC MOSFETs[2]. It is unnecessary for semiconductor doping or the growth of a gate oxide. Besidse, the number of needed photolithography steps is much smaller[2].

## 3. Summary

After the comparison above, we reach an answer of the winner on main technologies. SiC devices offer a higher breakdown voltage, while GaN devices offer a lower conduction loss (higher power efficiency) and lower cost. Besides, SiC do hold a great advantage on temperature endurance and reliability, and GaN have an overwhelming performance in high frequency and optoelectronic realm which are not specifically mentioned in the comparison.

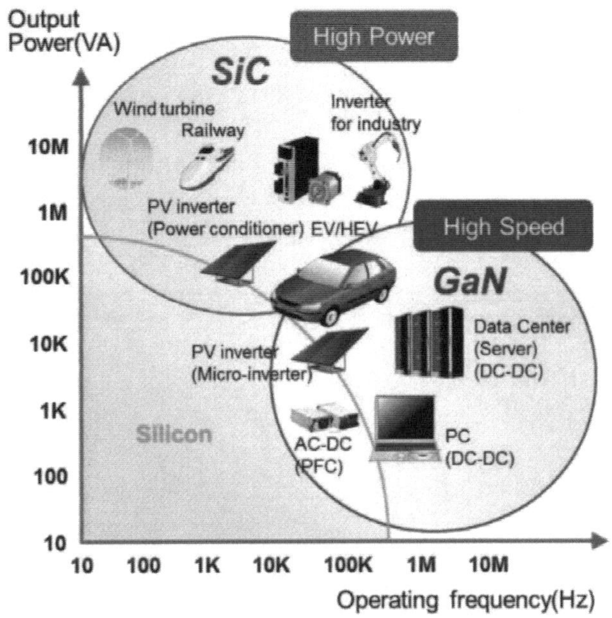


Figure 2. Possible applications of GaN and SiC from the aspects of outpower and operating frequency [8].

Figure 2 shows the way that the future markets of the wide bandgap semiconductors are shared by GaN and SiC[8]. As the increase of the breakdown voltage of GaN transistors is constricted up to 1000V by the existing GaN on Si substrates epitaxial growth technology, the target for GaN is the medium range of the output power with relatively high operating frequencies[8]. By taking advantages of the SiC's high breakdown voltages and better thermal conductivity, SiC should cover higher output power[8]. Therefore, at this late hour, coexistence between these two materials is the most realistic situation.

## 4. Conclusion

The capability of GaN and SiC and the devices based on two materials are analyzed and compared. Three key qualities—blocking voltage, conduction loss (power efficiency), cost—are compared. Though GaN provides a higher power efficiency and lower cost, SiC has its own advantage when it comes to the breakdown voltage. Both materials are important at present. Each of them can be applied to different ranges of voltage, power and operating frequency. That reflects a coexistence situation in the semiconductor market. However, the discussion above only referred to three factors, and a more meticulous analysis need further relevant qualities involved. Heat endurance, FOM, switching speed and erosion endurance, even size of the device should also be seriously considered.

IOP Conf. Series: Materials Science and Engineering **738** (2020) 012004 doi:10.1088/1757-899X/738/1/012004

**References**

[1] William S 1960 Introductory Remarks in Silicon Carbide, A High Temperature Semiconductor *Pergamon Press*

[2] Dimitrijev S and Han J and Moghadam H A and Aminbeidokhti A 2015 Power-switching applications beyond silicon: Status and future prospects of SiC and GaN devices *MRS BULLETIN* vol 40(05) p 43-46

[3] Kaminski N 2009 State of the Art and the Future of Wide Band-Gap Devices *Proceedings of the EPE 2009*

[4] Kaminski N and Hilt O 2012 SiC and GaN Devices-Competition or Coexistence? (ISBN 978-3-8007-3414-6) p 11

[5] Dimitrijev S 2012 *Principles of Semiconductor Devices* 2nd ed (New York: Oxford University Press)

[6] Lidow A and Strydom J and de Rooij M and Reusch D 2015 GaN *Transistors for Efficient Power Conversion* (New York: Wiley)

[7] Kaminski, Nando, and Hilt, O 2014 SiC and GaN devices – wide bandgap is not all the same *IET Circuits Devices & Systems.*

[8] Ueda T 2014 Reliability issues in GaN and SiC power devices *2014 IEEE International Reliability Physics Symposium* D.4.1. p 3

[9] Researchgate http://www.researchgate.net/publication/262881510_SiC_and_GaN_devices__wide_bandgap_is_not_all_the_same

IOP Conf. Series: Materials Science and Engineering **738** (2020) 012005 doi:10.1088/1757-899X/738/1/012005

# Analysis of Current Situation and Trend of Modern Coal Chemical Technology

**Du Jing, Xu Xujie**

National Quality Supervision and Inspection Center of Coal Chemical Products (Hebei), Handan, Hebei, China

**Abstract:** The rapid development of modern coal chemical enterprises and progress in the technical research and application demonstrate the great potential of the industry. For example, they have been strong performers in manufacturing substitute natural gas, coal-to-liquids and coal-based chemicals, boosting demands for its products. The scale of medium and large enterprises indicates a satisfied overall situation with promising future. The paper discussed the technical development of the modern coal chemical industry in general and the future of the industry.

## 1. Introduction

Modern coal chemical industry is based on the traditional coal chemical technology with a high demand for cleaning, utilization and efficient transformation of coal. It is a new industry featuring high efficiency, high yield and high quality. As early as in the "12th Five-year" period, China has proposed adjustment of external oil dependency and new standards on natural gas price and chemical cost, hoping to upgrade the diversified development paths of raw material industry by utilizing modern coal chemical technology. Meanwhile, conditions for industrialization were clarified to ensure effective transformation and upgrading of the coal chemical industry. Furthermore, it was also proposed that modern coal chemical technology should be eco-oriented, reflecting multiple considerations concerning all fields and areas.

## 2. Current development of modern coal chemical technologies in China

China's modern coal chemical industry is on the track of rapid development on a large scale. However, there are prominent problems along the way, including weak competitiveness of low oil price, increasing pressure of environmental protection and severe constraint of water resources, which seriously hinder the development of the industry and subject itself to criticism from the public. In the context of the new economic normal, new topics and new technological positioning have emerged centering on the development of modern coal chemical industry, offering guidance for the sustainable development of the nation's modern coal chemical industry.

### (1) Development of coal-to-liquids technology industry

During the 12th Five-year Plan period, China has built a complete coal-to-liquids chemical technology system, including direct and indirect liquefaction of coal as well as medium-low temperature hydrogenation of tar. Rich experiences have been accumulated in the construction, design and operation of large-scale CTL projects, and there has been significant improvement of technical level of domestic equipment. Under such circumstances, the scale of direct coal liquefaction industry has reached more than one million tons, making it an influential industry integrating fancy technologies, thus solving the problem of high concentration sewage treatment while meeting and the national standards of gas, liquid and solid waste discharge. Based on current technology utilization, oil products from direct coal liquefaction are very popular in the market, which are difficult to produce through traditional procedures. These oil products, with independent intellectual property rights, have taken a lead in the world. For example, domestic companies including Inner Mongolia Yitai Group, China Shenhua and Shanxi Jincheng Arthracite Mining Group have successfully established the indirect coal liquefaction system, integrating gasification, purification, air separation, fischer-tropsch synthesis and product processing technologies. The system, featuring rational design, high yield and high quality, ensures long-term stable and safe operation under full load.

**(2) Current development of substitute natural gas**

Compared with CTL, SNG is still in the early stage of development in China, with several SNG trials putting into practice. The trial results indicated that the SNG has been mature in each individual technology despite technical challenges, including inconsistency between gasification and coal quality in the gasification procedure when lignite is pressurized on the fixed crushed coal bed, gasifier wastewater treatment of fixed bed gasifier as well as difficulties in by-products utilization, etc. Generally, the SNG technology of China is still in the process of upgrading, and there is still plenty of space for optimization in terms of large-scale installations, high conversion efficiency, and emission reductions [1].

**(3) Current development of coal-based chemical industry**

Coal-based chemical industry is going through rapid development, with seven CTO (Coal to Olefin) trail projects in place, including four DMTO, two MTP and one DMTO-II, all of which have basically upgraded the core CTO technology based on local needs. In addition, school-enterprise cooperation further promoted R&D of such technologies. For example, the well-known Fluidized-bed Methanol to Aromatics (FMTA) adopts ZSM-5 as the catalyst system while the continuous reaction and regeneration system of fluidized bed is adopted in the aromatics structural chemical reactor. In addition, aromatics-based liquid products integrating trimethylbenzene, xylene and benzene could reach a purity of more than 95%. Taking Shaanxi Huadian Yuheng Coal Power Company as an example. The company manufactured the world's first 10,000-ton FMTA full-process industrialized test device, which met various standards with greater quality and efficiency, especially in terms of reducing energy consumption and pollutant emission, though there is still room for improvement [2].

## 3. Technical requirements for the development of China's modern coal chemical industry

Based on current situation, the coal chemical industry of China features large capacity of production and front-tier technologies, which will be analyzed from the following three main aspects.

**(1) Accelerated development of new products and technologies**

Domestic enterprises have never stopped the pursuit of new products and technologies, making reckless efforts to achieve this at present stage. After the introduction of numerous technologies from abroad, China is gathering momentum for innovation, as reflected in development of core technologies and engineering projects. However, there are still problems in the practical application of these technologies, such as inconsistency between procedures, device failures and defects in design, which have also become the bottleneck for the development and application of new products and technologies in coal chemical enterprises [3].

On the other hand, stability, energy consumption and economic rationality remain a challenge for large domestic equipment, which is far behind that of developed countries. For example, direct and indirect liquefaction requires huge investment with more complex process, especially in non-processing equipment while some equipment and devices still have relatively rich capacity.

Finally, supporting conditions of key equipment, materials and automatic control are not in place, not fully fulfilling the demands for technical innovation.

**(2) Constant fall in water and energy consumption**

In terms of water and energy consumption, the domestic enterprises keep searching for ways to minimize resources used for production, because the two resources are heavily consumed in traditional procedures. For example, CTL needs as much as 6.0 t/t of water while SNG takes up to 1.0t/1000m$^3$, and FMTO 28.0 t/t, a common phenomenon in northeast, northwest, and western China. Shortage of water resources in these regions prevents the development of coal chemical industry. Therefore, the overall development of innovation and conservation technologies for water and energy need to be improved. At present, it seems that the modern coal chemical engineering projects have been completed, with clear procedures and strong capability for energy conservation. For example, the oil consumption of direct million-ton coal liquefaction in Shenhua has decreased from 62.2MJ/t in 2014 to 48.1MJ/t in 2018, and the integrated energy consumption has decreased from 172.1MJ/t in the beginning to 160.0MJ/t. Combined with this key technology, its processing system has been further developed and improved [4].

**(3) Enhanced treatment and disposal of waste water and salt residue**

Wastewater produced by coal chemical projects is relatively complex in composition, so the water treatment generally takes long time. But at present, there is no domestic project preventing leakage of wastewater in the long run. In terms of high-salt wastewater, no enterprise has been able to separate and obtain relatively pure sodium sulfate and sodium chloride, with unclear destination of sewage discharge and more hazardous wastes with mixed salt content. After stabilizing and fixation treatment, landfill is also required, which is costly and demands a vast area of land. At present, the solution to this problem in China is not clear, requiring further improved experimental data verification. In particular, the research and development of environmental protection technology needs to be further strengthened along with integration and optimization of proprietary environmental protection technologies to fundamentally reduce the total emission and solve problems related to environmental protection [5].

## 4. Future development of China's modern coal chemical industry

At present, the development of modern coal chemical technology industry in China still needs to focus on three aspects: CTL, SNG and coal-based chemicals. New technologies should be developed in combination with coal chemical technologies for large-scale engineering projects from the following four aspects.

**(1) Draw experiences from advanced CTL technologies from abroad**

First of all, China's modern coal chemical industry needs to learn more from developed countries. For example, in the direct liquefaction of coal in CTL, the NEDOL technology of Japan could be a model to learn from, which is more suitable to produce bituminous coal combined with circulating solvent hydrogenation to improve production efficiency. The catalyst, made by replaceable natural pyrite and synthetic iron compounds, is environmentally friendly.

In addition, China also learned from IGOR+ solid-liquid separation technology from Germany, which adopted decompression distillation and red mud as the catalyst for coal chemical production. Then the HTI technology in the United States, a three-phase bubbling-bed reactor with full-back mixing of external circulation, uses highly dispersive nano-scale Fe as the catalyst for solid-liquid separation of critical solvent extraction.

While taking advantage of foreign technologies, Shenhua is working hard to upgrade its own procedures of direct coal liquefaction, where ultrafine hydrated iron oxide catalyst is used in the two-stage series full reverse-mixing suspended window reactor couple with super-fine hydrated ferric oxide as the catalyst. And this is the only successful technology in the world, meeting every requirement for optimization. The development of CTO has promoted the rapid progress of China's coal industry, whose products are widely used in the research and development of military low-freezing diesel, aviation kerosene and various rocket kerosene scarce products. For example, the wastewater treatment technology is still in the pilot stage, with the innovation process pressing ahead. As the core technology, it has dismantled the bottleneck hampering cyclist technology home and abroad, and successfully established a system of its own modern coal chemical technologies. By developing fischer-tropsch synthesis of alpha-olefin and synthetic wax, it has bridged technical gaps, paving way for further researches and development of innovative chemical technologies in China [6].

**(2) Trend and development of SNG technology**

SNG, as a mixture of renewable and unrenewable energy, has a promising future because of its technical features. In the early stage of domestic methanation industry, the R&D of new technologies is gathering momentum, with large-scale domestic projects of methanation catalyst in place at a scale of 5000m$^3$/d in trial projects, which offers an ideal solution to excessive energy consumption in the traditional cycle methane chemical process. At a scale over 1000m$^3$/h, it could fulfill the needs of the industry. Generally, SNG technology is still in the stage of experimental development, heavily reliant on independent development and introduction of key technologies. Taking indirect SNG as an example. Currently, the topso methanation technology, which is made up of 5 reactors and introduced from abroad, uses oxidized nickel-based catalyst. Also, it takes additional steam to ensure that the CH4 content can reach 99.17% and above. The Dalian Institute of Chemical Compounds in China discovered the non-cyclic methanation technology that adopts methanation catalyst, featuring hydrothermal stability under high temperature, strong resistance to carbon accumulation and good performance in catalyst activity and selectivity. After adopting the new non-cycle technology, its capacity of hydrogen-carbon ratio grading regulation has been upgraded. Meanwhile, the institute has independently developed a new high-temperature methane catalyst and methanation reactor, with the CH4 content reaching more than 95%. Moreover, in terms of direct SNG, China has also introduced the catalytic gasification technology of giant energy blue gas from Exxon of the United States, and the hydrogenation and gasification technology of coal from ARCH gas bed of Japan, etc. The products, highly active with strong capacity of high carbonization and deoxidization, can be directly hydrogenated to produce methane with high production efficiency [7].

**(3) Optimizing the FMTO technology**

During the 13th Five-year Plan period, China's coal chemical industry has developed the integrated C4 technology and DMTO-III based on the DMTO-II process, a new generation of FMTO processing technology. However, the R&D of downstream FMTO products in China are still in the initial stage of development. For example, in the process of $\alpha$ -olefin preparation, Yanshan Petrochemical has an olefin production capacity of more than 50,000 t/a, but it still cannot meet the current market demand. In the future, large-scale production of olefin copolymer and high-density polyethylene will be the key, requiring more efforts to be made in independent development and introduction of technologies, especially in the direct manufacture of aromatic hydrocarbons from syngas to improve thermal balance in the reaction process of methanol to aromatic hydrocarbons when applying direct syngas-based production of aromatics as well as the performance of xylene as a catalyst. At present, there have been in-depth researches on ethylene glycol in FMTO technology by domestic scholars. It has been a future trend to prepare ethylene glycol with oxalate and carry on large-scale production. [8].

## 5. Conclusion

Modern coal chemical enterprises have introduced various technologies from abroad along with independently researches in order to ensure further development of CTL, SNG, FMTO and MTO technologies. However, there are still a few technologies in domestic coal chemical enterprises that are still in the primary stage of R&D. Therefore, the modern coal chemical enterprises of China still has a long way to go, and the development and application of new products and technologies still need to be accelerated.

## References

[1] Jia Yanrong. Current Situation and Trend of Modern Coal Chemical Technology [J]. Chemical Enterprise Management,2017,0(35).

[2] Jia Zengqiang, Liu Chao, Hu Wenfeng, et al. Studies on Current Development and Trend of Modern Coal Chemical Industry [J]. Architectural Engineering Technology and Design,2018(8):4856.

[3] Chen Junjie, Wu Weihong. Studies on Current development and Trend of Modern Coal Chemical Technology [J]. Chemical Enterprise Management,2017(25):9.

IOP Conf. Series: Materials Science and Engineering **738** (2020) 012005 doi:10.1088/1757-899X/738/1/012005

[4] Wang Liguo. Thoughts on Rational Development of Modern Coal Chemical Industry from the Perspective of Overcapacity Prevention [J]. Macroeconomic Research,2012(1):3-12.
[5] Yang Hua, Jiang Hua, and Guo Sen. Construction of "Water-Coal-Carbon" System to Optimize the Development of Coal Chemical Industry [J]. Chemical Industry and Environmental Protection,2012,32(1):21-24.
[6] Gao Shitai. Thoughts on the Development of Modern Coal Chemical Industry (CTL) in China [C].// Chemical Editorial Office of National Coal Chemical Information Center. 2012 Collected Works of China Coal Chemical Industry Development Forum.2012:34-37,1-1.
[7] Xing Tao. *Studies on Current development and Trend of Modern Coal Chemical Technology* [J]. Chemical Enterprise Management,2017,0(20).
[8] Wang Kai. *Analysis of Wastewater Treatment Technology and Application in Modern Coal Chemical Enterprises* [J]. Architectural Engineering Technology and Design,2017(17):4093-4094.

IOP Conf. Series: Materials Science and Engineering **738** (2020) 012006 doi:10.1088/1757-899X/738/1/012006

# An Analysis of GaN-based Ultraviolet Photodetector

**Zitong Fan**

School of Physics and Electronics, Central South University, Changsha, Hunan, 410012, China

Corresponding E-mail: angela@cas-harbour.org

**Abstract.** The GaN-based UV photodetectors have been a hot topic in semiconductor field. To present the existing research, this paper introduces the development of GaN-based UV photodetectors and the common types of detectors. The difficulties and new technologies of UV photodetector are also introduced. About the difficulties, the whole paper would focus on the substrate of the GaN. To solve this problem, this paper have introduced the PSS, an enhancement of sapphire substrate. Also this paper have put a hypothesis on the shape of PSS. The surface plasmon, which is not a common technology in photodetectors is also introduced. It is hoped that this paper will provide research background and direction for researchers who want to enter this field.

## 1. Introduction

Ultraviolet detection technology is becoming more and more important in military and civilian fields nowadays, such as medical analysis, flame sensors, missile warning, ozone monitoring, and even UV astronomy. Compared with infrared detection technology, ultraviolet detection is a better choice.

Si-based ultraviolet detectors have been widely used in semiconductor ultraviolet detectors. However, due to its low band gap, it has a large limitation in practical applications. Therefore, ultraviolet detectors with wider band-gap materials are needed. Third-generation semiconductor materials such as GaN and their alloys AlGaN, are one of the hottest research topics in the UV detectors. They own the properties of small volume, low working voltage, low energy consumption, long lifetime, high temperature resistant, and good anti-irradiation stability [1]. Besides, the high chemical and thermal stability of GaN-based detectors have increasingly attracted fundamental research interest.

The study of photoconductive properties of GaN was started by Pankove and Berkeyheiser in1974[2]. In the 1990s, the first GaN-based UV photoconductor was developed by Khan et al. in 1992.[3] And then many fabricating techniques are invented, such as Lateral epitaxial Growth [4][5], Homogeneous epitaxial Growth [6]. Massive research have been done on the substrate of GaN nowadays. Considering many factors, sapphire substrate is widely used in industrial production. However, there are still some problems needed to discover.

The most important one is that we still have not found a perfect substrate of GaN. What's more, the performance of many devices are unsatisfactory. In this paper, I would like to make a review of the current research and discuss some new ways to enhance the devices' performance.

## 2. Types of photodetectors

### *2.1. Photoconductive photodetectors*

The photoconductive detector is one of the simplest structures. It is formed from a semiconductor thin film and two ohmic contact electrodes. When photons with energy greater than the width of semiconductor band-gap are incident into semiconductor materials, electron-hole pairs would be generated either by intrinsic transition and extrinsic transition. The intrinsic transition means the band-to-band transition while the extrinsic transition means transition involving forbidden-gap energy levels. Therefore, the conductivity of the semiconductors would change. When applied bias voltage, these changes in conductivity will form a directional photogenerated current. This is how the photoconductive photodetectors works.

However, there is a strong drawback of p.p called persistent photoconductivity (PPC), which means the photogenerated current would persist for a long time after the light is shut off[7], [8]. As a consequence of PPC, The measured responsivity is determined not just by the optical absorption in the semi-conductor, but also the time that the sample has been kept in the dark. Thus such a device is not suitable in real applications unless we make some change in the measuring technique.

In conclusion, photoconductive photodetector, for its low response speed and poor UV/contrast, is not an ideal device in real application, where the response speed and spectral contrast is required.

### 2.2. *p-i-n photodetectors*

In order to enhance the responsivity, the p-i-n diode has been fabricated. By adding an intrinsic layer in the p-n diode, the doping concentration in n region could be reduced. Therefore, the deletion region is widened. The width of depletion region could be changed by altering the thickness of intrinsic layer. This can improve the quantum efficiency and response rate of the device. Very high-speed GaN p-i-n photodiodes have been reported by Carrano et al. [9]

As Fig.1 shows, the cut-off shifts to shorter wavelengths and becomes smoother for increasing Al content. A visible rejection of more than four orders of magnitude is obtained in all the devices. III-nitride detectors provide an intrinsically sharp long-wavelength cut-off, which enables high resolution in detection.

In conclusion, the device has better performance than the photoconductive photodetectors. This type of device has a high UV/visible contrast, so it has a good UV resolution. In addition, due to the presence of potential barrier, the dark current of the device is low. Thus, the p-i-n photodetectors could be used in many situations.

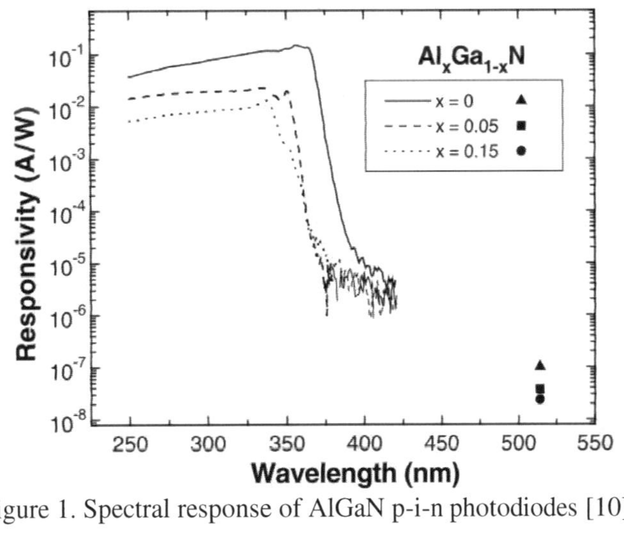


Figure 1. Spectral response of AlGaN p-i-n photodiodes [10].

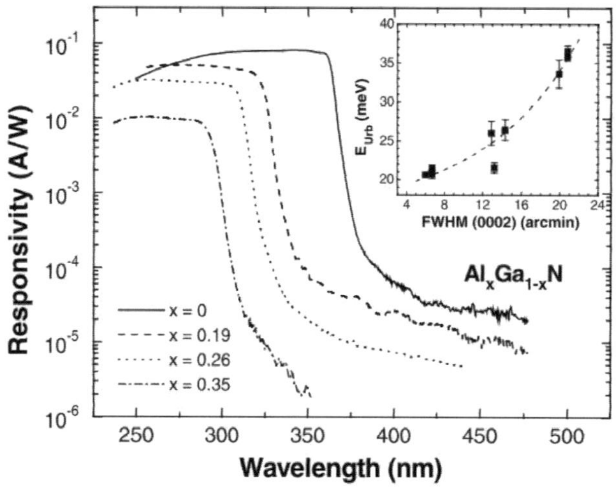


Figure 2. Pectral response of n-AlGaN/Au Schottky photodiodes.

*2.3. Schottky photodiodes*

The schottky structure can be divided into three parts, including semiconductor thin film, ohmic contact and schottky contact. Metals with high work function, such as Pt, Ni, Pd or Au, are commonly used for Schottky contact processing. The specific working principle of Schottky photodiode is similar to that of the p-n junction diode, the electron-hole pairs are generated by the incident photons and then form the photogenerated current with the help of electric field. The holes will then be pulled to the schottky barrier. As Fig.2 shows, when it is not approaching the cut-off wavelength, the response curve of the Schottky photodiode is relatively smooth.

Generally, the influence of incident light intensity and temperature can be ignored. This is mainly because the depletion region of schottky devices is usually located in the shallow layer of semiconductor materials. Even though the incident photon energy is increased, photocarriers can still form in the depletion region of the materials' surface and then form photocurrent. Compared to p-n photodiodes, Schottky photodiodes have more stable quantum efficiency, which makes it stable in most frequencies. Regarding device geometry, in theory, vertical structure could make the device to obtain better quality ohmic contacts, higher responsivity and lower capacitance.

To sum up, Schottky photodetector has the advantages of simple manufacturing process, high responsivity and smooth response curve, high quantum efficiency.

*2.4. Metal-semiconductor-metal photodetectors*

MSM photodiodes are particularly suitable for high-speed applications. It consists of two Schottky contacts, the two contacts are co-planar and opposite. Therefore, when one junction is forward conductive, another is reverse. MSM detectors has many advantages: low capacitance, high UV/vis contrast, and simple manufacturing process. The low capacitance decrease the RC delay, which makes the devices to work in a high responsivity. Due to the high UV/vis contrast, the device has high resolution.

The two Schottky contacts are usually interdigitated, which absorb more photons due to the clearance in the contacts. An enhancement of the MSM detector can be achieved by using semi-transparent electrodes, which enables photons pass through the electrodes, increasing the optical area. An increase of the responsivity by a factor of seven is observed when reducing the Au thickness of the electrodes from 60 to 20 nm [11].

In fact, the MSM photodetectors, compared with the p-i-n photodiode, the MSM structure is simpler, and its response is faster. In addition, it has the characteristics of small junction capacitance and high responsivity.

IOP Conf. Series: Materials Science and Engineering **738** (2020) 012006 doi:10.1088/1757-899X/738/1/012006

*2.5. Avalanche photodiodes*

Avalanche photodiodes (APDs) present the advantages of high speed, high sensitivity and large optical gain. APD is a special kind of pn junction structure or p-i-n junction detector working at avalanche breakdown point, it can also use the MSM structure or Schottky structure. An APD has the advantages of a PIN photodiode (cheap, small, high insensitivity) and those of a photomultiplier (high gain and speed).

An avalanche is formed as Fig 3[12]. The gain of APDs can be 1000 or more. The gain exponentially depends on the bias voltage and therefore the relative change of the gain with voltage is a linear function of the gain. At a gain of 100, the relative change is 10%/V and increases linearly to more than 30%/ V at gain 1000. Similar is the dependence of the gain on temperature changes. APDs must work at moderate internal gain and need low noise amplifiers because of the strong sensitivity to voltage stability and temperature [12].

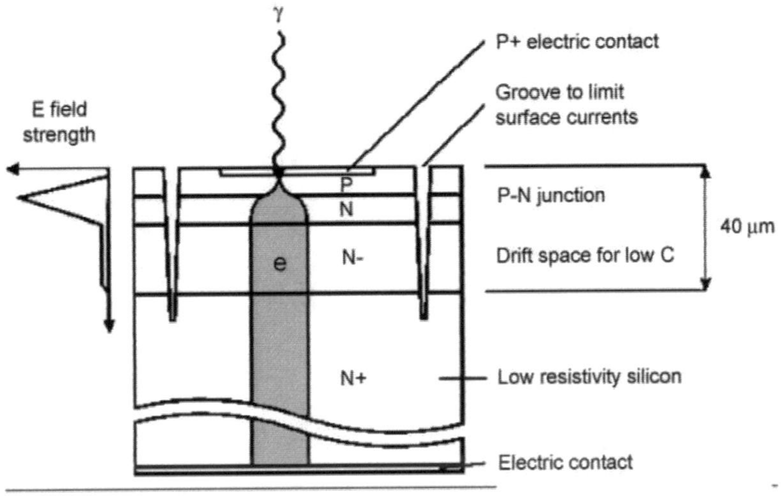


Figure 3. Structure and operation principle of an APD [12].

The advantages of APD are very significant, it is very suitable for detecting weak light and even single photon detection because of its extremely high internal gain. APD has applications in the military, aviation and life sciences. However, APD works under the high voltage, the noise will be amplified accordingly. Therefore, in the design of APD, both noise and gain parameters should be well balanced.

## 3. Discussion

*3.1. Technical Challenges*

Advances in WBG semiconductors have opened the possibility of developing high-performance UV photodetectors, capable of operating in harsh atmospheres. However, the technology of these materials is not mature yet. The development of high-quality substrates for homoepitaxial growth of GaN is a constant challenge, although the most serious difficulty lies in the achievement of high doping levels and reproducible ohmic and Schottky contacts. In industrial production, the development of high-quality substrates for homoepitaxial growth is the key point to improve the reproducibility and reduce costs.

The main problem is that high quality substrates have not yet been found. The GaN often grows on heteroepitaxy, and would face problems of high lattice mismatch, thermal mismatch, which would cause high dislocation concentration in materials. Among these dislocations, the spiral dislocation has the greatest influence. The main harm of dislocation is that it exists as a non-radiative composite center in forbidden band, which leads to a large loss of photogenerated electron-hole pair. Which has a negative effect on the optical properties of the device.

In addition to dislocation, there may be other non-ideal conditions in heteroepitaxial GaN, such as doping without activation, defect state and so on.These non-ideal conditions all require optimization of process technology. Therefore, how to improve the quality of GaN epitaxial layer becomes the key to study GaN-based devices.The key to solve this problem is to improve GaN heteroepitaxial substrate technology.

### *3.2. Patterned sapphire substrate*

During heteroepitaxial growth, there are serious problems in lattice constant mismatch and thermal expansion coefficient mismatch between GaN and sapphire substrate [13]. Aiming at these two problems, many methods have been proposed, such as Lateral Epitaxy Growth [4][5] and Homogeneous Epitaxial Growth [6]. However, these methods cannot improve both crystal quality and quantum efficiency.

Patterned sapphire substrate (PSS) technology developed in recent years has very good advantages. Using PSS as the substrate of GaN epitaxial material can effectively reduce the line dislocation density generated in the growth process of GaN epitaxial material, so as to improve the quantum efficiency of the device. On the other hand, PSS can scatter photons emitted from the optical region, which can improve the performance of LED devices.[14] The Fig. 4 shows a kind of PSS.

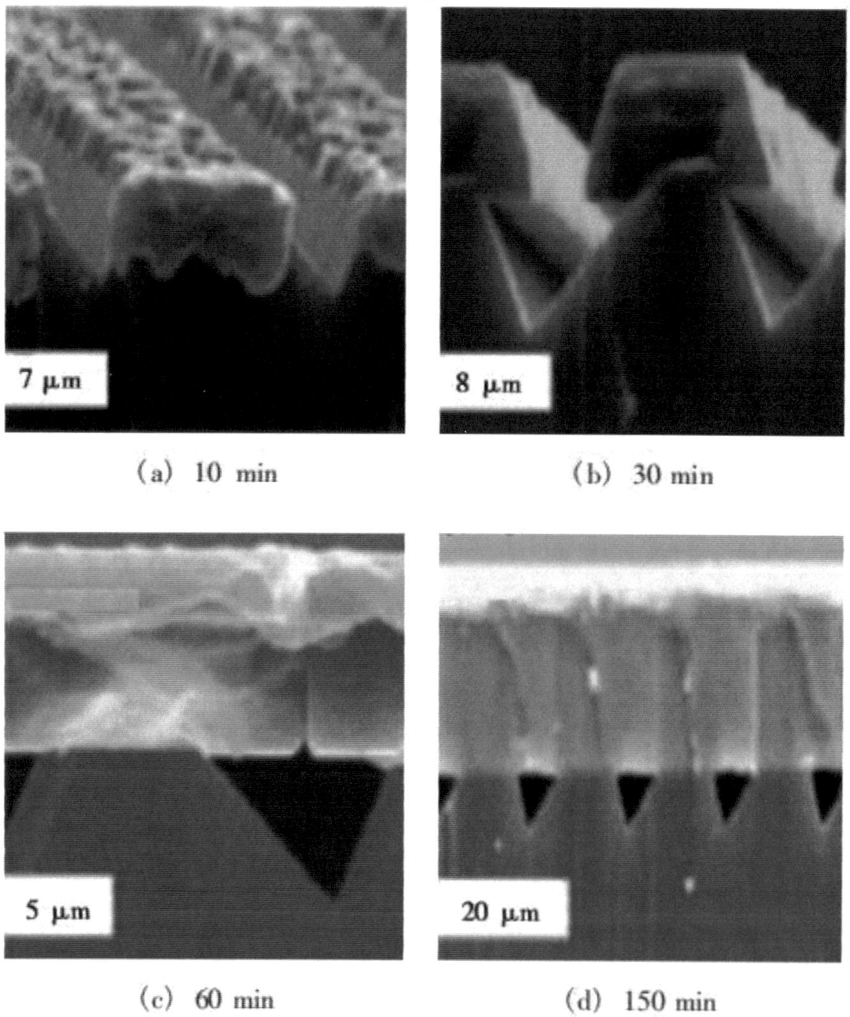


Figure 4. Cross-section SEM images of GaN exit-axial layer grown on the groove-shaped patterned sapphire substrate with different growth time [15].

PSS is mainly prepared by using materials with periodic pat-terns as mask on sapphire substrate, and etching the sapphire substrate with wet or dry etching methods to transfer the mask patterns to the sapphire substrate. Finally, remove the mask and then get the PSS.

At present, the graphic morphology of the sapphire substrate mainly includes groove shape, round hole shape, conical shape, trapezoid shape and hemispherical shape. There have been many studies on how PSS reduces material dislocation. However, there is still no agreement on which shape substrate works best because the exact mechanism of PSS is not completely clear. Based of the hypothesis, a conical shape substrate may be the best substrate. Due to the crystal structure of GaN, the formation of dislocation and the propagation mechanism of dislocation, the dislocation will produce inner triangular cone during lattice mismatch. When surface (0001) becomes the main growth direction, the dislocation will propagate upward. When the crystal grows in the substrate of the conical shape, the internal dislocation matches the slope of the conical shape, which may lead to the bending and annihilation of the dislocation growth in the subsequent lateral growth. However, more experiments are needed to confirm this hypothesis.

### *3.3. Surface plasmon*

With the improvement of theory and processing technology, surface plasmon can have many applications such as in optoelectronic devices, photocatalytic enhancement, biological sensor, and super resolution imaging. In recent years, surface plasmon has provided new approaches to fabricating high-sensitivity UV detectors. Due to the enhancement of surface plasmons, the highest responsivity of GaN-based UV photodetectors with annealed Ag NPs has increased more than 30 times[16].

The principle of SP enhanced photodetector is to use SP to enhance or regulate the light absorption of active layer. When the photon energy is larger than the band gap of the active layer, the active layer absorbs the photon and then generates electron-hole pair to form photocurrent. As shown in Fig.5 [17], SP enhanced light absorption can be explained in three aspects: the first is the scattering of metal nanoparticles, the second is the local field enhancement, and the third is the effective absorption of light enhanced by the propagation of SP.

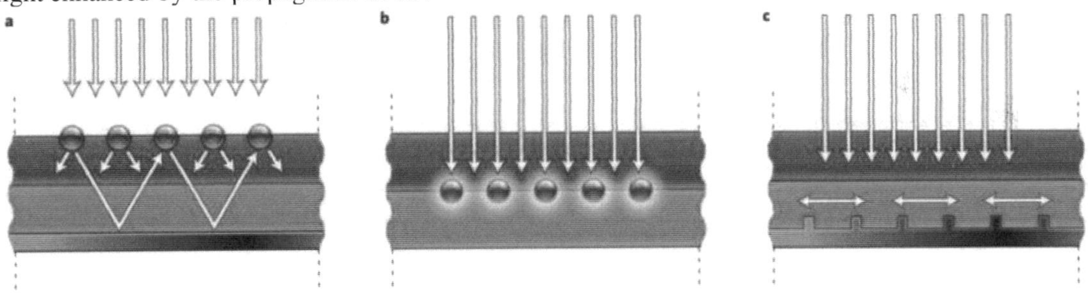


Figure 5. Plasmonic light-trapping geometries for thin-film. a, Light trapping by scattering from metal nanoparticles at the surface. b, Light trapping by the excitation of localized surface plasmons in metal nanoparticles embedded in the semiconductor. c, Light trapping by the excitation of surface plasmon polaritons at the metal/semiconductor interface [17].

Silver (Ag) nanoparticle is one of the best choices for near-UV photodetector due to its low parasitic absorption, high scattering efficiency, and easy preparation. There are many WBG semiconductors using Ag nanoparticles to form SP. In 2008, surface plasmon enhanced InGaN/GaN MQWs blueLEDs containing embedded Ag NPs layers were fabricated, whose optical output power was enhanced by 32.2% [18].

Aluminium (Al) is a highly promising material for commercial applications and it has the advantages of low price, high natural abundance and strong plasmon resonance. However, the disadvantages of Al as SP are the high parasitic absorption of Al nanoparticles and fabrication of Al nanoparticles.

In conclusion, the SP enhanced UV photodetectors has much potential to achieve remarkable progress, and there is also lots of technical problems need to be solved.

## 4. Conclusion

UV photodetector has very important practical value, among which, the third generation semiconductor material based UV photodetector has a good prospect. Various structures of photodetectors according to their different properties can be due to different practical conditions. However, there are still problems needed to be overcome, and the most important problem is to find a more suitable substrate. Meanwhile, the discovery of new technologies such as PSS and SP also provides good condition for the development of GaN UV photodetectors. The PSS is a more suitable substrate, because it could reduce the dislocation in the material. The conical shape may be the best shape, but it still need more experiments to prove it. And the SP is not a common technique, and it would be hard to utilize this technique in industrial production. Therefore, we should do more research on it to make enhancement. Except the substrate, the Ohmic contact problem in GaN devices is also important, which would also be focuses on in days to come.

## References

[1] Jain, S. C., Willander, M., Narayan, J. and Van Overstraeten, R. (2000) Iii– nitrides: Growth, characterization, and properties. *Journal of Applied Physics*, 87(3):965–1006.

[2] Pankove, J. I. and Berkeyheiser, J. E. (1974) Properties of zn-doped gan. ii. photoconductivity. *Journal of Applied Physics*, 45(9):3892–3895.

[3] Asif Khan, M., Kuznia, J. F. U. N., Olson, D. F. U. T., Van Hove, J. M., Blasingame, M. and Reitz, L. F. (1992) High-responsivity photoconductive ul-traviolet sensors based on insulating single-crystal gan epilayer. *Applied Physics Letters*, 60(23):2917–2919.

[4] Marchand, H., Wu, X. H., Ibbetson, J. P., Fini, P. T. and Mishra. U. K. (1998) Microstructure of gan laterally overgrown by metalorganic chemical vapor deposition. *Applied Physics Letters*, 73(6):747–749.

[5] Craven, M. D., Lim, S. H., Wu, F., Speck, J. S. and Denbaars, S. P. (2002) Threading dislocation reduction via laterally overgrown nonpolar (110) a-plane gan. *Applied Physics Letters*, 81(7): 1201–1203.

[6] Leszczynski, M., Beaumont, B., Frayssinet, E., Knap, W., Prystawko, P., Suski, T., Grzegory, I. and Porowski. S. (1999) Gan homoepitaxial layers grown by metalorganic chemical vapor deposition. *Applied Physics Letters*, 75(9):1276–1278.

[7] Qiu, C. H. and Pankove, J. I. (1997) Deep levels and persistent photoconductivity in gan thin film. *Applied Physics Letters*, 70(15):1983–1985.

[8] Li, J. Z., Lin, J. Y., Jiang, H. X. and Khan, M. A. (1998) Effects of persistent photoconductivity on the characteristic performance of an algan/gan heterostructure ultraviolet detector. *Applied Physics Letters*, 72(22):2868–2870.

[9] Carrano, J. C., Li, T., Brown, D. L., Grudowski, P. A. and Campbell, J. C. (1998) Very high-speed metal-semiconductor-metal ultraviolet photodetectors fabricated on gan. *Applied Physics Letters*, 73(17):2405–2407.

[10] Monroy, E., Hamilton, M., Walker, D., Kung, P., Sanchez`, F. J. and Razeghi, M. (1999) High-quality visible-blind algan pin photodiodes. *Applied Physics Letters*, 74(8):1171–1173.

[11] Palacios, T., Calle, F., Monroy, E. and Omnes`. F. (2010) Novel approaches for metal-semiconductor-metal gan uv photodetectors. *Physica Status Solidi A*, 194(2):476–479.

[12] Renker. D. (2007) New trends on photodetectors. *Nuclear Instruments and Methods in Physics Research* (Section A: Accelerators, Spectrometers, Detectors and Associated Equipment) 571(1-2 SPEC. ISS.):1–6.

[13] Yang, G., Chang, Jianli Zhao, J., Tong, Y., Feng, X., Jin, W., Zhang, Q., Huang, H. and Yan, D. (2015) Investigation of light output performance for gallium nitride-based light-emitting

diodes grown on different shapes of patterned sapphire substrate. *Ma-terials Science in Semiconductor Processing*, 33:149–153.

[14] Jeong, S., Kissinger, S., Kim, D., Lee, S., Kim, J., Ahn, H. and Lee, C. (2010) Characteristic enhancement of the blue led chip by the growth and fabrication on patterned sapphire (0 0 0 1) substrate. *Journal of Crystal Growth*, 312(2):258–262.

[15] Wang, J., Guo, L., Jia, H., Xing, Z., Wang, Y., Yan, J., Yu, N., Chen, H. and Zhou, J. (2006) Investigation of characteristics of laterally overgrown gan on striped sapphire substrates patterned by wet chemical etching. *Journal of Crystal Growth*, 290(2):398–404.

[16] Li, D., Sun, X., Song, H., Li, Z., Chen, Y., Jiang, H. and Miao, G. (2012) Realization of a high-performance gan uv detector by nanoplasmonic enhancement. *Advanced Materials*, 24(6):845–849.

[17] Berini, P. (2014) Surface plasmon photodetectors and their applications. *Laser & Photonics Reviews*, 8(2):197–220.

[18] Kwon, M., Kim, J., Kim, B., Park, I., Cho, C., Byeon, C. and Park, S. (2008) Surface-plasmon-enhanced light-emitting diodes. *Adv Mate,* 20(7):1253–1257.

IOP Conf. Series: Materials Science and Engineering **738** (2020) 012007 doi:10.1088/1757-899X/738/1/012007

# Recent research on ohmic contacts on GaN-based materials

**Yuan Liu**

Material Science and Engineering, Boston University, Boston MA 02215

**Abstract.** The most important components in most of the electronic products today, such as computers, mobile phones or digital recorders, are closely related to semiconductor. Semiconductors have a long history. Currently, there are some obstacles in the way of the further development of this material. Therefore, this paper mainly focuses on the ohmic contacts between metals and semiconductor with the help of recent research. In the end, a general conclusion is drawn on the electrical behavior of metal contacts on GaN for further advances in this field of GaN-based devices and materials.

## 1. Introduction

Semiconductors that can work at external temperatures over 150 °C without external cooling have broad application areas, such as automobiles, aircrafts, gas turbines, oil exploration, and space exploration industries[1]. However, there is a large voltage drop at the interface between GaN material and metal, which leads to poor electrical performance and stability of the devices [3]. Low resistance ohmic contact between metal and n-type and P-type GaN is one of the main solutions. Therefore, ohmic contact to GaN is the key technology for preparing commercial GaN devices, and is also the basis for further improving the performance of GaN devices.

This paper summarized some results achieved on ohmic contacts to GaN-based materials. In the first part, the basic principle and concepts of semiconductor-metal contacts are briefly introduced, and the most common methods to evaluate ohmic contacts will be presented. In the second part of the paper, the common approached for ohmic contacts to n-GaN and p-GaN will be presented. For n-GaN, particular emphasis will be given to Ti/Al series contacts and non-alloyed contacts. And for p-GaN, Ni/Au stacks will be discussed, and a new method which is InGaN/AlGaN superlattices to reduce resistance of p-GaN contacts as well as the effect of surface treatments will be presented.

## 2. Basic principle of semiconductor-metal contacts

Because of the rectification or excess power dissipation caused by the voltage thresholds, the low resistance of ohmic contacts leads to the easy flow of charge in both directions inside the two conductors[3]. Fig.1 shows the band diagram when metal contact to n-GaN and metal contact to p-GaN. Generally, metals with appropriate work functions are chosen to accomplish no barrier to the majority carrier [21]. According to the calculation, to achieve the ohmic contacts, the work function of metal should be equal to or smaller than that of the n-type semiconductor ( or bigger than the P-type semiconductor) [2]. As can be seen in fig. 2, Titanium (Ti) and Aluminium (Al) are in principle the materials of choice for ohmic contacts to n-type GaN[2]. Further more, in the terms of p-GaN, Ni, Au, Pt, Pd are the most suitable metals for the triggering of ohmic contacts.

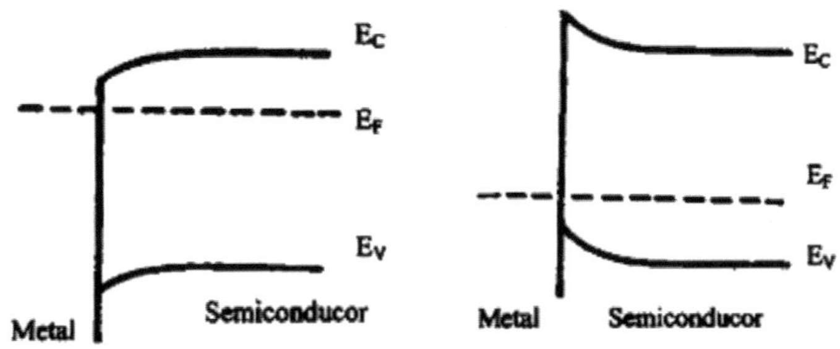


Figure 1. (a)metal contact to n-type semiconductor (b)metal contact to p-type semiconductor

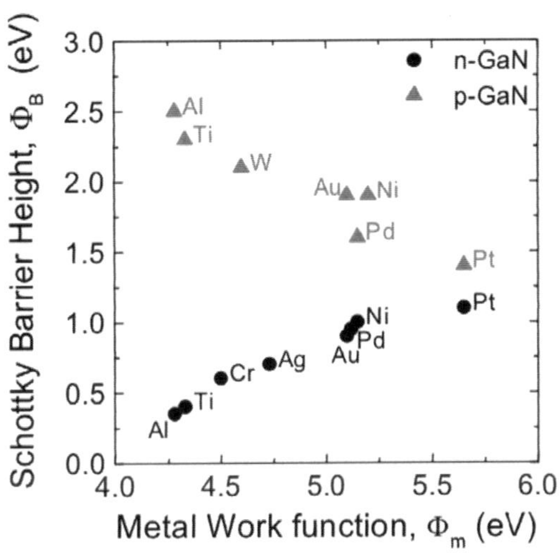


Figure 2. Survey of literature results of Schottky barrier height ΦB values as a function of the metal work function Φm for different metals on both n-type and p-type GaN. Data are taken from Ref. [4] for n-type GaN and Ref. [5] and [6] for p-type-GaN.

At present, the transmission line model (TLM) is commonly used to measure the specific contact resistance, which can evaluate the ohmic contacts. It includes rectangular, circular and circular transmission line models. It have proved that the rectangular and circular transmission line models have large errors, and the measured values are inaccurate. The circular transmission line model can eliminate the errors caused by terminal resistance, and has good operability, accuracy and repeatability.

## 3. Ohmic contacts to n-GaN

The valence bond of GaN shows obvious ionicity, and there is no high surface density of states at the interface between metal and GaN, so the height of contact barrier mainly depends on the work function of the metal. For n-GaN ohmic contact, as discussed in the last section, the preferred method is to select metal with small work function, of which Al (=-4.28eV) and Ti (=-4.33eV) are the two most commonly used metals.

### *3.1. Al, Ti Series Ohmic Contacts*

Al is one of the metals which was studied at first to used as contacts. This kind of ohmic contacts can be formed by single layer Al/n-GaN contact. The contact resistivity of n-GaN grown on sapphire substrates by ECR-MBE method is 1.2*10~4.4*10 cm$^2$ after 575°C and 10 min annealing. It is worthy

IOP Conf. Series: Materials Science and Engineering **738** (2020) 012007 doi:10.1088/1757-899X/738/1/012007

to note that long annealing at high temperature will decreases the contact properties, which may be because of the oxidation of Al to $Al_2O_3$, and formed a high resistance layer.

For single layer Ti/n-GaN ohmic contact, high temperature annealing is usually required. When annealing temperature rises to 900-950oC, the contact resistivity between Ti and n-GaN decreases from 2*10 to 7*10Ω *cm$^2$ [7]. This is because TiN is formed by the reaction of Ti and GaN at high temperature, and N vacancies are formed at the boundary between GaN and metal, which will cause heavy doping and formed tunneling current, which reduces resistivity.

However, Ti is more easily oxidized than Al at room temperature, so single-layer Al or Ti contact is not feasible for devices requiring stable and high performance. An improvement to the conventional Ti/Al/Ti/Au contact scheme has been developed to achieve better ohmic contact properties to *n*-GaN using a Ti/Al multi-layered structure [8]. Ti is used as the contact layer in the lower layer and Al as the cap layer. S. Ruvimov et al. reported [9] that the contact resistance of Ti/Al with n-GaN (grown on sapphire substrates by MBE method) reached 8*10cm$^2$ after 900 °C and 30 s annealing.

In order to further improve the contact performance of Ti/Al, a layer of less resistive and less oxidizable Au can be added to Ti/Al. In order to prevent the reaction between Au and Ti/Al, a barrier layer should be added. Recently, Ti/A1/Ti/Au, Ti/A1/Pt/Au, Ti/A1/Ni/Au and other multi-layer structures have also been extensively studied [10]. Fig3[2]. Shows a schematic of the multilayer used to fabricate ohmic contact to n-GaN.

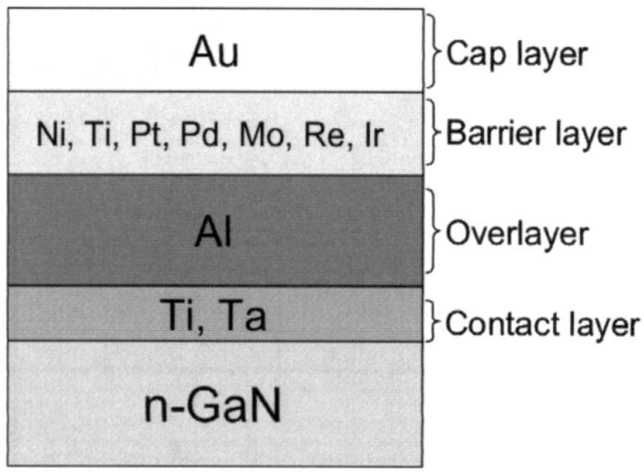


Figure 3. schematic of the multilayer used to fabricate ohmic contact to n-GaN [2]

*3.2. Non-alloy ohmic contact*

In order to effectively reduce the barrier height of gold semi-contact and obtain low ohmic contact, a thin epitaxial layer with narrow band-gap can be grown during crystal growth. InN, InGaN and InAlN are ideal narrow band-gap materials for GaN-based devices.

Although InN (Eg=1.9eV) can effectively reduce the contact barrier with metal, the abrupt heterojunction barriers of InN and epitaxy GaN still have adverse effects on ohmic contact. To solve this problem, a layer of InN/GaN short-period superlattice (sps) is grown between the top layer of InN and GaN, which can reduce the heterojunction barrier between GaN and InN. After depositing 200 nm Ti and 100 nmAl on the top layer of InN without annealing, the contact resistivity can reach 6*10-5cm$^2$[11].

## 4. Ohmic contact to p-GaN

*4.1. metal alloy*

The predicted value of work content of p-GaN is approximately 6.6 eV, which is higher than that of any other pure metals[13]. However, in the reality, the studies about the ohmic contacts between metal and p-GaN are still scarce and not under the consideration of all the conditions, in comparison to the conductivity of these contacts relies on the electronic properties of the materials[13].

In general, a p-type semiconductor like NiO, can be formed during the annealing at the contact/GaN interface region[14]. Therefore achieving proper conductivity can be realized by annealing the Ni/Au contact in an oxygen-containing ambient atmosphere[14]. And because of the binding of H atoms from the GaN lattice with help of the reaction with the ambient oxygen, which releases Mg dopant atoms that are passivated by forming Mg-H bonds in p-GaN. Hence, metals such as La in a Ni-La , which can absorb H contact, are put into use for p-GaN contact metal schemes[15].

According to He et al. (2016), the decrease of the ohmic contact resistance to p-GaN through the application of InGaN/AlGaN superlattice is obviously[19]. The strain-induced piezoelectronic field or the large valence-band discontinuity between GaN and InGaN can lead to an increase in the hole concentration, which can reach $1 \times 10^{19}$ $cm^{-3}$[16]. High concentration of two-dimentional hole gas(2DHG) not only helps to reduce contact resistance, but also serves as a current spreading layer. The application of InGaN/AlGaN superlattice can realize the positive ohmic contact of p-type GaN that is with the specific contact resistance of 7.27 *10-5cm$^2$[17]. The reason for that lies in the following two advantages of InGaN/AlGaN superlattice: one is that the InGaN/AlGaN superlattice can improve the ionization rate of Mg impurities to obtain higher void concentration. The other is the strong space restriction effect on voids and enhancement of the void expansion performance.

### *4.2. Metal oxide*

ITO (indium-tin oxide) with a relatively high resistivity ($10^{-1}$ Ω $cm^2$) when directly contact to p-GaN becomes the most frequently used transparent contact material[13]. As a result, the combination of ITO with other materials has been studied, including those with Ni/Au, Pt, Zn-Ni[13]. In contrast to the Ni/Au/ITO and Pt/ITO contacts, the Zn-Ni/ ITO contact has the lowest specific resistivity for the H-absorbing tendency of Zn-Ni[13]. With Ni layer between ITO and GaN, an ohmic contact with a specific resistivity of $8.6 \times 10^{-4}$ Ω $cm^2$ can be conducted[13]. This is required the proper conductivity, in addition to the formation of NiO in the contact[13].

Through the investigation of using other metals as p-GaN contacts it is found that the transparent conductor ZnO can be probably used as a p-GaN contact material for its lattice constant mismatch with the Wurtzite structure of GaN is small (1.8%) [18].

Compared to n-GaN, the preparation of effective ohmic contacts on p-GaN still remains a problem[20]. The main difficulties are as follows.

Firstly, there is a lack of metal with a high enough work function. GaN has a bandgap of 3.4 eV , and the work function of p-GaN was predicted to be about 6.6 eV,[12] but the work function of metal is generally less than 5eV, which is lower than p-GaN

The second is the difficulty in growing heavily doped p-GaN, which means the carrier concentration should over $10^{18}$. When doping p-GaN by Mg, the concentration of Mg is usually at the level of $10^{17}$ $cm^{-3}$ [16], which makes it difficult for the carrier concentration to exceed $10^{18}$ $cm^{-3}$.

Thirdly, the formation of N vacancies on GaN surface is easy to occur during the process, which makes the p-GaN surface possibly turn into n-type conductance.

## 5. Conclusion

The n-GaN ohmic contact using Ti/Al metal (alloy) and non-alloy as electrode has been studied in this paper. The contact resistivity can reach 10-5~10-8cm$^2$, which can meet the requirements of commercialization of devices. As being discussed before, through there are some achievements on making p-GaN ohmic contacts, it is still a practical and technological subjects to prepare ohmic contacts to p-GaN. Surface treatment, suitable metal or alloy system and suitable alloying process are the main methods to improve p-GaN ohmic contact performance.

Except the low resistance, a good ohmic contact should also have many other properties. Although many achievements have been made, there are some aspects on ohmic contacts that are not discussed in

IOP Conf. Series: Materials Science and Engineering **738** (2020) 012007 doi:10.1088/1757-899X/738/1/012007

this paper, such as the mechanism of action, the impact of various surface treatment technologies on contact, reliability, repeatability as well as antioxidation, which need to be further studied.

**Reference**

[1] Neudeck P G, Okojie R S and Chen L 2002 High-temperature electronics–a role for wide bandgap semiconductors? J. Proc. IEEE, 90(6):1065–1076.

[2] Fiorenza P, Greco G and Lucolano F 2016 Ohmic contacts to Gallium Nitride materials Applied Surface Science http://dx.doi.org/10.1016/j.apsusc

[3] Zhao S 2016 Developing Ohmic Contacts to Gallium Nitride for High Temperature Applications A Dissertation Presented in Partial Fulfillment of the Requirements for the Degree Doctor of Philosophy https://pdfs.semanticscholar.org/0d8a/2780f02dfe6fb788c97a5cf0912cd377260d.pdf

[4] Schmitz A C, Ping A T, Khan M A, Chen Q, Yang J W and Adesida I 1996 Schottky barrier properties of various metals on n-type GaN. *Semiconductor Science and Technology, 11: 1464* https://iopscience.iop.org/article/10.1088/0268-1242/11/10/002

[5] Ofuonye B, Lee J, Yan M, Sun C, Zuo J-M and Adesida I 2014 Electrical and microstructural properties of thermally annealed Ni/Au and Ni/Pt/Au Schottky contacts on AlGaN/GaN heterostructures. *Semiconductor Science and Technology* vol 29, chapter 9 [095005]. https://doi.org/10.1088/0268-1242/29/9/095005

[6] Rickert K A, Ellis A B, Kim J K, Lee J L, Himpsel F J, Dwikusuma F and Kuech T F 2002 X-ray photoemission determination of the Schottky barrier height of metal contacts to *n*–GaN and p–GaN. Journal of Applied Physics vol 92 p.6671. https://doi.org/10.1063/1.1518129

[7] Wu Y, Jiang W, Keller B P, Keller S, Kapolnek D, DenBaars S P, Mishra U K and Wilson B 1997 Low Resistance Ohmic Contact to n-GaN With a Separate Layer Method *Solid-State Electron* vol 41 p.165 https://doi.org/10.1016/S0038-1101(96)00151-7

[8] Pang L and Kim K 2015 Improvement of Ohmic contacts to n-type GaN using a Ti/Al multi-layered contact scheme. *Materials Science in Semiconductor Processing* vol 29 pp90-94 https://doi.org/10.1016/j.mssp.2013.10.011

[9] Ruvimov S, Liliental-Weber Z and Washburn J 1996 Microstructure of Ti/Al and Ti/Al/Ni/Au ohmic contacts for n-GaN *Applied Physics Letters* vol 69 chapter11 pp 1556–1558 https://doi.org/10.1063/1.117060

[10] Boturchuk I, Walter T, Julsgaard B et al. 2019 Structure and properties of Ta/Al/Ta and Ti/Al/Ti/Au multilayer metal stacks formed as ohmic contacts on n-GaN *Journal of Material Science: Materials in Electronics* vol 30 p 18144 https://doi.org/10.1007/s10854-019-02167-2

[11]Lin M, Huang F, Morkoc H 1994 Nonalloyed ohmic contacts on GaN using InN/GaN short-period superlattices *Applied Physics Letters* vol 64 chapter 19 p 2557 https://doi.org/10.1063/1.111573

[12] Nemanich R J, Benjamin M C, King S W, Bremser M D, Davis R F, Chen B, Zhang Z and Bernholc Z 1996 Gallium Nitride and Related Materials *In: Mater. Res. Soc. Symp. Proc. (Pittsburgh)* p 395

[13] Chen J and Brewer W D 2015 Ohmic contacts on p-GaN *Advanced Electron Materials* vol 1 p 1500113 https://doi.org/10.1002/aelm.201500113

[14] Ho J K, Jong C S, Chiu C C, Huang C N, Chen C Y and Shih K K 1999 Low-resistantce ohmic contact to *p*-type GaN J. *Applied Physics Letters* vol 74 p1275 https://doi.org/10.1063/1.123546

[15] Song J O, Leem D S, Kwak J, Lee S, Nam O, Park Y and Seong T Y 2004 Low resistance and transparent Ni–La solid solution/Au ohmic contacts to *p*-type GaN *Applied Physics Letters* vol 84 p 1504 https://doi.org/10.1063/1.1652238

IOP Conf. Series: Materials Science and Engineering **738** (2020) 012007 doi:10.1088/1757-899X/738/1/012007

[16] Kumakura K, Makimoto T and Kobayashi N 2000 Efficient Hole Generation above $10^{19}$ $cm^{-3}$ in Mg-Doped InGaN/GaN Superlattices at Room Temperature *Japanese Journal of Applied Physics* Part 2 vol 39 L195 https://doi.org/10.1143/JJAP.39.L195

[17] Adivarahan V, Lunev A, Khan M A, Yang J, Simin G, Shur M S and Gaska R 2001 Very-Low-Specific-Resistance Pd/ Ag/Au/Ti/Au Alloyed Ohmic Contact to p-GaN for High-Current Devices *Applied Physics Letters* vol 78 chapter 18 pp 2781-2783 http://dx.doi.org/10.1063/1.1353813

[18] Liu H Y, Li X, Liu S, Ni X, Wu M, Avrutin V Izyumskaya N Özgür Ü Yankovich A B Kvit A V Voyles P M and Morkoç H 2011 InGaN based light emitting diodes utilizing Ga doped ZnO as a highly transparent contact to p-GaN *Physics Status Solidi C* vol 8 chapter 5 p 1548 https://doi.org/10.1002/pssc.201000860

[19] He L, Yang, D, Ni, G. 2016 Technology for Advanced Focal Plane Arrays of HgCdTe and AlGaN. Springer, Berlin, Heidelberg.

[20] Retrieved on 03. February. 2019, from: http://www.galliumnitrides.com/Improvement-of-Ohmic-contacts-to-n-type-GaN-using-a-TiAl-multi-layered-contact-scheme.html

[21]Retrieved on 20. Dezember. 2019, from: https://pdfs.semanticscholar.org/0d8a/2780f02dfe6fb788c97a5cf0912cd377260d.pdf

IOP Conf. Series: Materials Science and Engineering **738** (2020) 012008 doi:10.1088/1757-899X/738/1/012008

# Power-switching beyond silicon: features of GaN devices

**Yonghan Liu**

Department of electronic engineering, Xidian University, Xi'an, Shaanxi, 710126, China

E-mail: angela@cas-harbour.org

**Abstract.** The generation, transmission, and use of electric power require different forms of power conversion. The commonly used electronic circuits for power conversion utilize switching techniques and, consequently, are referred to as switch-mode power electronic circuits. Two types of semiconductor-based switches are widely used in these circuits: two-terminal rectifiers (diodes) and three-terminal controlled switches (transistors). Currently, many integrated circuit devices use Silicon as the main substrate, because silicon is easy to purify and crystallize. Moreover, it is cheaper than nearly most of other semiconductor material. Silicon becomes main semiconductor material in today's developments of power-switching. However, silicon's energy gap is not large enough, which is the limitation of realizing more functions. With the requirement of new material to break the boundary of material limitation, the third-generation semiconductor rise up. This article will discuss the ideal requirements of power switching and how these requirements were realized by GaN through the method of deriving equations. Meanwhile, comparing with advantages of GaN, this paper will point out the silicon's shortages. Finally, discussing reasons why GaN did not substitute Si then become the main semiconductor material among diodes or transistors.

## 1. Introduction

Many scholars have done researches to explore the excellent qualities of GaN. Comparing to Silicon, the main semiconductor material nowadays, GaN is particularly suitable for integrated circuit devices, especially for power conversion. Generally speaking, a good kind of power conversion ought to have high-blocking voltage, high-power efficiency and high-switching speed. However, people do not put two materials together to compare materials' properties and analyzing why GaN is better than Silicon as a semiconductor material. Based on the purpose above, this essay will compare GaN and Silicon as a power switching material in three points: the blocking voltage, the power efficiency, and the switching speed. By comparing GaN power switching with Si power switching, people will know the reasons why GaN is better than Si as a material of power switching clearly.

## 2. Blocking voltage of GaN devices

The dielectric will lose its dielectric properties and become a conductor under the action of a strong enough electric field, which is called dielectric breakdown, and the corresponding voltage is called breakdown voltage. But for a switch, blocking voltage can be applied on the switch that is OFF mode, which is reverse-bias voltage. Blocking voltage need to interact with the atomic bonds. Nevertheless, blocking voltage relates to the strength of atomic bonds. The value of breakdown voltage also corresponds to planer's structures and to the semiconductor region's doping concentration. This region is drift region for GaN. But how do we get the blocking voltage for a planer structure?

According to one-dimensional Poisson equation

$$V_B = \frac{\varepsilon_S E_{Cr}^2}{2qN_D} \tag{1}$$

We can obtain breakdown voltage for a planar structure. In Equation 1, $V_B$ is the breakdown voltage, $E_{Cr}$ is the critical or breakdown electric field, which is related to the energy gap of the material. $N_D$ is the $n$ -type doping level, $\varepsilon_S$ is the semiconductor permittivity, and $q$ is the value of electron charge. For a specific material, $E_{Cr}$ and $\varepsilon_S$ are constants. Obviously, q is constant ($q = 1.6 \times 10^{-19} C$)also. So that if we want to get the maximum of $V_B$, we need to make $N_D$ as small as it can.

The breakdown electric field of Si is around $E_{Cr}$ =30V/$\mu m$ . And the semiconductor permittivity of Si is about $\varepsilon_S = 11.8 \times 8.85 \times 10^{-12} F\ /\ m$. Assuming the doping level $N_D$=5 × $10^{14} cm^{-3}$. The breakdown voltage is 600V.

As we have discussed, blocking voltage relates to the strength of atomic bonds, which can be indicated by the value of energy gap. The energy gap in Si is 1.12eV. However, the energy gap in GaN is 3.39eV, which means that $E_{Cr}$ of GaN is higher than Si. $E_{Cr}$ for GaN equals around 300V/$\mu m$, which is ten times bigger than $E_{Cr}$ of Si. The semiconductor permittivity for GaN is $\varepsilon_S = 10 \times 8.85 \times 10^{-12} F\ /\ m$. At the same doping level with Si ($N_D$=5 × $10^{14} cm^{-3}$). The blocking voltage of GaN is closer to 50,000V. From the results we can observe the huge gap between the two.

In the field of engineering applications, how do we get the maximum of breakdown voltage when the electric field is a specific value for GaN. According to (1) equation, $\varepsilon_S$ is constant. q is constant ($q = 1.6 \times 10^{-19} C$) too. The smaller the $N_D$ becomes, the bigger the breakdown voltage is.

## 3. Power Efficiency of GaN devices

The best efficiency of energy conversion, of course, is as close to 100 percent as possible. Due to the exist of resistance and current, parts of energy convert into heat then disperse into atmosphere. So, if we want to improve the efficiency of energy conversion, the first thing is to reduce the ratio of energy conversion to heat energy.

Energy losses can be classified into static and dynamic. Dynamic energy losses mean energy losses during charge and discharge of capacitors. The equation of dynamic energy losses is below.

$$P_d = fCV_C^2 \tag{2}$$

In this equation, $f$ means the switching frequency. C is the value of capacity and $V_C$ is the voltage of stray capacitance.

The static energy losses are decided by the voltage when switches are in on mode.

$$P_S = IV \tag{3}$$

According to the equation, the bigger voltage is, the more power converted into heat. In the case of both diodes and transistors, the main source of static power losses is the resistance of the low doped drift region, which is necessary for achieving high-blocking voltages.

IOP Conf. Series: Materials Science and Engineering **738** (2020) 012008 doi:10.1088/1757-899X/738/1/012008

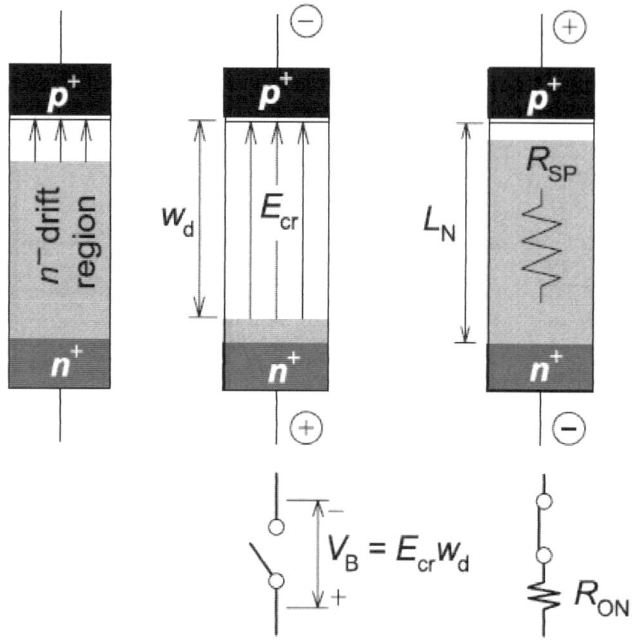


Illustration of the role of the drift region in power rectifiers/ switches and the relationship between the specific ON resistance ($R_{SP}$) and the maximum achievable breakdown voltage ($V_B$).[1] (Left) The fi rst schematic is for an unbiased *p–n* junction, whereas (middle) the second and (right) third schematics correspond to reverse-biased and forward-biased *p–n* junctions, respectively. The second schematic for the reverse-biased *p–n* junction (switch in OFF mode) shows that the reverse-bias voltage depletes the $n^-$ drift region, and the depletion-layer width ($w_d$) and the maximum electric fi eld ($E_{cr}$) determines the maximum voltage that can be applied ($V_B$). The third schematic shows that the $n^-$ drift region introduces a series resistance when the *p–n* junction is forward biased (the resistor $R_{ON}$ in series with the switch in ON mode). Note: $E_{cr}$, critical, or breakdown, electric field; $L_N$, length of the drift region; $R_{ON}$, ON resistance.

Figure 1. The role of the drift region in power rectifiers/switches and the relationship between the specific ON resistance and the maximum achievable breakdown voltage.

Defining specific ON resistance of the drift region as the ON resistance for a given area

$$R_{SP} = \rho L_N = \frac{L_N}{q\mu_n N_D} \tag{4}$$

As can be seen, the length of the drift region ( $L_N$ ) should not be smaller than the width of the depletion layer ( $w_d$ ) when the maximum reverse bias voltage, which is the breakdown voltage, is applied:

$$L_N \approx w_d = \sqrt{\frac{2\varepsilon_s v_B}{qN_D}} \tag{5}$$

In the diodes or transistors, the main resistance is the resistance of drift region (on mode). According to the equation below.

$$R_{sp} = \frac{4V_B^2}{\varepsilon_s \mu_n E_{cr}^3} \tag{6}$$

$V_B$ is breakdown voltage. $\varepsilon_s$ is semiconductor permittivity. $E_{Cr}$ is the breakdown electric field.

The relationship between the maximum blocking voltage and the ON resistance is illustrated in figure 1.

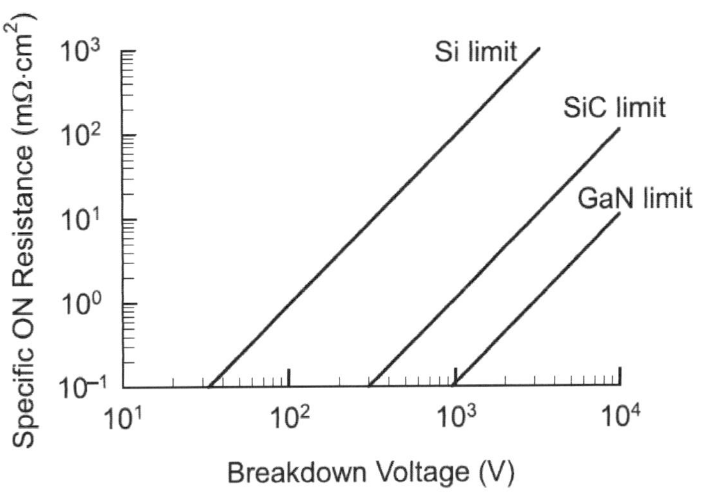


The relationship between the specific ON resistance of the drift region and the maximum achievable breakdown voltage for three different materials. These plots correspond to Equation 5 and the parameter values specified in the text. The plotted results show that, for a given specific ON resistance, the corresponding breakdown voltage is 10 times higher in SiC in comparison to Si, and even higher in the case of GaN. Reprinted with permission from Reference 6. © 2014 IEEE.

Figure 2. The relationship between the specific on resistance of the drift region and the breakdown voltage for Si, SiC, GaN.

Assuming that 20A of current have to be conducted by switch. The active area of device is 1 $cm^2$. For specific resistance of 0.1 $\Omega \cdot cm^2$, which corresponds to the breakdown voltage 1000V of GaN and corresponds to the breakdown voltage around 20V of Si. $R_{sp}$ equals to 0.1 Ω. Voltage quals to 2V, which below the breakdown voltage of GaN and the breakdown voltage of Si. And power loss is 40W. This analysis shows that the breakdown electric field limits the practical blocking voltages that can be achieved by GaN devices to values below 1000 V.

## 4. Switching speed of GaN devices

If a power switching device has a high-switching speed, this device will be widely used in integrate circuit, which avoids long switching delays. Meanwhile, comparing to the normal switch, the size of power switching device is much smaller. Therefore, many spaces in the chip are available. As we all known, spaces in chip are most important. By using these spaces, we can make chip become more and more small. Electric devices can be designed smaller and smaller. The size of a chip means the development of technology.

For using gallium nitride devices, power manufacturers can get the faster switching components. Most inefficiency of power conversion the result of switching losses. So that gallium nitride (GaN) devices have more efficiency advantages than slow MOSFET switching elements. Moreover, Faster switching speeds greatly reduce magnet and the size of capacitive circuit elements.

## 5. The potential of GaN

GaN is another wide energy gap material with a very high breakdown field which makes it attractive

IOP Conf. Series: Materials Science and Engineering **738** (2020) 012008 doi:10.1088/1757-899X/738/1/012008

for the development of power switches. GaN is usually grown on other substrates, most frequently on sapphire but also on SiC wafers. When grown on SiC wafers, the combined material is more expensive than just SiC, and it exhibits a much higher density of defects than SiC. Because of the higher defect density in comparison to SiC and a higher cost in comparison to GaN grown on other substrates, GaN grown on SiC is not used for power electronic devices; its use is for fabrication of blue (and consequently white) light emitting diodes that utilize the wide and direct energy gap of GaN. The very high density of defects does not cause fatal short circuits and recombination centers that would inhibit the light-emitting operation. Encouraged by the fact that these defects are not fatal, power transistors for microwave applications have been developed and commercialized with GaN grown on Si substrates. Although the defect density in this material is even higher than that for GaN grown on sapphire, the cost of this material is significantly reduced by the cheaper Si substrate.

Commercial power transistors using GaN are based on the HEMT structure. One disadvantage of the HEMT structure is that the current flow is lateral, and the power devices require both the source and drain contacts to be created at the surface. This results in less efficient utilization of the surface area in comparison to the vertical MOSFETs using SiC; however, the loss is smaller than the area gain that can be achieved by 100 times more electron mobility. The most important disadvantage of the HEMT structure is that the 2DEG is created by the polarization charge, which means the HEMTs are ON for zero gate voltage—a normally ON device. To turn the switch OFF, a negative gate voltage is applied. This feature appears to be an advantage when HEMTs are used as amplifiers for microwave applications. However, this is a undesirable feature, when HEMTs are used as power switches (the fourth key requirement for a power switch is the normally OFF operation). To resolve this problem, commercial GaN HEMTs are connected with a Si MOSFET in the configuration, which is typically referred to being as a cascode connection. As it can be seen, the Si MOSFET provides the normally OFF operation, whereas the GaN HEMTs provide high-blocking voltage and low ON resistance. The ON resistance of the Si MOSFET is added to the ON resistance of the HEMT; however, the Si MOSFET is a low-voltage device with a small ON resistance because it does not require a drift region.

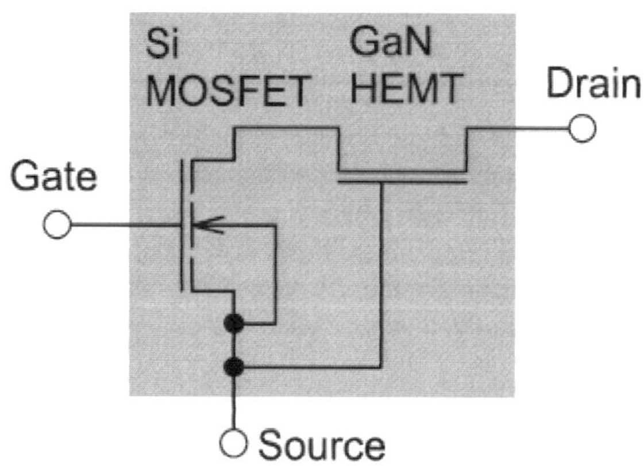


Cascode connection between a Si metal oxide semiconductor field-effect transistor (MOSFET), which provides normally OFF operation, and a GaN high-electron-mobility transistor (HEMT), which provides a high-blocking voltage.

Figure 3. Cascode connection between a Si metal oxide semiconductor field-effect transistor.

The cascode connection between a Si MOSFET and GaN HEMT enables the advantages of GaN HEMTs to be practically utilized; however, it is not the ideal solution. With this potential to significantly reduce the cost of power switches created in GaN, it becomes very attractive to work on the problem of normally OFF conduction and to create a normally OFF HEMT. Many research

IOP Conf. Series: Materials Science and Engineering **738** (2020) 012008 doi:10.1088/1757-899X/738/1/012008

activities are guided by this insight. 36 – 42 Some of the approaches have included the use of a gate material with a larger work function, while others have focused on introducing negative charge in the gate area, and yet another approach is to deposit a $p$ -type doped GaN or AlGaN layer on top of the AlGaN barrier. A gate dielectric between the gate metal and the AlGaN film is used to reduce the gate leakage. The structure created by utilizing a gate dielectric is usually referred to as a MOS-HEMT. If this effort is successful and a normally OFF HEMT or MOS-HEMT 43 – 47 is developed, it has the potential to become the dominant device because of the significant power dissipation and cost advantages.

## 6. Discussion

There are many applications in our live, which use many kinds of semiconductor devices. For instance, many electrical appliances in our life need to consume AC power. But if we obtain energy from solar cells, which is in the form of DC. Therefore, we need to convert DC power from solar cells to AC power that many appliances can use. Meanwhile, before AC power flood into DC circuit, AC-to-DC conversion is also needed. If we need low DC power, but we have high DC power. We need DC-to-DC conversion to convert high power into low power. Indeed AC-to-AC conversion is also needed. Conversing high AC power into low AC power, or low AC power into high AC power. Which devices can be used is determined by which kind of power we need. Based on the comparison I mentioned above, people might have question why GaN do not widely used nowadays?

The semiconductor device made of Si material has good resistance to not too high temperature and radiation. Because of the abundant storage of Si, silicon has become the most widely used semiconductor material. Actually, GaN has quality that limited widely use. Growing gallium nitride films on silicon carbide or sapphire substrates by heteroepitaxy is the main method of using GaN. The lattice matches between GaN and SiC is well, failure rate was only 3.5%. But SiC is expensive. So that SiC cannot be widely used. Sapphire and gallium nitride have a lattice mismatch of 14 %. And sapphire is also cheaper than SiC. But sapphire has high hardness, brittleness, high resistivity, and a much higher price than silicon. Si is cheap. however, the lattice mismatch between the GaN and Si is as high as 17 percent. Therefore, much research is needed to reduce epitaxial defects by using transition layers. Moreover, for a long time, the p-type effective doping concentration of GaN materials is not high, so it is impossible to realize p-type heavy doping. Currently, improving the technological level and finding ways to eliminate the barriers can make GaN devices more widely used than Si devices.

## 7. Conclusion

Through the comparations that are mentioned above, we can clearly know the advantages of GaN as a semiconductor material. However, these comparations just among aspects of properties. In order to realize the widely use of GaN as a material of power switching, we need to consider the price, the qualities of combination with other semiconductor materials and nature storage. Basically, if these obstacles are conquered, GaN will substitute Si soon. Silicon, now we used as main semiconductor material, have abandon sources and well combination qualities with most materials. More explorations are needed for GaN to improve its properties. In my reference, the alloy of GaN and other material might have surprise qualities. GaN is not as plenty as Si, what if the combination of Si and GaN. The combination might have unique properties. They may have complementation.

## Reference

[1] Q. Hua (2016) the material of GaAs and GaN's develop tendency and market applications. J. Integrate circuit application. 31(2): 30-33.
[2] F. Huang (1999) Wide gap semiconductor materials and devices. J. Word electronic components. 26(9): 24-26.
[3] Y. Cheng (2000) Development and application of third-generation semiconductor materials. J. Word products and technology.20(2):50-52.

IOP Conf. Series: Materials Science and Engineering **738** (2020) 012008 doi:10.1088/1757-899X/738/1/012008

[4] D. Li, L. Jia, Z. He, Z. Fan, X. Wang, F. Yang (2004) Research progress of GaN based SBD power devices. J. Micro-nano electronics. 51(5):277-296.
[5] S. Dimitrijev, J. Han, H. A. Moghadam, A. Aminbeidokhti (2015) power-switching applications beyond silicon: Status and future prospects of SiC and GaN devices. J. MRS BULLETIN. 40.10:399-405.
[6] N. Zhang, B. Moran, S. P. DenBaars, U. K. Mishra, X. W. Wang, T. P. Ma (2001) Effects of surface traps on breakdown voltage and switching speed of GaN power switching HEMTs. https://ieeexplore_ieee.xilesou.top/abstract/document/979575/authors#authors

# Comparison between The ultra-wide band gap semiconductor AlGaN and GaN

**Jiajie He**

College of Science, Changchun University of Science and Technology, Changchun, Jilin, 130000, China

Corresponding E-mail: angela@cas-harbour.org

**Abstract.** The emergence of the ultra-wide band gap semiconductor, whose band gaps considerably wider than 3.4 eV (GaN), opens a new era in many fields because of their many superior property. The major material of the UWBG semiconductor is high Al-content AlGaN and the studies find these material have stunning strengths in ultra-high power and ratio-frequency electronics, deep-ultraviolet optoelectronics, quantum information and electronics for extreme-environment applications. In this article, by comparing GaN and ultra-wide band gap AlGaN, people can know the strengths of AlGaN, lacks of GaN and realize why AlGaN will replace GaN in some applications and why GaN will continually be used by people in some applications.

## 1. Introduction

GaN is a popular and widely used semiconductor that can be operated under a temperature over 600 °C [3]. However, when the density of temperature-induced intrinsic carriers in GaN surpasses the concentration of dopant-induced carriers at this temperature, the performance of device will degrade. Compared with GaN, ultra-wide band gap semiconductors like AlGaN have a wider band gap, higher levels of breakdown field, better temperature stability and irradiation robustness, so that they can be operated above 1000 °C and tolerate high radiation. Under this background, this article will introduce the properties of GaN and AlGaN and their performance. By analyzing the limitations of GaN, the reasons why AlGaN will replace GaN in some specific fields are discussed.

## 2. Material

### *2.1. GaN*

Firstly, GaN as the third semiconductor is a hexagonal wurtzite [11] under the barometric pressure and has a stable chemical property even under high temperatures or in the alkaline solution. At the same time, GaN has a small atomic volume and therefore has a good hardness. Secondly, GaN has a low permittivity so it has high RF power level. Thirdly, it has high saturation electron mobility [7] (maximum 3x107 cm / s), so it has good carrier transport capability. Finally, GaN has a band gap of up to 3.4 eV, so its inherent carriers are relatively low at high temperatures and high radiation [10].

However, GaN material has still some drawbacks. For example, the heteroepitaxy defect density is high due to the unsolved substrate single crystal problem. There are lots of drawbacks in GaN epitaxial thin films, including points defects caused by growth technology or doping, buffer layer defects and dislocation caused by substrate mismatch. The residual impurities in epitaxial GaN mainly come from H, O, C and N vacancies, while H forms high resistance in the complex compensation holes formed by

Mg.O is the shallow donor of GaN and affects carrier concentration. Dislocation limits the thickness of nitride materials with high Al component required for deep uv light.

### *2.2. AlGaN*

As a frontier material, AlGaN has a crystal structure and they are also ternary alloy [12]. Just like other ternary alloys, the major exploit of AlGaN is focus on power of hetero structures and band gap engineering [6]. Compared to GaN, firstly, AlGaN own wider band gap up to 6.0 eV while the GaN wide only 3.4 Ev; secondly, its breakdown field higher than 4 MV $cm^{-1}$ of GaN; thirdly, AlGaN has higher saturation velocities than $10^7$ cm $s^{-1}$ ; fourthly, it has a high electron mobility [7] up to1000 $cm^2$ $V^{-1}$ $s^{-1}$ ; fifthly, it's easier for AlGaN to be doped n-type with Si. Above these are standpoints from electronic perspectives. From the optoelectronics perspective, AlGaN can directly access to emission wavelengths shorter than 365 nm into UV-A; -B; -C.In this area, AlGaN is stronger than GaN.However, just only a few AlGaN has been applied in applications and devices because there are two major problems [9].

The first problem is the controlling over doping. Firstly, when the band gap increase, most dopant ionization energies grow up. But if the ionization energies grow up, the fraction of carriers will decrease so under the room temperature, it is difficult for many UWBG devices to reach the special free carrier concentration. Secondly, the wide band gap make themselves incline to self-trapping carriers because the localized nature of valence band state form easily small hole polarons and this occurrence happens easily to self-trapping because of their lattice dislocation. Thirdly, dopants in semiconductor has a tendency called compensation via impurities, native defects and so on. For instance, n-type doping AlGaN by Si, if the proportion of Si increase continually, the free carrier concentration will increase and it will decrease as it reaches the turning point.This phenomenon is called compensation and lower compensation is better. Considerably, AlGaN/AlN own spontaneous and piezoelectric polarization, people can consider new method such as polarization-induced doping, which can generate limited carriers at the hetero interface but because of the deep-level flaws of the wide band-gap layer, this change is relative.

The second problem is the absence of single-crystal used for epitaxial growth and the lack of scientific understanding of controlling heteroepitaxy on such substrates. The high price and low ability of AlN constrain the development of UWBG electronic and optoelectronic devices. Except for high crystalline quality and large diameter, low lattice bow is necessary. Usually people need a large bow radius but it's difficult to obtain one. Traditional method is not used because to obtain AlN boules from melt calls for a high pressure, which is difficult to realize in lab. So now the alternative method is vapor-phase and solution growth approach, which can let the microscopic seed regrowth again and again and finally expand the suitable size. However, the expansion speed of this method is slow, as a result, the production is low, while the wafer is expensive. Therefore, the actual production doesn't take place. Now, some scientists come up with a mixing method such as ammonothermal method, which is validated that it can produce high quality homoexpital HVPE- grown GaN.

## 3. Devices and applications

The high-power electronics is the widest application for UWBG technology because each application controlled by power switch need high power. When the switch is off, it helps these applications block high voltage and when the switch is on, it helps these applications monitor the current. As people demand more sustainable energy, high-power electronics will play more important role in our lives.

Now GaN-on-Si is the major material used in high-power electronic. Current WBG technology can reduce power switching to a low level, on which the heat dissipation and heat sinks will not encumbered by system design, and reduce waste and cost, so that people will focus on the cost and performance of system instead of that of the device. This semiconductor device can realize a stronger doping under the given break voltage, the reduction of electric conduction and switching and the increase of the efficiency of power transformation. For high voltage, the first consideration is vertical SiC and for low voltage, the first consideration is GaN. Compared to traditional semiconductors, UWBG can applied in higher or lower voltage. Then, UWBG semiconductor has a higher FOM's ( $= \frac{1}{4}\varepsilon\mu EC^3$ ) Based on these properties, it will be applied in electronics and thermal fields.

From continuous and pulsed power, especially in some highest power and voltage application in pulsed power, the UWBG semiconductor will have well performance than WBG semiconductor.

As to continuous power, power grid and some applications such as used in the electric- car and ferry can be impacted by UWBG semiconductor. In general, if UWBG semiconductor want to be applied in these application such as mobile and electric-applications in airline and so on, they should be designed as a vertical structure similar to SiC. The design of vertical AlGaN need a processing technology depended on lateral selective area doping [5]. Compared to traditional WBG semiconductor, a high quality gate insulator can form on the UWBG material because of the stable properties. And another application is Pulsed power including electromagnetic rail guns, electric reactive armor, pulsed microwave sources, directed energy weapons, high-power pulsed lasers, and fusion energy. With the remarkable development of semiconductor, more and more scientists start to investigate high-voltage semiconductor. GaN and SiC are limited to dealing with $\approx 10^5$ W. However, the high power applications required $\approx 10^6$ w, the UWBG semiconductor which has lower carrier density,robust radiation resistance, and good thermal stability,will be a factor to accomplish this problem.

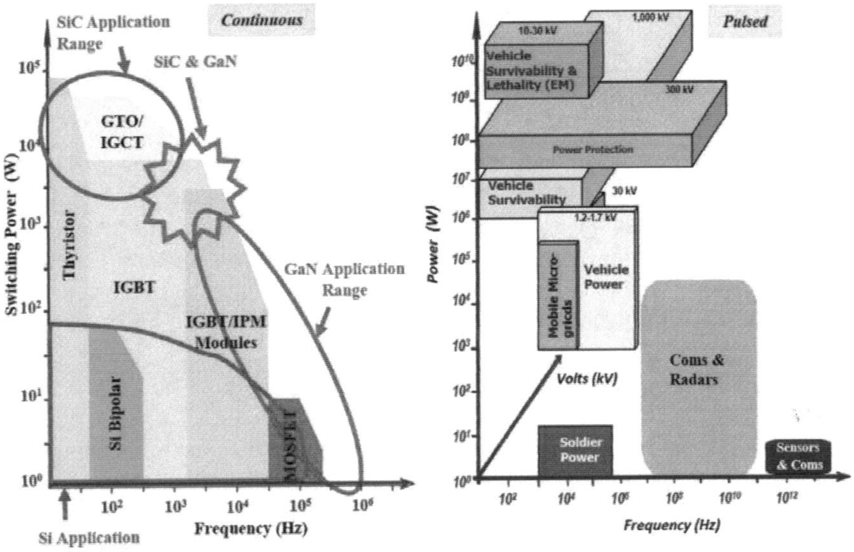


Figure 1. Left: Map of current device families used in continuous power switching applications, organized by switching frequency and switching power and Map of applications for pulsed power devices [2]

In the continuous-power application, UWBG can make power switch work under higher frequency then traditional silicon. Because of this, it allows smaller passive element used in the switch and this method can increase integration. This way can increase dramatically property of power switch and finally UWBG can come up more small foot script, lighter weight and operating under high temperature. These advantages will have a huge impact on transportation, military system and aerospace and other fields. In pulsed power, improvement of passive elements need them to follow the potential system of UWBG but the challenge is how to keep passive element good properties under the high-power switching.

Except for high-power electronics, UWBG are widely used in extreme environment and in this section, The following two parts will be discussed: the surrounding environment is extreme such as the presence of radiation and high temperature or pressure or corrosive chemical. However, under this extreme environment, the sensors and electronic- applications are necessary. For some equipment we have talked about before, WBG and UWBG technology possess better robustness than under the high radiation environment. And this kind of high voltage, huge current and radiation resistance mainly utilized as electronic thrust control actuators of rockets and missiles. WBG and UWBG have robust for high operation, another advantage for space application. Currently, almost all useful commercial GaN power devices are not sufficiently rated for high-voltage power conversion envision by NASA [4]. However, UWBG can replace it because application which UWBG is applied in possess operate in high radiation and high temperature. And UWBG devices may become the important factor in next space generation. The reasons of UWBG has wider band gap which need higher energy of photons or ion clusters to come up ionization of electron- hole in semiconductor. And another reason is UWBG can use material between two different band gaps to form hetero-structure and then form 2DEG area and form channel in HEMT, which is not sensitive to hard radiation [1].

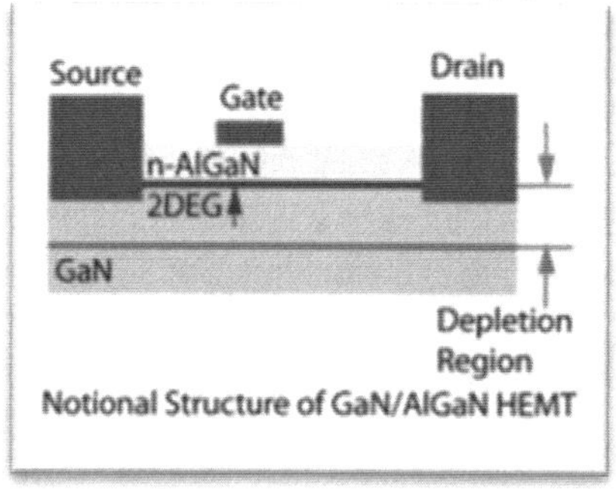


Figure 2. a notional GaN HEMT (right), biased in the off state with 100 V bias on the drain

UWBG material will improve dramatically immunity of HMET transistor to hard radiation and it will get small devices and reduce the volume for collecting radiation-induced charge. Next, for some simple sensors, they can by used in extreme environment, but now we have no devices such as microprocessor and wireless electronics and powering devices can be used in 250°C, even above. Today, acting cooling is used in high temperature for complex electronic devices and in lower temperature, we can not use acting cooling and we can only expend the time of heating. Although, SiC and GaN-based devices can operate under above 600°C but the shortcoming of them is the temperature-induced intrinsic carriers in these materials surpassed the concentration of dopant-induced carriers at these very high temperature (beyond 800°C for SiC and 1000°C), reducing the device performance [3]. Based on the situation, scientists have more incentive to develop UWBG to adept the extreme environment. Using UWBG need to integral electronic-sensor system and there are so many challenges such as material diffusion, formation between metals and temperature gradient and so on. Another major challenge is how to delivery power to system.

## 4. Suggestion

AlGaN is one of the important representatives of the third generation of new wide band gap semiconductor materials, whose band gap width can be continuously adjustable in the range of 3.4eV (GaN) to 6.2eV (AlN). AlGaN has more flexibility than GaN in band gap, people can change percentage of Al content in AlGaN and find a percentage of Al that reflects AlGaN's performance or a performance optimality [8]. With the increase of Al component in AlGaN, the receptor position deepens and the activation energy increases, leading to the decrease of activation efficiency and hole concentration. Scientists can discover the percentage of Al components under different optimal performance conditions. For example, when Al content increase a certain percentage with high break field, high electron mobility [8], this AlGaN can be used in high electronic power.

## 5. Conclusion

Although GaN still dominant the semiconductor field, scientists have found scientists have found some flaws in the application process. These problems concentrate on doped hetero structures [6], and operating temperatures. However, scientists have found that UWBG with a wider band gap and higher breakdown voltage can solve some problems. But so far, only a handful of laboratories are studying

IOP Conf. Series: Materials Science and Engineering **738** (2020) 012009 doi:10.1088/1757-899X/738/1/012009

UWBG, and UWBG is rarely used because ultra-wide band gap semiconductors still face many problems such as high cost and material properties. When these problems are solved, AlGaN will eventually replace GaN in some specific fields.

**Reference**

[1] Patrick H. Carey, Fan Ren, Albert G. Baca, Brianna A. Klein, Andrew A. Allerman, Andrew M. Armstrong, Erica A. Douglas, Robert J. Kaplar, Paul G. Kotula, Stephen J. Pearton, "Extreme Temperature Operation of Ultra-Wide Bandgap AlGaN High Electron Mobility Transistors", Semiconductor Manufacturing IEEE Transactions on, vol. 32, no. 4, pp. 473-477, 2019.

[2] Tsao, J. Y., Chowdhury, S., Hollis, M. A., Jena, D., Johnson, N. M., Jones, K. A., Kaplar, R. J.,et al. (2018) Ultrawide-band gap Semiconductors: Research Opportunities and Challenges. J. Advanced Electronic Materials, 49: 3-5,19-23,35-37.

[3] Beechem, T. E., Mcdonald, A. E., Fuller, E. J., et al. (2016) Size dictated thermal conductivity of GaN. J. Journal of Applied Physics, 120(9): 095104.

[4] Lauenstein, J. M., Casey, M. C., LaBel, K. A., Ikpe, S., Topper, A. D., Wilcox, E. P., Kim, H., Phan, A. M. (2015) Single-Event Effects in Silicon Carbide Power Devices, NEPP Electronics Technology Workshop, Greenbelt, MD, June 23–26, 2015, pp. 1–15

[5] Kimoto, T., Cooper, J. A. (2014). Fundamentals of Silicon Carbide Technology (Growth, Characterization, Devices, and Applications) || Physical Properties of Silicon Carbide. John Wiley & Sons Singapore Pte. Ltd, pp. 11-38.

[6] Bajaj, S., Akyol, F., Krishnamoorthy, S., Zhang, Y., Rajan, S. (2016). Algan channel field effect transistors with graded heterostructure ohmic contacts. J. Applied Physics Letters, 109(13): 133508.

[7] Tˇapajna, M., Hilt, O., Bahat-Treidel, E., et al. (2016) Investigation of gate-diode degradation in normally-off p-GaN/AlGaN/GaN high-electron-mobility transistors. J. Applied Physics Letters, 107(19): 193506.

[8] Wunderer, T., Northrup, J. E., Johnson, N. M. (2016) Technology and Applications, In: Michael, K, Jens, R., (Eds.), III-Nitride Ultra- violet Emitters. Springer, Cham, 227: 193–218.

[9] Baca, A. G., Armstrong, A. M., Allerman, A. A., Douglas, E. A., Sanchez, C. A., King, M. P., et al. (2016). An aln/al\r, 0.85\r, ga\r, 0.15\r, n high electron mobility transistor. J. Applied Physics Letters, 109(3): 033509.

[10] Liao, M., Shen, B., Wang, Z. (2019) Wide Bandgap Semiconductor Materials, Matthew Deans, India pp.1-110

[11] Sze, S. M., Lee, M. K. (2012) Semiconductor Devices Physics and Technology, In: Quad Graphics, Versailles, United States pp.15-22, 223-224

[12] Xu, Q. (2019) MOCVD Growth and P-Substitution of High Al-AlGaN Thin Films. D. University Nanjing, 164: 4-5.

# Dynamics and strain-based design of the large-diameter pipe under the impact of a falling rock

**Bing Shao [1,3], Yifei Yan[1,3], Xiaomei Wang [2,3], Xiangzhen Yan[3]**

[1]College of Mechanical and Electronic Engineering, China University of Petroleum (East China), Qingdao, 266580;

[2]Qingdao energy and Huarun gas Co., Ltd., Qingdao, 266580;

[3]Oil and Gas Technology Research Center, China University of Petroleum (East China), Qingdao, 266580

corresponding author: Bing Shao, upcshaobing@163.com

**Abstract:** Falling rocks are one of the essential causes to pipeline failures in the most frequently occurring geological disasters and the pipes in a thin-shell structure can hardly resist the huge impact energy. The traditional design are conducted by keeping the maximum stress smaller than the material strength which is too conservative to make a full use of the pipeline. In this work, codes for the explicit dynamics based on the Lagrange algorithm are adopted to numerically simulate the impact of falling spherical stones on the large diameter pipe with the consideration of nonlinearity of big deformation. Dynamic responses of the pipe as well as strains on pipe's cross section are discussed in detail. Influences on the pipe strains from factors like impact speed, stone size and moving direction are investigated. Results show that, most significant effect is from the diameter-thickness ratio of the pipe and it shows the resistant capability of the pipe drops exponentially with the decreasing of the thickness and the ratio is recommended to be chosen in a range of 40~60. Eccentricity also effects the strains exponentially and the impact extent is greatly reduced with the eccentricity being smaller than 50%. Safety check is carried out by an ultimate-strain criterion and so, the ultimate size and velocity of the falling stone are got. What is done in this paper is expected to be of referential values for geological disaster resistant design of pipeline.

## 1. Introduction

The damages on the pipelines from the falling rocks in the geological disasters including landslides and debris flows are mainly caused by the transient impacts which can easily result in failure of the pipe. Both laboratory tests and computations have been conducted. Qixin Yang & Shubao Guan adopted a method by a gravity hammer to fall freely onto the soil groove to investigate the influence of the falling stone impulsive force to the gallery structure[1]. By the test, the experimental formula of calculating falling stone impulsive force was given. Di Prisco & Galli conducted a small scale plane-strain text in the laboratory and the lateral response curves of pipe in the varying buried depth under motivations in different directions were obtained[2]. Although many tests have been conducted, imperfections always exist for the laboratory tests. Due to the complex factors in the falling rocks in the real happening geological disasters such as huge impact energy, transient interacting time and complicated interactions, it is hard to be simulated by the laboratory tests. Thereupon, numerical simulations are being adopted more and more widely and of course, it is no exception for the impacts of the falling rocks in the geological disasters. Plassiard & Donze used the discrete element method (DEM) to simulate the impact process of the falling stones on the protective embankment[3]. Xuejing Deng et al. numerically simulated the impacts of falling rocks on the buried pipelines by the DEM software 3DEC and investigated the factors effecting the surface soil pressure and the pipe information[4]. Hongyuan Jing did some research work on the stress responses of the gas pipeline under the impact of falling stones by the ANSYS/DYNA[5]. Min Lou & Haiqin Ming also used the ANSYS/DYNA to simulate the damage of the subsea pipeline under the impact of heavy objects[7].

In the design of oil & gas pipeline, the traditional stress-based standards appear to be conservative compared with the strain-based standards. The deformation resistant capability of the pipe materials is more fully used and meanwhile, the structural stability is remained in the stain-based standards which could be applied in the areas like landslides and frost regions where large deformations may more likely occur. For now, dozens of pipelines have been designed by the strain-based methods and applied in the areas containing potential geological disasters[7-13]. However, focus on the impact of falling rocks on the pipe has not been attracted enough.

## 2. Explicit dynamic FEM Modeling

### *2.1. Lagrange explicit algorithm*

By the principle of the virtual work, the discrete control equations in Lagrange, given as

$$u(x,t) = N(X)u(t) = \sum_I N_I(X)u_I^e(t) \tag{1}$$

$$\varepsilon = \sum_I \frac{\partial N_I}{\partial X}u_I^e, \quad \sigma = [D]\varepsilon \tag{2}$$

$$[D] = \frac{E(1-\mu)}{(1+\mu)(1-2\mu)}[\mu] \tag{3}$$

$$f_e^{\text{int}} = \int_\Omega B^{\mathrm{T}}\sigma \mathrm{d}\Omega \tag{4}$$

$$f_e^{\text{ext}} = \int_\Omega \rho N^{\mathrm{T}} b \mathrm{d}\Omega + \int_{\partial b} N^{\mathrm{T}} t \mathrm{d}s \tag{5}$$

$$M_e = \int_{\Omega} {}_e\rho N^{\mathrm{T}} N \mathrm{d}\Omega \tag{6}$$

$$M\ddot{u} + f^{\text{int}} = f^{\text{ext}} \tag{7}$$

Equation (1) ~ (3) respectively are the node force in the cell, node force out of the cell and cell mass matrix. Equation (4) is the momentum equation where the *u(x, t)* is the trial function of the node displacement, the $N_I(X)$ is the interpolating function, $u_I^e(t)$ is the node displacements. ε is the node strain and σ is the cell stress. $f^{int}$ is the node force in the cell and $f^{ext}$ is the node force out of the cell. *B* is the strain matrix and *b* denotes the volume force. $\int_{\partial b} N^{\mathrm{T}} t \mathrm{d}s$ is the node force caused by the boundary conditions, $M_e$ is the lumped mass matrix and $\ddot{u}$ is the node acceleration.

### *2.2. FEM simulation model*

In this section, we use the explicit dynamic FEM codes to simulate the impact process of the falling rocks on the bare pipe and investigate the large deformation and stress distribution of the pipe. Furtherly, the applicability of the pipe is verified by the utmost strain rule and the factor influences of the falling rock such as speed, size and impacting direction on the resulting strain of the pipe is discussed in detail. In the simulation, the falling rock is simplified to be an ideal sphere in the diameter of 0.6m and the diameter and the wall thickness of the X80 pipe are 1219mm and 22mm, respectively. Other parameters used in simulations are given in table 1. The falling sphere is taken as a rigid body and the rock is falling to the pipe with its centroid moving perpendicularly to the pipe axis. Figure 1 shows the simulation model and grids.

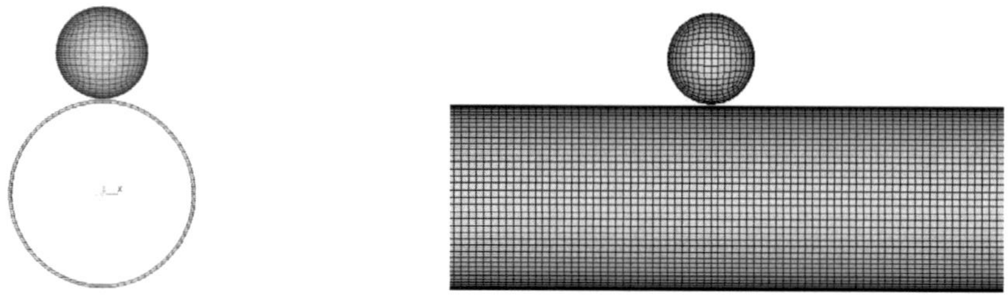

Figure 1. The FEM model of the pipe and the falling rock

Table 1. Material parameters for the FEM model

| parameters | rock | pipe |
|---|---|---|
| density/(kg·m³) | 2000 | 7850 |
| Young's modulus /Pa | $1.0\times10^9$ | $2.06\times10^{11}$ |
| Poisson's ratio | 0.25 | 0.30 |

## 3. Influences of parameters during the impact

In this section, the impacting speed of 10m/s of the rock with diameter of 0.6m are adopted as the initial speed of the falling rock. Five moments in the impacting process are taken and shown in figure 2. From the stresses and deformations in figure 2, the point contact is happening in the initial moment and in the second moment, the stress sharp rises from 0 to 27.2MPa in 3ms with an unconspicuous deformation. However, significant deformation happens to the pipe wall at the impacting point in next moment and the stress spreads. From the fourth and fifth moment, the pipe wall recovers to some extent and the stress goes down responsively, however, permanent plastic deformation has been produced.

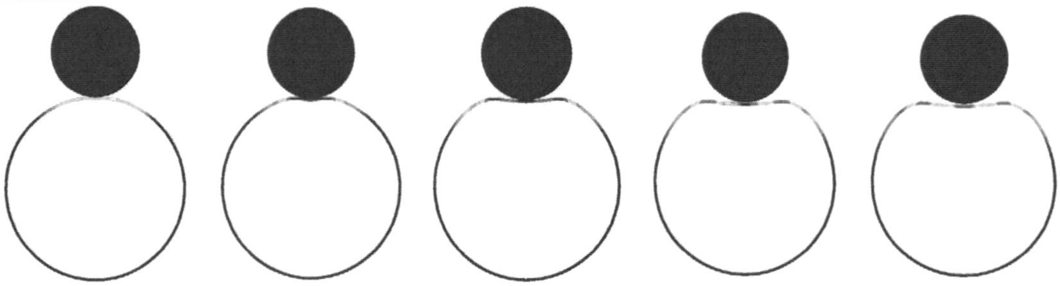

Figure 2. The stress contour and deformation of pipe during the impact process

In this case, the maximum stress of the pipe during the impact does not reach the yield strength, but by the stress-based standard, the minimum safety factor is 1.10, smaller than the design value 1.25, which says the pipe can be no longer competent for service. However, for the strain-based standards, the maximum radial strain is 0.001, the maximum axial strain is 0.00225 and the maximum effective strain is 0.00213, all of which are lower than the permissible value of 0.0054[14].

### *3.1. Ratio between pipe diameter and wall thickness $R_{wt}$*

According to the requirement for $R_{wt}$ for being over 20, values of 20, 40, 60, 80 and 100 are chosen for simulations and the corresponding wall thickness will be 60.95mm, 30.48mm, 20.32mm, 15.24 and 12.196mm respectively. The maximum effective strain is extracted. Figure 3 shows the displacements of the contacting cells along the impacting direction with varying $R_{wt}$, which reflects the subsidence depth of the pipe wall at the impacting point. Maximum subsidence depths are used in figure 4 and the fitting curve is obtained. From figure 3 and 4, it can be easily seen that, with an increasing $R_{wt}$, the subsidence depth rises quickly which means, with a small $R_{wt}$, the pipe owns a better capability to resist the impact of a falling rock. From the curves, $R_{wt}$ is suggested to be chosen during the range of 40~60.

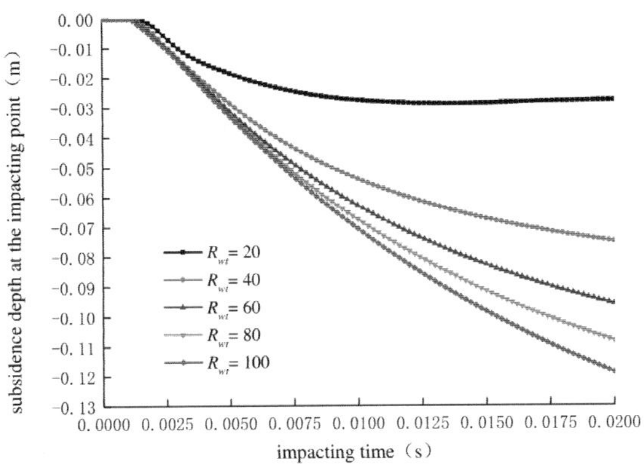


Figure 3. Subsidence depth at the impact point versus impacting time

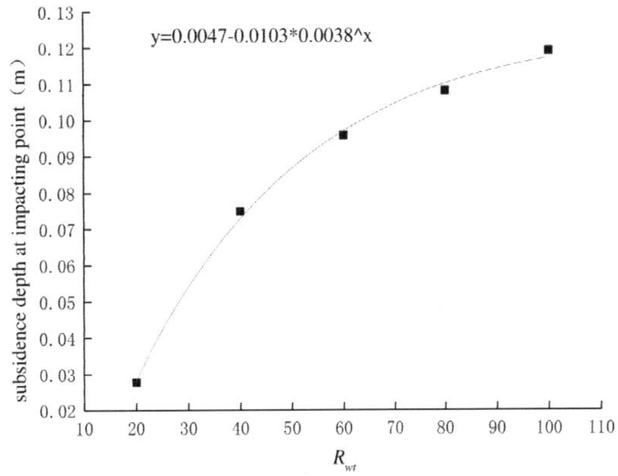


Figure 4. Final subsidence depth at the impact point versus $R_{wt}$

*3.2. Impacting direction*

The impacting of the falling rock can cause different consequences in different directions. In our work, the eccentricity ratio is taken to describe the impact direction which is the ratio of actual eccentricity to the maximum one, given as:

$$H_R = \frac{H}{r+R} \times 100\% \tag{8}$$

where $H$ denotes the eccentricity which is the horizontal projected distance between the centers of the rock and the pipe cross section; is $r$ the radius of the rock and the $R$ denotes radius of the pipe. In this case, five values of 10%、30%、50%、70% and 90% are chosen.

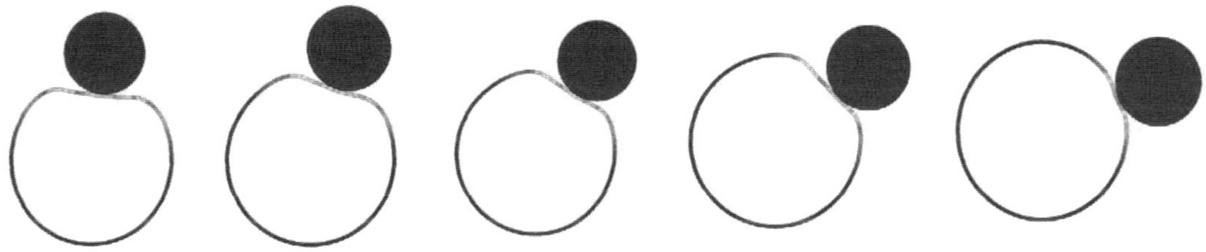


Figure 5. Sectional deformations of the pipe versus impact directions

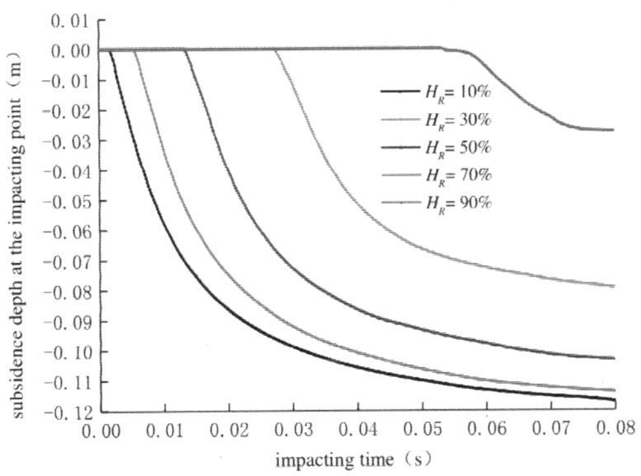


Figure 6. subsidence depth versus time with varying eccentricity

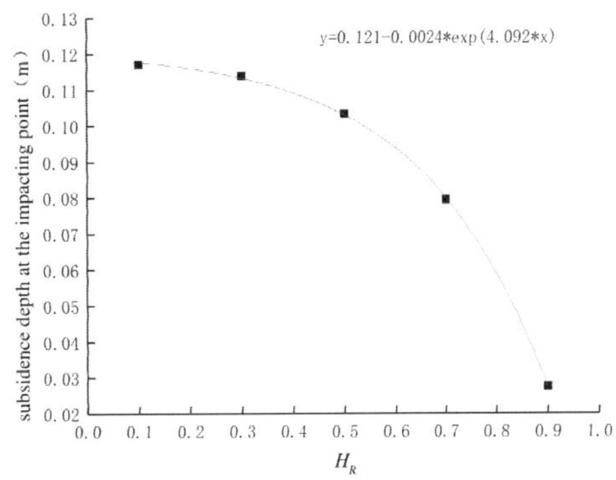


Figure 7. Final subsidence depth with varying versus eccentricity

Figure 5 shows the stress and deformations of final moments of the impacts with different $H_R$. With a bigger $H_R$, which means the rock laterally contacts the pipe, the impact will cause a smaller impact because only partial impacting energy is transferred to the pipe. Figure 6 shows the displacements of the contacting cell under different $H_R$. By extracting the maximum subsidence depth, figure 7 and corresponding fitting curve are obtained. It can be seen that, the subsidence depth decreases exponentially with the increasing $H_R$. The 50% of $H_R$ is more likely to be a watershed because a sharp decrease happens when is $H_R$ smaller than 50%.

### *3.3. The size and speed of the falling rock*

For a pipe with specific $R_{wt}$ and other parameters, there are critical values of the size and speed of the falling rock that the pipe can resist. In this section, the 0.0054 axial strain is chosen to be the permissible maximum value to judge the resistant capability of the pipe to the falling rock. Parameters are same to those in section 2.2 except the varying diameter and falling speed of the rock.

From figure 8 obviously, a faster moving and bigger rock results in a lager axial strain and curves are linear. Extracting the permissible axial strain, figure 9 is got. From figure 9, the pipe can resist a fast impacting speed of the rock when its diameter is smaller than 1.0m. A fitted equation is given in figure 9 and for specific pipe and falling rock, the permissible impacting speed of the rock can be determined by the equation.

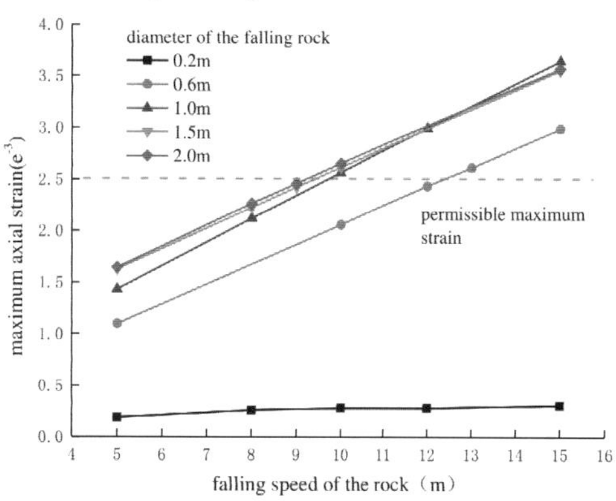


Figure 8. Maximum axial strain of pipe with varying diameter and speed of the falling rock

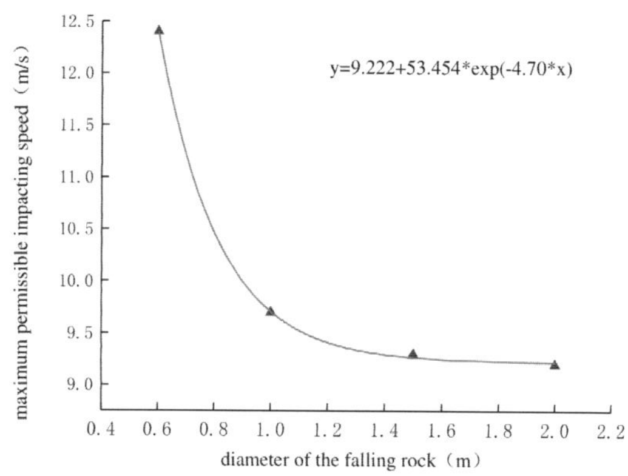


Figure 9. Maximum permissible speed and the diameter of the rock by the strain based design guidelines

## 4. Conclusions

In this work, we discussed the dynamics and parameters effects during the impacting process of the falling rock onto a pipeline. From the strain-based standards, the permissible speed and the diameter of the falling rock can be determined by the fitted equations. The results and corresponding advices may provide some engineering reference.

## Acknowledgments

The authors are grateful to the financial support of China Postdoctoral Science Foundation (grant number 2018M640665), National Natural Science Foundation of China (grant number 51804330 and 51875579).

## References

[1] Yang Qixin, Guan Shubao. Test and research on calculating method of falling stone impulsive force [J]. Journal of the China Railway Society, 1996, 18(1): 101-106.

[2] Di Prisco C, Galli A. Soil-pipe interaction under monotonic and cyclic loads; experimental and numerical modelling. Proceedings of the First Euro Mediterranean Symposium in Advances on Geomaterials and Structures [J]. Hammamet, Tunisia. 2006: 3-5.

IOP Conf. Series: Materials Science and Engineering **738** (2020) 012010 doi:10.1088/1757-899X/738/1/012010

[3] Plassiard J P, Donzé F V. Rockfall impact parameters on embankments: a discrete element method analysis [J]. Structural Engineering International, 2009, 19(3): 333-341.
[4] Deng Xuejing, Xue Shifeng, Tong Xinghua. Numerical simulation on response of buried pipeline induced by rock-fall transverse impaction [J]. Journal of China University of Petroleum, 2009, 33(6): 111-115.
[5] Jing Hongyuan. Analysis and numerical simulation on dynamic response of buried pipeline caused by rockfall impaction [D]. China University of Geosciences, 2007.
[6] Yan Xiangzhen, Zhang Lisong, Yang Xiujuan. Strain response study of oil-gas pipeline crossing earthquake fault based on pipeline-soil coupling and large deformation shell model [J]. China Civil Engineering Journal, 2010, 43(8): 132-139.
[7] Lou Min, Ming Haiqin. Dynamic response analysis of the submarine suspended pipeline impacted by dropped objects based on LS-DYNA [J]. Marine Science Bulletin, 2015, 34(01): 113-120
[8] Liu Bing, Liu Xuejie, Zhang Hong. Pipeline design criteria based on strain and the control factors [J]. Journal of Southwest Petroleum University (Science & Technology Edition), 2008, 30(3): 143-147.
[9] Yu Zhifeng, Shi Hang, Tong Lei, et al. Application of strain-based design method in West-to-East Parallel Gas Transmission Pipline Project [J]. Oil & Gas Storage and Transportation, 2010, 29(2):143-147.
[10] Han Bing, Wang Zhiyin, Wu Zhangzhong, et al. Application of strain-based theory in failure analysis of pipeline subjected to mining collapse areas [J], Journal of China University of Petroleum, 2012, 134-138.
[11] Yi Liqi, Din Keqing, Qian Caifu. Research on calculation method of long-distance pipeline's deformation based on strain [J]. Chinese Journal of solid mechanics, 2011, 32: 310-313.
[12] Liu Bing, Liu Xuejie, Zhang Hong. Pipeline design code based on morphotropy strain [J]. Natural Gas Industry, 28(2): 129-131.
[13] Shao Bing, Yan Xiangzhen, Yang Xiujuan. Nonlinear finite element analysis of dynamic response of suspended submarine pipeline with continuous spans [J]. Science Technology and Engineering, 2012, 12(23):49-53.
[14] DZ/T 0219-2006, Specification of design and construction for landslide stabilization[S].

# China's Energy Consumption and Economic Growth under the Constraints of Carbon Emissions

**Ji chang-lin, Liu yi-ning and Zhang Ya-ning***

Northeast Electric Power University, Jilin, Jilin, PR China

*Zhang Ya-ning (Corresponding author)522707763@qq.com

**Abstract:** This paper was based on the provincial unbalanced panel data of 31 administrative regions in China from 2005 to 2016, investigated the effects of carbon emission constraints and economic growth on China's energy consumption, energy consumption structure and energy policy changes. The empirical results showed that a higher level of economic development would increase regional carbon emissions. Active energy policy and consumption structure adjustment will have a significant inhibitory effect on carbon emissions. This study provided micro-theoretical support for the optimization of regional energy consumption structural adjustment strategy, spatial prediction of emission reduction, and resolving the balance between energy supply and demand. It was helpful for local governments to analyze the energy consumption situation and the overall structural change trend according to the energy consumption structure. It could further tap the energy-saving potential of the region and formulate reasonable long-term development strategies.

## 1. Introduction

Carbon cycle is one of the basic driving forces of global climate and environmental change. With the continuous development of social economy, the exchange scale and rate of matter energy in land surface systems are increasing. The asymmetry development of global has brought about significant negative effects on global climate and environment. In recent years, with the rapid development of China's economy, carbon emissions have attracted wide attention due to huge energy consumption and environmental pressures. In the past two decades, energy standard coal consumption has increased from 1.46 billion tons in 2000 to 3.25 billion tons in 2010, and then to 4.36 billion tons in 2016. The emission reduction strategy has also developed from simple low-carbon and new energy to more specific and instructive capacity removal and gradually adapted to the new normal economy. In the government report of 2017, Premier Li Keqiang clearly proposed expanding the work of cutting overcapacity into the field of coal and electricity, and formulated the energy consumption target of 2017, the coal consumption target and the specific targets of eliminating, suspending and delaying the construction of coal and electricity production capacity. The above-mentioned central decision-making and government actions fully reflect the central government's determination to implement the emission reduction strategy and provide a good institutional environment basis.

## 2. Theoretical Analysis and Literature Review

In the study of the influencing factors of the change of energy consumption structure, Gabriel et al. (2001), Tol (2007) and other scholars built a complete macro-model earlier, and analyzed the influencing factors of the change of energy consumption structure. Nakata (2004) extends the study on the external environmental impact of energy consumption structure, and considers that the adjustment of energy consumption structure should take into account the impacts on economic, population growth and environmental, and points out the importance of renewable energy development for green economy. Li et al. (2012) further assessed the development potential and sources of green energy in a region in two scenarios, and on this basis analyzed the contribution potential of green energy in achieving carbon intensity targets. There are many variables affecting the structure of energy consumption, so far, the academic community has not reached a consistent conclusion. From the direct impact factor analysis, regional population, industrial structure, resident consumption level and other aspects will affect energy consumption demand, indirect factors mainly include domestic or regional gross domestic product, proportion of industrial output value, national or regional gross energy product, total energy import value, energy efficiency and so on. In the further study of the influencing factors, Lin Boqiang et al. (2009, 2010) constructed a co-integration model based on energy consumption and GDP, urbanization level, industrial structure, energy price and energy efficiency, and used scenario analysis method to calculate the energy consumption structure under various scenarios, pointing out the key influencing factors and feasible paths for the development of low-carbon economy. Fan Decheng et al. (2012), Ye Yuyao et al. (2014) and Li Shuang et al. (2015) measured the impact of various external factors on energy consumption structure under the low-carbon economic target from a more macro perspective. As far as research methods are concerned, considering the complex conduction mechanism of energy consumption structure change, its changing law is affected by internal and external factors, and the interaction of various factors may also form a variety of complex influence paths and make the variables appear multiple collinearity. In order to eliminate the impact of collinearity on regression results, Zeng Fanhua (2013), Dai Hongjun (2016) uses ridge regression analysis method to fit, and achieves a more realistic and reliable estimation of morbid data by losing a small amount of information. Generally speaking, the adjustment of energy structure is a systematic project, which needs to take into account the economic development, resource endowment, population status, industrial structure and policy environment of the region in an all-round way.

In summary, scholars have used various methods to study the optimization of energy consumption structure. They have gone through the analysis of influencing factors based on theoretical model to the feasibility demonstration of structural adjustment based on empirical level, from the trend prediction of energy structure change at the policy level to the evaluation of low-carbon effect closer to the people's livelihood. The existing achievements have maturity in both theoretical research and empirical support, and have relatively stable conclusions in many research units, and have the conditions for expanding application in regional socio-economic development.

## 3. Empirical Analysis and Discussion

### *3.1. Research methods and modeling*

In order to improve the reliability of the model, incorporate the total energy, energy consumption structure, industrial structure, GDP, total population and policy variables into the general relationship between energy structure and carbon emissions. Construct a state-space variable parameter model of the impact of energy structure adjustment on carbon emissions as follows:

$$\ln C = a_0 + a_1 \ln toen + a_2 \ln enst + a_3 \ln engdp + a_4 \ln inst + a_5 \ln pop$$
$$+ a_6 \ln gdp + a_7 \ln gdp2 + a_8 \ln gdp3 + a_9 policy + u \quad (1)$$

Formula (1) is the measurement equation of the model, which reflects the relationship between the explained variable and all explanatory variables. Toen represents the total energy consumption, enst represents the adjustment of energy consumption structure, engdp represents the energy intensity of the region, Inst represents the industrial structure, GDP represents the economic development level of the region, pop represents the total population, policy represents the policy variable, u is the random perturbation term. In order to measure the quantitative relationship of percentage changes among variables, all variables (except policy variables) are in the form of natural logarithms. Considering the possible inverted U-shaped effect of economic growth on carbon emissions, incorporate the square term lngdp2 and cubic term lngdp3 of GDP to the numerical value into the formula.

### *3.2. Data Sources and Processing*

This paper chooses the provincial unbalanced panel data from 2005 to 2016 as a sample to study the relationship between energy consumption structure and emission reduction, which removes the administrative regions such as Tibet, Hong Kong and Macao where data are difficult to obtain or lack seriously. The starting date of the sample is 2005, which is due to the significant changes in the world energy pattern in 2005. The frequent rising oil prices have led many countries to adjust their energy policies substantially and adopt energy-saving or expansion of energy development structure to solve the problem of energy prices. On the other hand, in the context of the domestic coal, oil and electricity shortages, the State Council established the National Energy Leading Group in May 2005 as the highest-level deliberative and coordinating body for energy work. As a result, the State Council's "large-ministerial system reform" has merged the nuclear power management functions of the State Development and Reform Commission's Energy Bureau, the State Energy Leading Group and the former National Defense Science and Technology Commission, and the State Energy Administration has been officially listed and put into operation. Therefore, 2005 is an important turning point in China's energy development. After 2005, various regions began to carry out phased energy reform. The data mainly come from China Energy Statistics Yearbook, China Statistics Yearbook and the websites of provincial statistical bureaus. The measurement and data processing of variables are described as follows:

Carbon emissions are the explained variable in this study. The commonly used method is based on the calculation method of Volume II (Energy Volume) of IPCC Guidelines for National Greenhouse Gas Inventories. The formula is as follows:

$$C=\sum C_i=\sum E_i\times NCV_i\times CC_i\times COF_i\times 44/12$$

Among them, i is a positive integer, indicating the type of energy, C represents the total amount of carbon dioxide emissions, E represents energy consumption, NCV is the average low calorific value, CC represents the carbon level per unit calorific value, COF is the carbon oxidation factor, that is, the carbon oxidation rate when energy burns, 44 and 12 respectively represent carbon dioxide and the molecular weight of carbon .

This study mainly focuses on carbon emissions at the national level. China Energy Statistics Yearbook statistics primary energy consumption in four types of energy. The data show that the carbon dioxide emission coefficients of coal, oil, natural gas and water, nuclear and wind power respectively are 2.745, 2.146, 1.629 and 0. Therefore, from the availability and reliability of data, it is easier to use another calculation method to achieve the research purpose. The specific formula of calculation is as follows:

$$C=\sum E\times\left(e_i/E\right)\times c_i$$

In this expression, E represents China's total primary energy consumption (converted into 10,000 tons of standard coal), ei/E is the proportion of energy in category i, and ci is the carbon dioxide emission coefficient of energy in category i.

### *3.3. Empirical test and Result analysis*

Table 1 shows the unit root test results of explanatory variables, explained variables and their corresponding differential variables. The statistical value of the final test is less than the critical value to indicate that the data is stationary, and greater than the critical value to indicate non-stationary. From the test results, it can be found that most of the original explained variables in the unit root test, the test statistic z(t) value is greater than the critical value at any significant level, indicating that the original variable data sequence is not stable, therefore, conduct first-order difference of variables. The results show that the statistic value of the lnC first-order difference variable dlnC is -3.013, which is greater than the critical value of -3.709 at the level of 1% significance and less than the critical value of -2.983 at the level of 5% significance. It shows that the first-order difference variable of lnC is stable at the level of 5% significance. Conduct other variables to similar stationarity tests,except for the lnpop second-order difference variable ddlnpop, other variables are all achieved that the data sequence of first order difference is stable, which means that the second-order difference sequence of all variables is stable.

Table 1 Unit Root Test Results of explanatory variables and explained variables

| variable | statistic z(t) | p-value |
|---|---|---|
| lnC | 0.397 | 0.9723 |
| dlnC | -3.013 | 0.0289 |
| dlntoen | -6.287 | 0.0000 |
| dlnenst | -3.512 | 0.0097 |
| dlninst | -3.133 | 0.0198 |
| ddlnpop | -7.126 | 0.0000 |
| dlnengdp | -5.973 | 0.0000 |
| dlngdp | -2.973 | 0.0683 |
| dlngdp2 | -5.379 | 0.0000 |
| dlngdp3 | -11.637 | 0.0000 |

From the regression results, we can find the following conclusions. After the adjustment of energy structure, the explanatory coefficients of total energy consumption and energy structure variables are all decreasing. Although the range is small, it can be explained to some extent that the supply of energy structure policy at the national level makes the energy structure of different regions converge to a certain extent, and the relationship between carbon emissions and the two variables in different provinces is no longer as significant as before. The coefficient of industrial structure variables is not significant, which indicates that the increase of the proportion of tertiary industry in the region has no significant impact on the decline of carbon emissions. The reason may be that with the development of economy, the economic contribution efficiency of the tertiary industry is increasing. In this process, the carbon emissions of the primary and secondary industries may not necessarily decrease, or even increase, and the tertiary industry itself may also increase regional carbon emissions. For the relationship between energy intensity and carbon emissions, the coefficient is significant at the level of 10%, which shows that better energy intensity can reduce regional carbon emissions to a certain extent. However, after the policy adjustment after 2012, all regions pay more attention to emission reduction, and the emission reduction effect of energy intensity decreases, which is also reflected in the coefficient test results of the policy. Population and economic growth also show a significant positive correlation with carbon emissions, which is consistent with the conclusions of previous studies.

Table 2 Fixed-effect model

| explanatory variable | total sample | sample before adjustment | adjusted sample |
|---|---|---|---|
| lntoen | $-0.613^{***}$ | $-0.627^{***}$ | $-0.497^{***}$ |
| | (-8.732) | (-9.120) | (-6.533) |
| lnenst | $-0.136^{***}$ | $-0.145^{***}$ | $-0.113^{**}$ |
| | (-3.569) | (-3.964) | (-2.071) |
| lninst | -0.012 | -0.013 | -0.012 |
| | (-0.430) | (-0.451) | (-0.427) |
| lnengdp | $-0.137^{*}$ | $-0.142^{*}$ | -0.135 |
| | (-1.671) | (-1.713) | (-1.591) |
| lnpop | $0.259^{**}$ | $0.262^{**}$ | $0.237^{*}$ |
| | (2.017) | (2.237) | (1.933) |
| policy | $0.031^{*}$ | 0.021 | 0.025 |

| | | | |
|---|---|---|---|
| | (1.813) | (1.260) | (1.293) |
| lngdp | 2.132*** | 2.390*** | 2.017*** |
| | (5.337) | (6.189) | (5.033) |
| lngdp2 | 0.667* | 0.531 | 0.445 |
| | (1.732) | (1.370) | (1.113) |
| lngdp3 | 0.009 | (0.009 | 0.007 |
| | (0.023) | (0.019) | (0.018) |
| cons | 0.782*** | 0.749*** | 0.643*** |
| | (4.620) | (4.122) | (3.428) |

Explanation: *, ** and *** respectively indicate that the statistical values of coefficients have passed the significance test at the levels of 10%, 5% and 1%.

## 4. Conclusions and Policy Recommendations

This study examines the impact of energy consumption, energy consumption structure changes and energy policy changes on carbon emission reduction through empirical testing of provincial panel data of 31 administrative regions in China from 2005 to 2016. The empirical results show that the reform of energy policy has achieved remarkable results in the short term, and the cleaner energy consumption structure can significantly affect carbon emissions. After 2012, after the unified energy consumption reform at the national level, this remarkable degree has declined slightly as a whole due to the improvement of areas with poor energy consumption structure before, but it can still fully explain carbon emissions. Therefore, we can conclude that although carbon emissions are affected by many factors, the reform of energy consumption structure can play a significant emission reduction effect.

## References:

[1] Kiran Chand T R, Badarinath K V S, Elvidge C D, et al. Spatial characterization of electrical Electricity consumption patterns over India using temporal DMSP-OLS night-time satellite data[J]. International Journal of Remote Sensing, 2009, 30(3): 647-661.
[2] Amaral S, Câmara G, Monteiro A M V, et al. Estimating population and energy consumption in Brazilian Amazonia using DMSP night-time satellite data[J]. Computers, Environment and Urban Systems, 2005, 29(2): 179-195.
[3] Amaral S, Monteiro A M V, Câmara G, et al. DMSP/OLS night-time light imagery for urban population estimates in the Brazilian Amazon[J]. International Journal of Remote Sensing, 2006, 27(05): 855-870.
[4] Xueliang Yuan,Jian Zuo. Transition to low carbon energy policies in China: From the five-year plan perspective [J]. Energy Policy, 2011, 39: 3855-3859.
[5] Cools M, Brijs K, Tormans H, et al. Optimizing the Implementation of Policy Measures Through Social Acceptance Segmentation[J]. Transport Policy, 2012, 22: 80-87.

IOP Conf. Series: Materials Science and Engineering **738** (2020) 012012 doi:10.1088/1757-899X/738/1/012012

# Development of $Fe_3O_4$-modified-carboxylic MWCNTs for extraction and determination of pigments in beverage

**Zufei Feng*, Sijin Li, Yan Lu, Yingjuan Zhao and Xinyue Zhai**

Department of Applied Chemistry, School of Science, Xi'an University of Technology, Xi'an 710061, P.R. China

Email: zufeifeng@xaut.edu.cn

**Abstract.** A nanomaterial comprising $Fe_3O_4$-modified-carboxylic multi-walled carbon nanotubes ($Fe_3O_4$-MWCNTs-COOH) was prepared by a co-precipitation method. Combined with HPLC-photodiode array detector (DAD), $Fe_3O_4$-MWCNTs-COOH was used to determine pigments in beverage. Some important parameters that could influence extraction efficiency were optimized. The results indicated that $Fe_3O_4$-MWCNTs-COOH combined with HPLC-DAD is a promising solid-phase extraction material for the sample pretreatment in the determination of pigments.

## 1. Introduction

Edible pigment is a broad category of the most common food additives, which can improve the sense of food and stimulate consumers' purchasing desire. Many foods suffer from oxidation, fading and other phenomena in the process of manufacturing, transportation or storage, which seriously affecting their appearances. Therefore, it is essential to add the pigments to regulate and control this problem [1]. Many countries have made corresponding regulations and strict provisions on the amount and type of synthetic pigments added to food [2]. Up to now, many analytical techniques have been developed for the determination of the edible pigments, such as high performance liquid chromatography (HPLC) [3], SPE combined UV-visible spectrophotometry [4], liquid chromatography/tandem mass spectrometry (LC-MS) [5] and so on.

In this study, $Fe_3O_4$ magnetic nanoparticles (MNPs) were prepared by the classical liquid phase synthesis-chemical precipitation method, and the carboxyl modified multi-walled carbon nanotubes (MWCNTs) were magnetically functionalized with the MNPs [6]. The obtained magnetic material was used to adsorb pigment molecules in the beverage samples. After desorption, HPLC-PAD was used for qualitative and quantitative analysis of the sample solution.

## 2. Experimental methods

### *2.1. Reagents and materials*

Ferric chloride was purchased from Tianjin No.3 Chemical Reagent Factory (Tianjin, China). Ferrous chloride was purchased from Fuchen Chemical Reagents Factory (Tianjin, China). Ammonium hydroxide was provided by Xi`an SanPu Chemical Reagent Co., Ltd. (Shaanxi, China). Acetic acid was bought from Tianjin Beilian Fine Chemicals Development Co., Ltd. (Tianjin, China). All the above reagents are of analytical grade purity. Acetonitrile and methanol (HPLC grade) were supplied by Guanghua SciTech Co., Ltd. (Guangzhou, China). Carboxylated MWCNTs (*o.d.*, 30-50 nm; -COOH content, 1.06 wt%; length, ~20 μm; purity >95%) were purchased from Chengdu Organic

Chemical Co., Ltd. (Chengdu, China). The beverage (orange Fenda) was obtained from Shaanxi Taigu Coca Cola Beverage Co., Ltd. (Shaanxi, China). Ultrapure water was prepared by an ultra-purification system. Food additives Sunset yellow and Tartrazine were from Shanghai Dyestuff Research Institute Co., Ltd. (Shanghai, China).

### *2.2. Instrumentation*

The HPLC analysis was performed in a WATERS Series (WATERS Technologies, Milford, MA) LC system, which was equipped with an e2695 Alliance Quaternary Pump, a 2998 Photodiode-Array Detector (PAD), an Alliance Col Heater column oven and an automatic sampler. The system was controlled by an Empower 2 Personal Single System. Chromatographic separation was performed with an Innvoal ODS-2C18 reverse separation column (5 μm, 250 mm × 4.6 mm) (Tianjin Agela Technologies Co., Ltd., Tianjin, China). Transmission electron microscopy (TEM) images were taken on a Hitachi H-7650 transmission electron microscope (Hitachi, Japan). The PALL ultrapure water systems were supplied by ELGA Lab Water Instrument Co., Ltd. (Bucks, UK).

### *2.3. Preparation of $Fe_3O_4$ Nanoparticles*

Firstly, 0.1257 g ferric chloride hexahydrate and 0.0470 g ferrous chloride tetrahydrate were accurately weighed into a 250 mL three-necked flask containing 80 mL of deoxygenated water followed by continuously stirring to dissolution. When the temperature of the solution reached 80°C, dropwise added 60 ml of 5% ammonium hydroxide with intensive stirring for 60 min. The oxygen must be prevented from the entering system during the whole reaction process. After the reaction was completed, $Fe_3O_4$ deposition was separated by an external magnetic field and washed with deoxygenated water until the pH value was nearly neutral. After static settlement, the supernatant was decanted. The target product $Fe_3O_4$ nanoparticles were obtained after dried at room temperature. The optimized chemical coprecipitation method was used to prepare magnetic multi-walled carbon nanotubes.

### *2.4. Preparation of Magnetic Multi-walled Carbon Nanotubes*

Firstly, 0.1257 g ferric chloride hexahydrate and 0.0470 g ferrous chloride tetrahydrate were added into a 250 mL three-necked flask containing 80 mL of deoxygenated water, and were continuously stirred to dissolution. Then 0.2041 g MWCNTs-COOH was added to the above solution. When the temperature of the solution reached 80°C, dropwise added 60 ml of 5% ammonium hydroxide with intensive stirring for 60 min. Oxygen must be prevented from entering the system during the whole reaction process. After the reaction was completed, the sediment $Fe_3O_4$-MWCNTs-COOH was separated by an external magnetic field, washed with deoxygenated water, and the pH value was detected to be about 7 after static duration. By this time, the supernatant was poured out, dried and ground at normal temperature to obtain the magnetically functionalized multi-walled carbon nanotube.

### *2.5. Experimental Process*

Pigment extraction was completed in four steps: extraction, separation, desorption and analysis.

*Extraction* Firstly, 20 mg of magnetic MWCNTs-COOH solid powder were accurately measured as the adsorbent, and then 4 mL of beverage sample solution were added and shaken for about 30 s. The magnetic carbon nanotube particles were uniformly dispersed in the solvent. After being shaken gently for 25 min and then kept undisturbed for a period of time to achieve the balance of target analyte between magnetic multi-walled carbon nanotubes and the sample solution.

*Separation and Desorption* A strong magnet was placed at the bottom of the micro centrifuge tube. Then magnetic MWCNTs-COOH and the sample solution were separated by an external magnetic field, and the supernatant was sucked off by a syringe. A proper amount of desorbent was added into the micro centrifuge tube for desorption. Then the desorption solution was collected.

*HPLC anylysis* The conditions of HPLC were set as follows: column temperature 30°C, mobile phase (methanol: water = 40:60), flow rate 1.0 $ml/min^{-1}$, wavelength 450 nm, injected volume 20 μL.

## 3. Results and discussions

### *3.1. TEM characterization*

Transmission electron microscope (TEM) was used to characterize $Fe_3O_4$ nanoparticles, carboxylated multi-walled carbon nanotubes and carboxylated multi-walled carbon nanotubes modified by magnetic nanoparticles. Figure 1a showed that the $Fe_3O_4$ nanoparticles were spherical particles. It was observed that the carboxylated multi-walled carbon nanotubes were tubular substances with small diameter and long length, as shown in Figure 2b and 2c. The particle size distribution of $Fe_3O_4$ nanoparticles could be observed between 10 nm and 15 nm.

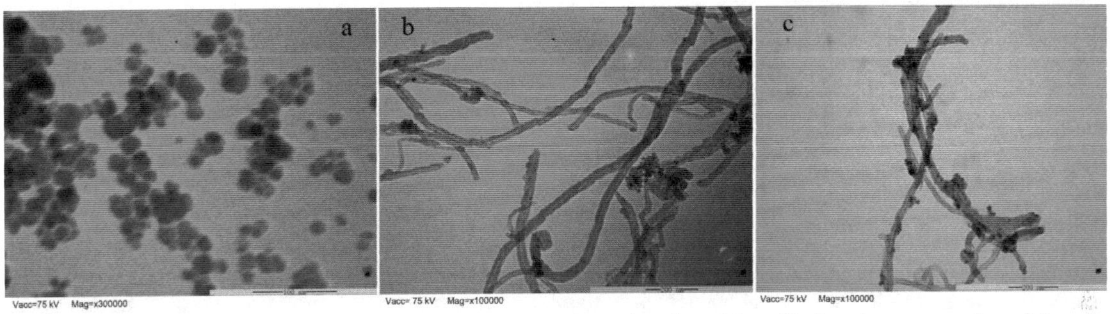


Figure 1. TEM image of $Fe_3O_4$ nanoparticles (a), carboxylated multi-walled carbon nanotubes (b) and carboxylated multi-walled carbon nanotubes modified by magnetic nanoparticles (c).

### *3.2. VSM characterization*

The VSM hysteresis loops of pure $Fe_3O_4$ nanoparticles and magnetic carboxylated multi-walled carbon nanotubes were illustrated in Figure 2. The maximum saturation magnetization (Ms) of $Fe_3O_4$ nanoparticles and magnetic carboxylated multi-walled carbon nanotubes were 67.5 emu/g and 7.5 emu/g respectively, which proved that the carboxylated carbon nanotubes modified with $Fe_3O_4$ nanoparticles had certain paramagnetism and could be separated quickly by the external magnetic field. The maximum saturation magnetization of magnetic multi-walled carbon nanotubes was much lower than that of pure $Fe_3O_4$ nanoparticles, because the content of $Fe_3O_4$ nanoparticles attached to the surface of the modified multi-walled carbon nanotubes was very few. Consequently, this might reduce the magnetic characteristics of $Fe_3O_4$, but do not affect the magnetic utilization of the whole compound.

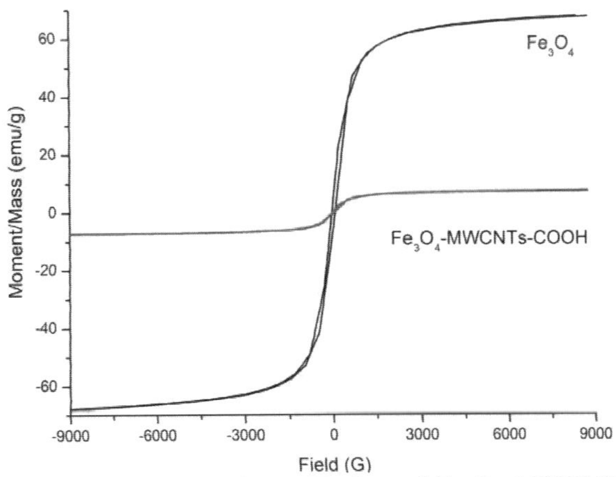


Figure 2. VSM magnetization curves of pure $Fe_3O_4$ and $Fe_3O_4$-MWCNTs-COOH sorbents.

### *3.3. Optimization of adsorption condition*

In order to optimize the conditions for extraction of pigment from beverages by the magnetic material, this study investigated the extraction efficiency under different conditions.

*Effect of adsorbent dosage* The amount of magnetic MWCNTs would have a greater impact on the extraction effect. Carboxylic radicals which generated after ionization of carboxylated MWCNTs could enhance the adsorption with pigment molecules in beverages by forming hydrogen bonds. Figure 3a showed the adsorption effect when the different amounts of adsorbent were used. When 10 mg adsorbent was used, the adsorption effect was the best.

*Effect of extraction time* Theoretically, the longer the adsorption time was, the more pigments would be obtained during the final desorption. However, the best adsorption time was 25 min (Figure 3b). It was likely because that after reaching the adsorption equilibrium, part of the adsorbed pigments were redissolve into the solution. Or the adsorbent did not have specificity for this substance. With the increase of adsorption time, other substances were adsorbed in large quantities, resulting in a decrease in the extraction content of the pigment.

*Effect of desorption agent types* After the adsorption, four commonly used organic solvents-methanol, acetonitrile, methanol/acetic acid (9:1) and acetonitrile/formic acid (9:1) were selected as the desorption agents. As shown in Figure 3c, methanol had the best desorption effect.

*Effect of desorption agent dosage, time and number of desorption times* Four dosages were investigated for the same desorbent. The result showed the amount of 2 mL desorbent had the best desorption effect (Figure 3d). Then, the desorption time was evaluated. It was clearly observed that the peak area was the largest when the desorption time was 20 min (Figure 3e). However, with the increase of desorption time, some of the desorbed pigments would be adsorbed again, which would lead to a decrease of adsorbed pigment contents. Three levels of desorption times were optimized. The result showed that almost all the pigments adsorbed on the magnetic MWCNTs can be desorbed at the first time (Figure 3f).

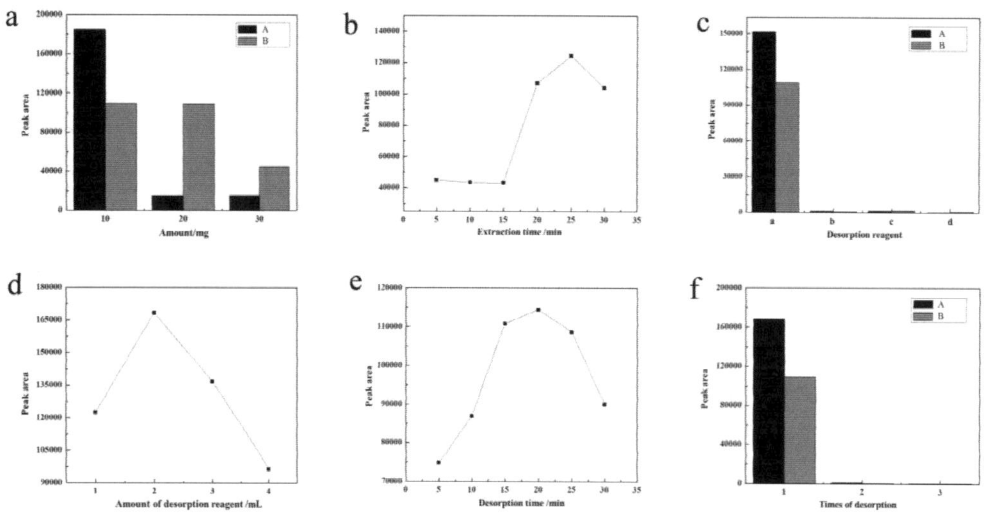


Figure 3. The optimization of extraction process. A-Tartrazine; B-Sunset yellow. (a) $Fe_3O_4$-MWCNTs-COOH amount; (b) extraction time, (c) desorption reagent (a-methanol, b-acetonitrile, c-methanol/acetic acid (9:1) and d-acetonitrile/formic acid (9:1)); (d) amount of desorption solvent; (e) effect of desorption time; (f) number of desorption times.

*Determination of pigment content in beverages* Under the optimal conditions, a series of analytical parameters were optimized, including linear range and correlation coefficients. The linear range of the calibration curve was 0-30 μg/mL both for Tartrazine and Sunset yellow; the correlation coefficients ($r^2$) were 0.9922 and 0.9190, respectively; the regression equation of Sunset yellow is $y$=4615$x$+12985; the curve regression equation of Tartrazine is $y$=13010$x$+29754. The results indicated that the developed analytical method was reliable.

The national standard of China (GB 2760-2007, "Hygienic Standard for the Use of Food Additives") stipulates that the maximum usage limits of Tartrazine and Sunset yellow in food are 0.05 and 0.025 g/kg [7], respectively. The experimental result (Figure 4) showed that, the contents of two synthetic pigments in Fenda beverage are 0.0973 and 0.022 mg/kg, respectively, which are not beyond the national standard.

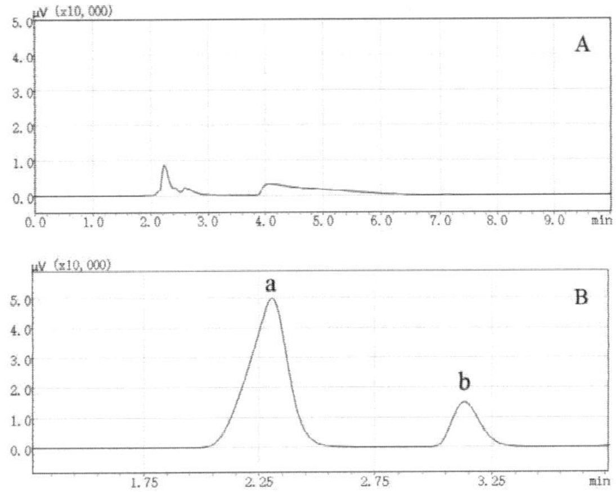


Figure 4. HPLC chromatogram of the extraction process: A-beverage sample without extraction; B-injected sample solution after extraction and desoroption. a-Tartrazine; b-Sunset yellow.

## 4. Conclusion

In the present study, $Fe_3O_4$-MWCNTs-COOH was successfully developed as an ideal magnetic nanomaterial for solid phase extraction in beverage analysis. Under the optimized condition, the sorbents exhibited satisfactory affinity to target analyte. This proposed method also has several advantages, such as rapidity, simplicity, ease-of-use, and environmentally friendly. Therefore, $Fe_3O_4$-MWCNTs-COOH sorbents are potential materials for solid phase extraction in the sample pretreatment of pigment analysis in beverage sample.

## Reference

[1] Zheng Q and Kuang H M 2004 *Beverage Industry* (Chinese journal) 7 (04) 16-19
[2] Li J Y, Wang H B and Lin Z H 2009 *Chinese Horticulture Abstracts* (Chinese journal) 10 165-167+178
[3] Zor S D, Asci B and Donmez O A 2016 *J. Chromatogra. Sci.* 54 (6) 952-957
[4] Bisgin A T 2018 *J. AOAC Int.* 101 (6) 1850-56
[5] Tsai C F, Kuo C H and Shih D Y C 2015 *J. Food & Drug Anal.* 23 (3) 453-462
[6] Meng X X and Gao M L 2002 *J. Binzhou Teachers College* (Chinese journal) 18 (04) 69-72
[7] GB2760-2007 2007 *Hygienic Standards for Uses of Food Additives* Standards Press of China Beijing

# Ubiquitin binding domains – from structure to application

**Ruofan Yang**[1]

[1]College of Chemistry and Chemical Engineering, Xiamen University, Xiamen, Fujian, 361000, People's Republic of China

**Abstract:** Ubiquitin is one of the most important signal molecular in biological process and ubiquitination plays essential roles in many signal transduction pathways. In most cases, they function as a post transcription modification on target protein as a monomer or a chain, which could recruit other protein with ubiquitin binding domains (UBDs) to enable signal transduction. The UBDs are variant on structure and recognition site on ubiquitin. For the variant function the UBDs gives, UBDs are good tools pool for material design such as ubiquitin pathway inhibitor, ubiquitin enrichment material and manufactural specific ubiquitin chain antibody. In this review, we summarize some recent work on UBDs characterization and application.

## 1. UBIQUITIN AND UBIQUITINATION

Ubiquitin consists of 76-amino-acid residues, which is highly conserved from yeast to man. Reversible PTM on proteins play important role in signal sensing and intracellular signal transduction. The ubiquitination of target proteins is a widespread and conserved regulatory post-translational modification in eukaryotes during protein degradation, signal transduction and DNA repair pathways[1]. Conjugation of ubiquitin to a target protein as a PTM or to itself resulting in polyubiquitin chains is regulated by activity of ubiquitin-activating (E1), ubiquitin-conjugating (E2) and ubiquitin-ligating (E3) enzymes. The addition of single ubiquitin or ubiquitin chains can alter protein activity and its localization such as target to 26 S proteasome[2]. So far, many cellular progresses are known to involve ubiquitination, including DNA repair[3], protein degradation[4], endocytosis[5, 6], vesicular trafficking and gene silencing. In these processes, ubiquitin acts as a reversible signaling tag that can trigger biological events in cells by recruiting ubiquitin-binding domains (UBDs) in cellular proteins.

So far, most UBDs characterized well interact with ubiquitin through three surfaces on ubiquitin including hydrophobic patch including and surrounding isoleucine 44 (Ile44), acidic hydrophilic Asp58-centred face (Asp58) and flexible glycine tail (Gly76)[7] as figure 1 shows. Helical UBDs share a common attraction to the same binding surface on the ubiquitin moiety, formed by the hydrophobic patch including and surrounding isoleucine 44 (Ile44). In contrast, some of ubiquitin binding zinc fingers (ZNFs)[8], such as the A20-ZnF and the ZnF-UBP, prefer binding to acidic hydrophilic Asp58-centred face[9]. In this review, we look into molecular basis of ubiquitin recognition by UBDs mainly through the structure properties of different UBDs and comprehend how different types of UBDs recognize corresponding ubiquitin or polyubiquitin chains as well as their roles in cellular signaling transduction.

| | | Binding affinity | | Structure | | |
|---|---|---|---|---|---|---|
| Ub-binding domain | Source protein | $K_d$ ($\mu$M) | Method | PDB ID | Method (resolution, Å) | Note |
| UBA | Dsk2 | 14.8 ± 5.3 | SPR | 1WR1 | NMR | |
| | hHR23A | 400 ± 100 (mono-Ub) | NMR | 1Z06 | NMR | UBA2–two Ub |
| | Mud1 | 390 ± 50 (mono-Ub) | SPR | 1Z96 | X-ray (1.8) | PDB for UBA alone |
| | Ede1 | 83 ± 9 | NMR | 2G3Q | NMR | |
| CUE | Vps9 | 20 ± 1 | ITC | 1P3Q | X-ray (1.7) | Dimeric CUE |
| | Cue2 | 155 ± 9 | NMR | 1OTR | NMR | Monomeric CUE |
| GAT | GGA3 | 181 ± 39 | ITC | 1YD8 | X-ray (2.8) | Two Ub-binding sites; Ub binds to site 1 in the crystal structure |
| | GGA3 | | | 1WR6 | X-ray (2.6) | |
| | TOM1 | 409 ± 13 | SPR | 1WRD | X-ray (1.75) | |
| UEV | Vps23 | | | 1UZX | X-ray (1.85) | |
| | Tsg101 | 510 ± 35 | SPR | 1S1Q | X-ray (2.0) | |
| Ubc | UbcH5 | ~300 | NMR | 2FUH | NMR | |
| UIM | Vps27 | 277 ± 8 (UIM1)<br>177 ± 17 (UIM2) | NMR | 1Q0W | NMR | UIM1–Ub |
| | Vps27 | 246 ± 1 (UIM1)<br>1690 ± 40 (UIM2) | SPR | 1O06 | X-ray (1.45) | UIM2 only; no Ub |
| | S5a | ~350 (UIM1) | NMR | 1YX5 | NMR | |
| | S5a | 73 (UIM2) | NMR | 1YX6 | NMR | |
| DUIM | Hrs<br>491 (site 1)<br>543 (site 2) | 190 (wt) | SPR | 2D3G | X-ray | Two Ub-binding sites |
| MIU | Rabex-5 | 29 ± 4.8 (Y25A, SPR)<br>29 ± 1 (Y26A, ITC) | SPR, ITC | 2FID, 2FIF | X-ray (2.5) | |
| | | 28.7 | ITC | 2C7N | X-ray (2.1) | |
| NZF | Npl4 | 126 ± 26 | SPR | 1Q5W | NMR | |
| A20 ZnF | Rabex-5 | 22 ± 0.4 (A58D, SPR)<br>21 ± 1 (A58D, ITC) | SPR, ITC | 2FID, 2FIF | X-ray (2.5) | |
| | Rabex-5 | 12 | ITC | 2C7N | X-ray (2.1) | |
| ZnF UBP | Isopeptidase T | 2.8 | ITC | 2G45 | X-ray (2.0) | |

Figure 1. complex structure and affinities of different UBDs[32]

## 2. UBDs interact with ubiquitin hydrophobic patch including Ile44

The interaction of UBDs and ubiquitin hydrophobic patch is based on hydrophobic interaction. The hydrophobic α-helical regions of UBDs enable their interaction with ubiquitin hydrophobic face. The UBDs fold into α-helical based structures constituting the largest class of ubiquitin-binding domains including UBA [10](ubiquitin associated), UIM (ubiquitin-interacting motif), MIU (motifs interacting with ubiquitin), CUE (coupling of ubiquitin conjugation to endoplasmic reticulum). All of them above are known to interact with the single hydrophobic region on ubiquitin including Ile44. Both UBA and CUE domains contains a common three-helical bundle architecture thus they bind to the Ile44 patch in similar pattern. The UIM and MIU domain is also helical and interact with hydrophobic patch.

### *2.1 UBA*

The UBA domain was the first reported and well-characterized ubiquitin-binding domain from Rad23 and R23A, Dsk2[11] and NBR1 which are key protein in proteasome targeting, kinase regulation and autophagy process[12]. UBA domains were discovered as a homology region in many proteins involved in ubiquitination cascades[13]. UBA domains consist of three compact helix bundles. Two of the helical, α1 and α3 helices, form a conserved hydrophobic patch which is able to bind hydrophobic Ile44 patch on mono-ubiquitin[14] as figure 2 shows. UBA domains are able to bind mono-ubiquitin in vitro with affinity in mM magnitude. What's more, they have been found to play a role in a variety of other protein–protein interactions. Through some orientation variation of the UBA domain relative to ubiquitin exist in some reports, but the identity of the binding sites on both allows of no dispute.

Besides, the UBA domains is essential in polyubiquitin recognition[13] because its ability to interact with two ubiquitin hydrophobic Ile44 faces simultaneously through two opposite hydrophobic faces formed by three bundle helix[15]. The UBA domains of hHR23A[16] is a good example to illustrate how two surface patches on the UBD allow it to be sandwiched into the hydrophobic binding cleft formed by Lys48-diubiquitin using the Ile44 faces of both proximal and distal ubiquitin moieties. NMR studies of the hHR23A UBA domain bound to Lys48-linked di-ubiquitin in the centre of a 'sandwich' between the two mono-ubiquitin moieties. More interesting, some UBA domains show preference to the ubiquitin chain in specific linkage form[17]. Taking hHR23A, A human Rad23 orthologue, for example, hHR23A effectively bind to chains containing up to eight ubiquitin molecules with its C-terminal UBA domain sandwiching between the ubiquitin moieties of Lys48-linked diubiquitin[17-19], contacting closely with the linker peptide bonds connecting two moieties. These facts give reason to the ability of R23A to inhibit substrate deubiquitylation thus influence binding and activation of Uch37 by Rpn13. They probably make Rad23 proteins better shuttling factors for passing substrates to the proteasome because substrates are more likely to remain ubiquitylated during transport. Cecile Pickart group pioneered studies on interaction between 30 different UBA domains and polyubiquitin chains in different types of linkage and concludes that the UBA domains can be sorted into four major specificity classes. The class 1 selectively binds to Lys48-linked polyubiquitin including two known members, hHR23A UBA and Mud1 UBA while class 2 prefers Lys63- linked polyubiquitin[18]. In contrast, class 3 and class 4 show no selectivity on polyubiquitin linkages. However, with the fact that this classification does not show patterns of sequence conservation and most of the discrimination mechanistic basis remains to be elicited, this result must be considered provisional.

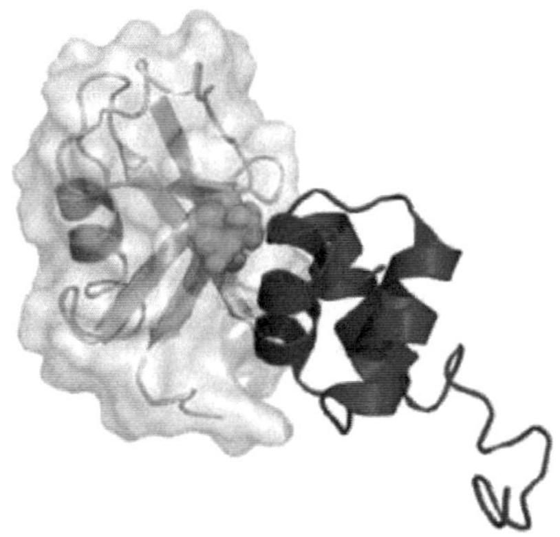

Figure 2. Structure of UBA-mono ubiquitin complex[32]

*2.2 UIM*

The UIM is found in S5a[20], Vps27, STAM[21], epsins and RAP80, most of which are trafficking progress related proteins. The UIM contains a single α-helix, centred around a conserved alanine residue and binds to mono-ubiquitin with relatively low affinity in mM magnitude. They mediate the recognition and direction of some protein complex to corresponding ubiquitinated cargo. For example, s5a, the first characterized ubiquitin-binding site in a proteasome subunit having only one ubiquitin interacting surface, is necessary for proteasome to recognize their substrates. The NMR structures of ubiquitins and S5a UIMs complex show that the UIM helix interact with a shallow hydrophobic groove on ubiquitin. Another UIM containing protein, receptor associated protein 80 (RAP80)[22], targets BRCA1 through its double UIMs which binds Lys63-linked ubiquitin chains.

*2.3 MIU*

The MIU contains a single helix and seems to be unique to one protein, the Rab5 exchange factor Rabex-5, so far[23, 24]. The Rabex-5 is necessary in endocytosis and kinase regulation[25, 26]. The MIU in Rabex-5 is in tandem with A20 ZnF (zinc finger) domain and attach to the helical C-terminus of the

A20 ZnF domain. The MIU consists a functionally essential alanine residue, which contacts Ile44 hydrophobic face of ubiquitin in similar mechanism with UIM. However, the MIU helix lies on the same hydrophobic groove that binds UIM but in the opposite orientation, which gives MIU another name, invert UIM. The N-terminus part of the UIM show similarity with MIU C-terminus and vice versa. Different from UIM, the MIU contains one more turn of helix that contacts with ubiquitin thus its affinity for ubiquitin is relatively higher to $\mu$M magnitude.

*2.4 CUE domain*

The CUE domain was digged by bioinformatics analysis of proteins related to Endocytosis and kinase regulation pathways including Vps9, TAB2 and TAB3. CUE domains share structurally similarity to the UBA domains as both of them consists of three-helix bundles and bind ubiquitin via conserved hydrophobic residues at the C-terminus of the $\alpha 1$ helix. With screens for mono-ubiquitin interactors in yeast, the CUE function in ubiquitin binding was uncovered. Ubiquitin binding ability seems to be the universal property of CUE domains as all the domains in CUE family show binding affinity to mono ubiquitin. Interestingly, one member of CUE domains, CUE domains in Vps9, shows special property as it bind to mono-ubiquitin with much higher affinity than the other CUE domain. The researcher find that the Vps9 CUE domain forms a domain-swapped dimer which is not observed in other CUE domains. In the dimeric structure, one of the α-helices in each domain is swapped, which forms a structure that is kinked to allow the two equivalent CUE domain surfaces to contact different portions of the same ubiquitin molecule. This structural contortion expands the CUE domain-ubiquitin interface from 550 $Å^2$ to 870 $Å^2$, such that it includes residues near the C terminus of ubiquitin. This type of binding interaction probably explains the relatively high binding affinity of this particular CUE-domain-containing protein for ubiquitin.

## 3. UBDs interact with ubiquitin acidic hydrophilic Asp58-centred face

ZnFs are the second biggest family of ubiquitin binding domains including NZF domains, A20 ZnF domains and ZNF UBP domains. The ZnF ubiquitin-binding domains offer much more diversity in recognition and binding affinity than the helical domains and both NZF fomains and A20 ZnF domains recognize Asp58-centred polar face. Like other UBDs, they were also discovered initially through domain dissections of known ubiquitin-binding proteins. In studies on Ufd1 Npl4, a ubiquitin binding adaptor in endoplasmic reticulum-associated degradation pathway, the NZF (Npl4 Zn-F) domain was established. subsequently, NZF domains in Vps36, TAB2 and TAB3 were discovered. The A20 ZnF domain was found in the NF-κB signalling cascade.

*3.1. NZF domains*

NZF domains contain approximate 30-residue and built around a single zinc-binding site. For the NZF domains tested to bind ubiquitin including Npl4, TAB2, TAB3 and Vps36, their affinities are about 100 $\mu$M or weaker. Two residures, Thr and Phe, lie in the first ‘zinc knuckle’ and a hydrophobic residue in the second knuckle, the TF fingerprint, appear to be the most important position for ubiquitin binding.

*3.2. A20 ZnF domains*

A20 ZnF domains were first discovered in ubiquitin binding protein, A20 protein, which function as a Lys63-linked polyubiquitin chain remover from its substrate, RIP (receptor-interacting protein), using another domain, OTU family de-ubiquitination domain and subsequently attaches Lys48-linked polyubiquitin chains to the substrate protein via its A20 ZnF domains. Another protein, Rabex-5, which also contains a A20 ZnF domain is a homologues of yeast Vps9vin mammals. Unlike Vps9, Rabex-5 binds to ubiquitin with a N-terminal A20 ZnF domain fused to a MIU instead of CUE domain in Vps9 (Figure 3). The Rabex-5 A20 ZnF domain binds to ubiquitin with Kd about 15 $\mu$M. The A20 ZnF domains use a pair of aromatic residues and several polar residues to bind to a predominantly polar patch on ubiquitin centred on Asp58. The A20 ZnF binds to ubiquitin surface epitope which does not overlap with the Ile44 patch.

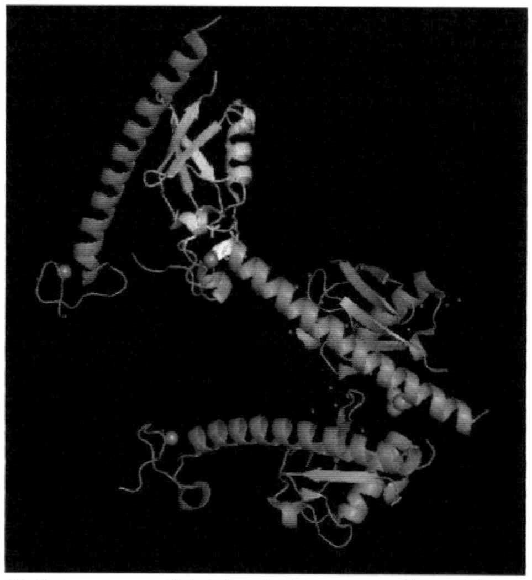

Figure 3. Structure of A20_ZNF-mono ubiquitin complex

### *3.3. ZnF UBP*

The ZNF UBP (ubiquitin-specific processing protease) ZnF domain was discovered originally in Isopeptidase T (USP5) via an exploration of the ubiquitin-binding sites of histone deactylases. The ZnF UBP domain is wildly distributed in different sorts of deubiquitinating enzymes including the ubiquitin ligase IMP/BRAP2 and the microtubule deacetylase HDAC6. The ZnF UBP has also been referred to as the PAZ (polyubiquitin-associated zinc binding) domain, even though it has no relation with the widely studied RNAbinding PAZ domain. The ZnF UBP domain is approx. 130 residues in length, much larger than other ubiquitin-binding ZnFs. The domain is built around a single zinc-binding site in its Nterminal half which is fused to an α/β fold. The ZnF UBP is one of the ubiquitin binders with highest affinity among known binding domains: Kd = 3 μM for the human isopeptidase T ZnF UBP. The free C-terminal glycine residue of ubiquitin is required for ZnF UBP binding. The structure of the isopeptidase T ZnF UBP domain–ubiquitin complex shows that the extended C-terminus of ubiquitin insert deep into a tube-shaped cavity in the ZnF UBP domain, which tightly binds and enclose the free C-terminal ubiquitin tail (the sequence: RGG) through an aromatic pocket.

## 4. Application of UBDs

### *4.1. Tandem UBDs as enrichment materials or inhibitor*

When the research on ubiquitin has accumulated[27-31], some are inspired by the series arrangement of UBD in nature, trying to construct multiple repetitive UBD modules in the same protein chain to enhance their binding ability[32]. This idea does not require much thought, it is very easy to think of, because nature in the process of evolution is doing the same. However, this constructed element, or tandem ubiquitin binding element (TUBEs), has achieved quite good results.[33] According to reports, most of these elements contain more than five UBD repeats, separated by flexible amino acid linkers, and contain commonly used protein tags in the sequences so that they can be immobilized.

This kind of component has been put into use and has shown quite good performance. For example, TUBEs of UBA1 domain based on hHR23A and UQ1-UBA domain based on ubiquilin-1 exhibit high affinity for tetramer ubiquitin chains linked by Lys48 and Lys63, which is 100-1000 times higher than that of single UBA domain[33]. These two non-specific domains are most widely used.

### *4.2. Isolation free mono ubiquitin and ubiquitin chains*
This form of UBD is bound to the carrier, and its affinity effect depends on its concentration on the carrier. These UBDs contain Znf-UBP domain and A20 Znf domain of ZNF216 protein etc. Their binding sites include targets from C terminal[34], Asp58[35] to Ile44. Among them, the Znf domain is used to capture the polyubiquitin chains that do not bind to proteins in the samples.

In summary, although these types of UBDs are unconventional and relatively few, they are still used in many special situations and may be extended in the future.

### *4.3. UBD as endo-cellular monitor or indicator*[36]
In this case, UBD with fluorescent labels, such as GFP, RFP and so on, can be recombined to obtain UBD with fluorescent effect. These UBD tools can bind to target proteins like fluorescent antibodies. [37]When observed by fluorescent microscopy, they can indicate their location with fluorescent signals, this method can be used to monitor the ubiquitination of proteins, as well as the level and process of ubiquitination of various parts of cells.

There is also a large family of enzymes in cells, called deubiquitinating enzyme (DUBs), which hydrolyze ubiquitin molecules from ubiquitin-linked proteins or precursor proteins by hydrolyzing ester, peptide or isopeptide bonds at the carboxyl end of ubiquitin. The human genome encodes nearly 100 deubiquitinating enzymes, making them the largest family of ubiquitin system enzymes. DUBs genes in humans can be divided into two categories: cysteine protease family and metalloproteinase family. Ubiquitin can be produced in areas where DUB is active in cells, so UBD can be used as indicator the activity of DUBs. However, fusion proteins are only a fairly basic method. In order to improve its specificity, some people have designed fluorescence-quenching mechanism or fluorescence recovery mechanism. Both of them are targeted at specific ubiquitin junction types and use multiple junctions to improve their specificity.

The reason is very simple. If a lock core has only one identifying position with the key, the lock will also be easily cracked, and even may be unlocked by the key that does not match the lock core by accident. This will be a disaster. If a lock requires more sophisticated key, and unlocking them "by chance" becomes difficult, or even, impossible.

Facing problem how to ensure that only when all matches will phenomena occur, FRET come into people's sight. Two UBDs targeting the adjacent ubiquitin junctions are recombined with fluorescent and quenching atom groups respectively. Only by observing the phenomenon of fluorescence appearing and quenching, can the existence of a specific ubiquitin junction structure be explained, otherwise, there will be false positive possibilities. Or two kinds of UBD can be recombined to easily stimulated substances and fluorescent substances that are difficult to stimulate respectively, so that the existence of specific ubiquitin junction structure can be judged by observing whether there is "no fluorescence at first, then fluorescence gradually appears"[38]. Examples of this are Venus proteins. It is expected that more of these kits will be available in the future, and this approach may be used to monitor ubiquitinated proteins in the body.

### *4.4. Inhibitors of living activities*
UBD can bind to ubiquitinated proteins to act as an indicator protein, but from another point of view, when proteins in a metabolic or signaling pathway are combined with engineered UBD, it will not perform its normal physiological function, at least will be more or less affected. So UBD can also be directly used to bind to a protein, it can be a signal molecule, it can be a metabolic precursor. UBD can inhibit ubiquitin-dependent pathways in cells by binding these proteins.[39]

In addition, the ability of UBD as an inhibitor goes far beyond that. UBD plays a recognition role by binding with ubiquitin chains. DUBs also rely on its ubiquitin-binding domain to recognize and decompose ubiquitin chains. UBD and DUBs normally compete with each other for polyubiquitin chains. In this way, UBD with high affinity can be designed to protect ubiquitinated proteins from being deubiquitinated by DUBs or from being decomposed by proteases. Such a mechanism has been widely observed in organisms such as yeast.

Based on this mechanism, we can design some inhibitors to regulate the process of life activities, or to protect certain target proteins from decomposition, which can play a role in controlling the symptoms of some incurable genetic diseases.

**Reference**

[1] Komander, D., Rape, M., The Ubiquitin Code. **2012,** *81* (1), 203-229.
[2] Hicke, L., Schubert, H. L., Hill, C. P., Ubiquitin-binding domains. *Nature Reviews Molecular Cell Biology* **2005,** *6* (8), 610-621.
[3] Huang, T. T., D'Andrea, A. D., Regulation of DNA repair by ubiquitylation. *Nature Reviews Molecular Cell Biology* **2006,** *7* (5), 323-334.
[4] Meyer, H.-J., Rape, M., Enhanced Protein Degradation by Branched Ubiquitin Chains. *Cell* **2014,** *157* (4), 910-921.
[5] Hicke, L., A New Ticket for Entry into Budding Vesicles—Ubiquitin. *Cell* **2001,** *106* (5), 527-530.
[6] Raiborg, C., Rusten, T. E., Stenmark, H., Protein sorting into multivesicular endosomes. *Current Opinion in Cell Biology* **2003,** *15* (4), 446-455.
[7] Dikic, I., Wakatsuki, S., Walters, K. J., Ubiquitin-binding domains — from structures to functions. *Nature Reviews Molecular Cell Biology* **2009,** *10* (10), 659-671.
[8] Sato, Y., Fujita, H., Yoshikawa, A., Yamashita, M., Yamagata, A., Kaiser, S. E., Iwai, K., Fukai, S., Specific recognition of linear ubiquitin chains by the Npl4 zinc finger (NZF) domain of the HOIL-1L subunit of the linear ubiquitin chain assembly complex. *Proceedings of the National Academy of Sciences* **2011,** *108* (51), 20520.
[9] Husnjak, K., Dikic, I., Ubiquitin-Binding Proteins: Decoders of Ubiquitin-Mediated Cellular Functions. **2012,** *81* (1), 291-322.
[10] Reyes-Turcu, F. E., Shanks, J. R., Komander, D., Wilkinson, K. D., Recognition of Polyubiquitin Isoforms by the Multiple Ubiquitin Binding Modules of Isopeptidase T. **2008,** *283* (28), 19581-19592.
[11] Chen, L., Shinde, U., Ortolan, T. G., Madura, K., Ubiquitin-associated (UBA) domains in Rad23 bind ubiquitin and promote inhibition of multi-ubiquitin chain assembly. *EMBO reports* **2001,** *2* (10), 933-938.
[12] Elsasser, S., Finley, D., Delivery of ubiquitinated substrates to protein-unfolding machines. *Nature Cell Biology* **2005,** *7* (8), 742-749.
[13] Wilkinson, C. R. M., Seeger, M., Hartmann-Petersen, R., Stone, M., Wallace, M., Semple, C., Gordon, C., Proteins containing the UBA domain are able to bind to multi-ubiquitin chains. *Nature Cell Biology* **2001,** *3* (10), 939-943.
[14] Bienko, M., Green, C. M., Crosetto, N., Rudolf, F., Zapart, G., Coull, B., Kannouche, P., Wider, G., Peter, M., Lehmann, A. R., Hofmann, K., Dikic, I., Ubiquitin-Binding Domains in Y-Family Polymerases Regulate Translesion Synthesis. *Science* **2005,** *310* (5755), 1821.
[15] Hirano, S., Kawasaki, M., Ura, H., Kato, R., Raiborg, C., Stenmark, H., Wakatsuki, S., Double-sided ubiquitin binding of Hrs-UIM in endosomal protein sorting. *Nature Structural & Molecular Biology* **2006,** *13* (3), 272-277.
[16] Raasi, S., Pickart, C. M., Rad23 Ubiquitin-associated Domains (UBA) Inhibit 26 S Proteasome-catalyzed Proteolysis by Sequestering Lysine 48-linked Polyubiquitin Chains. **2003,** *278* (11), 8951-8959.
[17] Reyes-Turcu, F. E., Horton, J. R., Mullally, J. E., Heroux, A., Cheng, X., Wilkinson, K. D., The Ubiquitin Binding Domain ZnF UBP Recognizes the C-Terminal Diglycine Motif of Unanchored Ubiquitin. *Cell* **2006,** *124* (6), 1197-1208.
[18] Trempe, J.-F., Brown, N. R., Lowe, E. D., Gordon, C., Campbell, I. D., Noble, M. E. M., Endicott, J. A., Mechanism of Lys48-linked polyubiquitin chain recognition by the Mud1 UBA domain. *The EMBO Journal* **2005,** *24* (18), 3178-3189.
[19] Varadan, R., Assfalg, M., Raasi, S., Pickart, C., Fushman, D., Structural Determinants for

Selective Recognition of a Lys48-Linked Polyubiquitin Chain by a UBA Domain. *Molecular Cell* **2005,** *18* (6), 687-698.
[20] Deveraux, Q., Ustrell, V., Pickart, C., Rechsteiner, M., A 26-S PROTEASE SUBUNIT THAT BINDS UBIQUITIN CONJUGATES. *J. Biol. Chem.* **1994,** *269* (10), 7059-7061.
[21] Fisher, R. D., Wang, B., Alam, S. L., Higginson, D. S., Robinson, H., Sundquist, W. I., Hill, C. P., Structure and Ubiquitin Binding of the Ubiquitin-interacting Motif. *J. Biol. Chem.* **2003,** *278* (31), 28976-28984.
[22] Sims, J. J., Cohen, R. E., Linkage-Specific Avidity Defines the Lysine 63-Linked Polyubiquitin-Binding Preference of Rap80. *Molecular Cell* **2009,** *33* (6), 775-783.
[23] Penengo, L., Mapelli, M., Murachelli, A. G., Confalonieri, S., Magri, L., Musacchio, A., Di Fiore, P. P., Polo, S., Schneider, T. R., Crystal Structure of the Ubiquitin Binding Domains of Rabex-5 Reveals Two Modes of Interaction with Ubiquitin. *Cell* **2006,** *124* (6), 1183-1195.
[24] Lee, S., Tsai, Y. C., Mattera, R., Smith, W. J., Kostelansky, M. S., Weissman, A. M., Bonifacino, J. S., Hurley, J. H., Structural basis for ubiquitin recognition and autoubiquitination by Rabex-5. *Nature Structural & Molecular Biology* **2006,** *13* (3), 264-271.
[25] Mukhopadhyay, D., Riezman, H., Proteasome-Independent Functions of Ubiquitin in Endocytosis and Signaling. *Science* **2007,** *315* (5809), 201.
[26] Huang, F., Kirkpatrick, D., Jiang, X., Gygi, S., Sorkin, A., Differential Regulation of EGF Receptor Internalization and Degradation by Multiubiquitination within the Kinase Domain. *Molecular Cell* **2006,** *21* (6), 737-748.
[27] Layfield, R., Tooth, D., Landon, M., Dawson, S., Mayer, J., Alban, A., Purification of poly-ubiquitinated proteins by S5a-affinity chromatography. *PROTEOMICS* **2001,** *1* (6), 773-777.
[28] Kirkpatrick, D. S., Weldon, S. F., Tsaprailis, G., Liebler, D. C., Gandolfi, A. J., Proteomic identification of ubiquitinated proteins from human cells expressing His-tagged ubiquitin. *PROTEOMICS* **2005,** *5* (8), 2104-2111.
[29] Schwertman, P., Bezstarosti, K., Laffeber, C., Vermeulen, W., Demmers, J. A. A., Marteijn, J. A., An immunoaffinity purification method for the proteomic analysis of ubiquitinated protein complexes. *Analytical Biochemistry* **2013,** *440* (2), 227-236.
[30] Phu, L., Izrael-Tomasevic, A., Matsumoto, M. L., Bustos, D., Dynek, J. N., Fedorova, A. V., Bakalarski, C. E., Arnott, D., Deshayes, K., Dixit, V. M., Kelley, R. F., Vucic, D., Kirkpatrick, D. S., Improved Quantitative Mass Spectrometry Methods for Characterizing Complex Ubiquitin Signals. *Molecular &amp, amp, Cellular Proteomics* **2011,** *10* (5), M110.003756.
[31] Low, T. Y., Magliozzi, R., Guardavaccaro, D., Heck, A. J. R., Unraveling the ubiquitin-regulated signaling networks by mass spectrometry-based proteomics. *PROTEOMICS* **2013,** *13* (3-4), 526-537.
[32] Hurley, James H., Lee, S., Prag, G., Ubiquitin-binding domains. *Biochemical Journal* **2006,** *399* (3), 361.
[33] Hjerpe, R., Aillet, F., Lopitz-Otsoa, F., Lang, V., England, P., Rodriguez, M. S., Efficient protection and isolation of ubiquitylated proteins using tandem ubiquitin-binding entities. *EMBO reports* **2009,** *10* (11), 1250-1258.
[34] Xia, Z.-P., Sun, L., Chen, X., Pineda, G., Jiang, X., Adhikari, A., Zeng, W., Chen, Z. J., Direct activation of protein kinases by unanchored polyubiquitin chains. *Nature* **2009,** *461* (7260), 114-119.
[35] Scott, D., Oldham, N. J., Strachan, J., Searle, M. S., Layfield, R., Ubiquitin-binding domains: Mechanisms of ubiquitin recognition and use as tools to investigate ubiquitin-modified proteomes. *PROTEOMICS* **2015,** *15* (5-6), 844-861.
[36] Melvin, A. T., Woss, G. S., Park, J. H., Dumberger, L. D., Waters, M. L., Allbritton, N. L., A Comparative Analysis of the Ubiquitination Kinetics of Multiple Degrons to Identify an Ideal Targeting Sequence for a Proteasome Reporter. *PLOS ONE* **2013,** *8* (10), e78082.
[37] Altun, M., Kramer, Holger B., Willems, Lianne I., McDermott, Jeffrey L., Leach, Craig A.,

Goldenberg, Seth J., Kumar, K. G. S., Konietzny, R., Fischer, R., Kogan, E., Mackeen, Mukram M., McGouran, J., Khoronenkova, Svetlana V., Parsons, Jason L., Dianov, Grigory L., Nicholson, B., Kessler, Benedikt M., Activity-Based Chemical Proteomics Accelerates Inhibitor Development for Deubiquitylating Enzymes. *Chemistry & Biology* **2011,** *18* (11), 1401-1412.

[38] Chen, Z., Zhong, Y., Wang, Y., Xu, S., Liu, Z., Baskakov, I. V., Monteiro, M. J., Karbowski, M., Shen, Y., Fang, S., Ubiquitination-Induced Fluorescence Complementation (UiFC) for Detection of K48 Ubiquitin Chains In Vitro and in Live Cells. *PLOS ONE* **2013,** *8* (9), e73482.

[39] Raasi, S., Pickart, C. M., Rad23 Ubiquitin-associated Domains (UBA) Inhibit 26 S Proteasome-catalyzed Proteolysis by Sequestering Lysine 48-linked Polyubiquitin Chains. *J. Biol. Chem.* **2003,** *278* (11), 8951-8959.

IOP Conf. Series: Materials Science and Engineering **738** (2020) 012014 doi:10.1088/1757-899X/738/1/012014

# The approach to sustainable space mining: issues, challenges, and solutions

**Fengna Xu**

Law School, Xi'an Jiaotong University, 28 Xianning West Road, Xi'an, Shaanxi Province, China

E-mail: fengna.xu@foxmail.com

**Abstract.** Harvesting and using space resources and energy has always been a dream of mankind for centuries as well as a popular topic for science fiction novels. In recent years, technological development makes space mining possible or even a reality in the near future. Accordingly, not only governments but also ambitious private actors are engaged in space mining activities. In addition to increasing investment, the USA and Luxembourg have erected their national space law to answer industrial needs. However, existing international law fails to address these activities in an effective manner, and many issues remain unresolved. Without further regulation, a laissez-faire approach would be detrimental to the growth of this nascent industry. Some challenges are likely to happen, such as conflicts between multiple States, inequalities in benefits sharing, and environment contamination. As a result, for promoting sustainable exploitation and use of space resources and energy, this article suggests some solutions, including: 1) guaranteeing legal certainty; 2) ensuring sufficient investment; 3) strengthening international cooperation; 4) promoting environment preservation and protection.

## 1. Introduction

As the resource and energy supplies of the Earth continue to dwindle, outer space is the only place we could find fresh supplies of both energy and raw materials. An asteroid belt between Mars and Jupiter, and near-Earth asteroids, with an abundance of resources, are capable of exploitation. For example, the Moon contains vast amounts of natural resources, like uranium, titanium, silicon, etc. And it's estimated that Helium-3, a light and non-radioactive fusion fuel that is abundant on the lunar surface while rare on Earth, is enough to power human energy needs for up to 10,000 years [1]. At present, the extraction and use of space resources and energy is not technically feasible, but is expected to happen in the near future. Many believe that space mining could support commercial application, stimulate technological innovation, produce economic return and give important contribution to create and redistribute wealth in the world. Consequently, not only governments but also ambitious private actors are engaged in space mining activities. For example, National Aeronautics and Space Administration (NASA) launched OSIRIS-REx spacecraft in 2016 to explore the asteroid Bennu before collecting a sample to return to Earth, which will improve our understanding of the resources in near-Earth space. Similarly, China is devoting more effort to the space-based economic opportunities just on the horizon. On January 3, 2019, China soft-landed the Chang'e-4 robotic lander and rover on the far side of the moon, and released Yutu-2 to explore the lunar landscape. Besides, Planetary Resources embarking on the world's first commercial deep space exploration program has announced plans to mine asteroids for gold, platinum and rare minerals and transform asteroid water into rocket fuel. Hence, as 'money time' has come, the exploitation and use of space resources and energy require proper governance now.

Space sustainability has become an important concern at modern times, aiming for the balance between economic growth, care for the environment and social well-being. Although a multifaceted approach is being adopted to promote space mining, the current legal framework suffers various inadequacies in addressing this issue effectively. When the *Treaty on Principles Governing the Activities of States in the Exploration and Use of Outer Space, Including the Moon and Other Celestial Bodies* (19 December 1966) (the Outer Space Treaty) and the *Agreement Governing the Activities of States on the Moon and Other Celestial Bodies* (24 October 1970) (the Moon Agreement) were concluded, there was no urgency to regulate it. In particular, there was no sufficient scientific information available regarding space resources and the possibility of their economic use. Consequently, the two treaties do not regulate the exploitation and use of space resources directly. Nevertheless, the Outer Space Treaty, as the cornerstone of international space law, could provide guidance for space mining, but their contents are not at all clear. The Moon Agreement further emphasized and developed the Outer Space Treaty, which were not accepted by the major space-faring countries.

Thus, so far there are no international agreed rules to govern the exploitation and use of space resources and energy. A laissez-faire approach would be detrimental to the growth of this nascent industry. This article focuses on the issue of promoting sustainable exploitation and use of space resources and energy. Part 2 analyses the latest developments in space resources and energy and summarizes deficiencies of the current legal framework in addressing this issue effectively. Part 3 discusses possible challenges due to lack of proper regulation at the dawn of space mining. Part 4 suggests solutions for sustainable space mining.

## 2. Latest Developments in Space Resources and energy

As shown in Figure 1, many modifying factors need to be considered to convert Mineral Resources to Ore Reserves. There include mining, processing, metallurgical, infrastructure, economic, marketing, legal, environment, social and government factors. Hence, in recent years, a multifaceted approach is being developed to advance the technological, economic, and legal readiness for the harvest and utilization of space resources and energy.

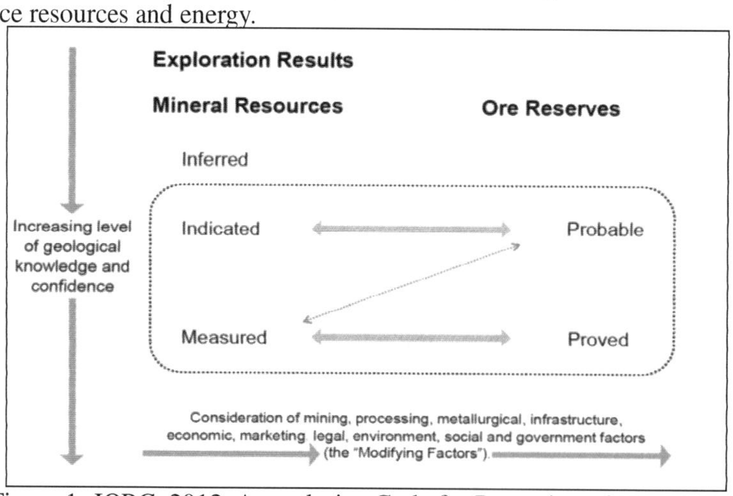


Figure 1. JORC, 2012. Australasian Code for Reporting of Exploration Results, Mineral Resources and Ore Reserves (The JORC Code).

### *2.1. Advances in technology and hardware*

Many technologies and hardware have been developed to promote space mining, such as Remote Sensing, Visible Imaging, Neutron Spectrometer, Lunar Reconnaissance Orbiter, Cosmic Ray Telescope for the Effects of Radiation (CRaTER). And a substantial breakthrough has been made to

IOP Conf. Series: Materials Science and Engineering **738** (2020) 012014 doi:10.1088/1757-899X/738/1/012014

achieve the benefits of In Situ Resource Utilization (ISRU) for a reasonable cost, mass, and risk. As we all know, space resources can be used in situ or transported to Earth. ISRU means any hardware or operation that harnesses and utilizes 'in-situ' resources to create products and services for robotic and human exploration [2]. And ISRU will change the way we explore the outer space. For example, Moon Direct would be the most effective transportation system to access to the lunar surface. By making using of LOx/H2 propellant produced at a lunar polar base, it could support the operation of a lightweight Lunary Excursion Vehicle (LEV) flight system. These space-sourced propellants can dramatically reduce the cost of every other activity in cislunar space. Table 1 provides some examples. Besides, ISRU could also support lunar missions in other ways. For example, a repairable lunar power system can be generated founded on the fabrication of silicon solar cells by thin film growth technology in an ultra-high vacuum environment, resulting in an energy-rich environment for the Moon.

Table 1. Benefits of space-sourced propellants [a].

| Cislunar Activity | Space-Sourced Propellant Benefit |
|---|---|
| Transportation from Earth to Geosynchronous orbit | 10-20% lower cost |
| Transportation from Earth to Lunar surface | 70% lower cost |
| Cost of a human mission to Mars | 2-3 times reduction |
| In-space transportation | Essentially the cost of space-sourced propellant |

[a] The data are collected from the Lunar Polar Prospecting Workshop: Findings and Recommendations.

### 2.2. *Economic support from public and private entities*

With a potential customer base, space mining is considered to be promising and interests States as well as private actors recently. In addition to USA and China, Luxembourg is also taking part in the race by setting up a $227 million fund to entice space mining companies to open offices in the country and make it the hub of space mining innovation. Similarly, the United Arab Emirates is making a multipronged effort to establish a space mining industry, such as an investment of more than $5 billion. At the same time, governments are turning to private companies increasingly and providing confidence for investors in order to generate revenues and open new markets. Accordingly, several commercial initiatives have started for asteroids and the moon, by Planetary Resources, Deep Space Industries, Google Lunar X-Prize, Lunar Missions Ltd, Moon Express and so on.

### 2.3. *Inadequacies of the current legal framework*

To answer industrial needs, the US has adopted *Commercial Space Launch Competitiveness Act (CSLCA)* to facilitate the commercial exploration for and commercial recovery of space resources by U.S. citizens. Luxembourg followed suit and erected its national space law. Concerns have been raised about whether the domestic legislation conflicts with international law. Under the Outer Space Treaty, States shall carry on activities in the exploration and use of outer space in accordance with international law. However, the current legal framework suffers various inadequacies in addressing space mining activities effectively.

The Outer Space Treaty provides that outer space shall be free for exploration and use by all States; such freedom shall be non-discriminatory, equal and in conformity with international law; all areas of celestial bodies shall be freely accessible; and scientific investigation in outer space shall be free in Article I. It's worth noting that the Outer Space Treaty restricts such freedom to science investigation, without referring to commercial exploitation and use of space resources. It is argued that the freedom to engage in scientific exploration of outer space does not justify large-scale exploitation of extraterrestrial resources for commercial purpose. Meanwhile, Article I of the Outer Space Treaty also provides that exploration and use of outer space shall be carried out for the benefits and in the interests of all countries, regardless of their degree of economic or scientific development, and shall be the province of all mankind. Under this provision, the obligation of countries to share benefits is legally binding. However, States Parties hold opposing positions on what constitutes 'the interests of all countries' and how to share benefits might rely largely on their good faith.

Besides, Article II of the Outer Space Treaty embodies the non-appropriation principle, which proscribes the national appropriation of outer space 'by claim of sovereignty, by means of use or occupation, or by other means'. From article VI of the Outer Space Treaty, it can be inferred that 'national appropriation' covers both States appropriation and private entities appropriation. Although this principle precludes the possibility of appropriation of outer space and celestial bodies, whether it extends to natural resources therein is uncertain as the Outer Space Treaty never makes a distinction between them, and the term 'outer space' does not suffice to include both outer space broadly considered and its natural resources.

What's more, Article IX of the Outer Space Treaty has laid the basis for environment protection of the outer space. It requires that States pursue studies and conduct exploration of outer space so as to avoid their harmful contamination and also adverse changes in the environment of the Earth. And States are obliged to consider the environmental aspects for the authorization and supervision of national activities in outer space and adopt appropriate measures when necessary. However, the provisions contained in this principle are rather vague and broad.

## 3. Possible challenges at the dawn of space mining

### *3.1. Conflicts between multiple States*

Space resources, as *res communis* [3], can be appropriated to some extent on the basis of freedom of exploration and use of the outer space. However, it is likely to follow a 'first come, first served' approach to space resources activities. In fact, the 'first come, first served' approach drove early and rapid development of oil industry of the US in the 19th century, although a frenetic race among surface owners followed and led to an extraordinary waste of oil and gas. Given that so far there are no agreement or property rights on space resources, they are essentially in a 'state of nature'. Allocation by the 'first come, first served' approach is simple and requires very little government involvement to deter another one (called a 'junior') from displacing the rightful first comer (called a 'senior'). However, overprotecting the senior by priority rights could run the risk of disorder, waste, inequality, and even monopoly. The Outer Space Treaty, requires State parties to conduct all their activities in outer space 'with due regard to the corresponding interests of all other States Parties'. Without specific coordinating rules, conflicts between multiple States are likely to happen. Private entities may choose to arm themselves to safeguard their own interests. In extreme cases, States may also protect them by placing weapons of mass destruction in outer space if necessary [4]. As a result, priority rights should not be absolute but subjected to some arrangements.

### *3.2. Inequalities in benefits sharing*

Basically, the exploitation and use of space resources would bring a wide-range of benefit and interests to all countries. Space-faring countries who have invested much on such activities are certainly entitled to profit from the work of their own labor. Developing countries, despite with limited capabilities for space exploration, could also benefit from such activities. Nevertheless, the common benefit and interests of all countries are not equivalent to the common benefit and interests of every country. It's conceivable that, with different level of economic or scientific development among States, the benefits and interests they gain from space mining vary. However, under current space law, developing countries hardly benefit from space mining. Because of the competing views and interests among different countries, substantial disagreement also exists with regard to benefits sharing. This issue has become more tangible and pressing especially given that space mining would become a reality soon. Specifically, space-faring countries are reluctant to sacrifice the 'hard earned' benefits to those who don't have the capacity to invest or gain them on their own, while developing countries are pressing for equitable distribution of benefits.

### *3.3. Environment contamination*
Space resources activities are *per se* ultra-hazardous activities, which may be harmful to both the outer space and the Earth environment. Forward contamination arising from Earth affects the environment of outer space. This type of contamination may include: all forms of debris found in outer space of nonhazardous nature; hazardous waste which is chemically or physically dangerous; radioactive waste which is the residue of nuclear powered space objects; biological material from Earth to a planetary body with space probes or human space missions, etc. [5]. While backward contamination arising in space adversely affects the surface or atmosphere of Earth. In particular, returning spacecraft may also spread pollution or bring back waste, such as radioactive debris and extraterrestrial material. For this reason, we should carry out these activities with a high standard of care and due diligence, in spite of uncertainties about specific contamination at present.

## 4. Solutions for sustainable space mining
The World Commission on Environment and Development (WCED) defines the concept of sustainable development as 'development that meets the needs of the present without compromising future generations to meet their own needs' in its report *Our Common Future* in 1987. After continuous development, some principles are incorporated into sustainable development concept, such as sustainable use and integration, intragenerational equity and intergenerational equity, precautionary approach, international cooperation. For sustainable space mining, these principles could provide a useful reference for us.

### *4.1. Guaranteeing legal certainty*
The negotiating history of the Outer Space Treaty demonstrates that the term 'appropriation' in Article II is primarily concerned with the expansion of State territory. But the legal status of natural resources is not necessarily the same as the surface of land, drawing from three different types of property over natural resources of the subsoil in Earth: i) the regime of accessio; ii) the national property model; and iii) the model of absolute property of the State. Consequently, the land in outer space not under national or private dominion doesn't mean that exploitation of natural resources in the subsoil is unlawful under Article II of the Outer Space Treaty. And many examples confirm that the exploitation and use of resources to some extent can occur in areas beyond national jurisdiction or of global interest. For example, the protection of property rights over resources that private actors may recover from the deep seabed, doesn't necessarily amount to a sovereignty claim over the territorial area. Mineral resources could become the property of miners based on their labor or effort rather than their sovereignty over the deep seabed. Hence, considering the recent resurgence in public and private interest in the mineral exploitation in space, an international regime should be formulated for the purposes of clarity and legal certainty. Otherwise, the investment pool would shrink and the process of the exploration and use of outer space by mankind would be hindered.

### *4.2. Ensuring sufficient investment*
The financial support for the research and application of related technology is essential. In particular, commercial incentives could promote investment in the exploration of space resources and enable the development of the most efficient use of resources. With the prospect of property rights, private sectors would make every effort to invest and develop technology for profits. For instance, Deep Space Industries (DSI) is pioneering a new capability for space exploration with low cost and great performance. However, without government interference, the commercial approach does indeed generate extreme distributions of wealth. Due to huge economic and technological disadvantages,

developing countries may be left far behind by space-faring countries. As a result, the international community needs to carefully balance the interests of developed and developing countries which are reconcilable and not mutually exclusive.

### *4.3. Strengthening international cooperation*

To avoid the risk of conflicts, the international community should not vest an absolute monopoly in the senior and deprive a whole neighborhood or community of his rights. By international cooperation, rights therein can be distributed equitably and waste and abstract claims be avoided. Besides, a dual system may be another good choice to coordinate international competition for space resources. Due to the vastness of outer space and the uncertainty of resources, it's more advisable to establish reserve zones on the moon and Mars for those countries with limited capabilities for space mining.

Additionally, space mining is beneficial in a general sense, as it is sufficient to achieve 'the greatest good for the greatest number'. But, to make these benefits truly accrue to all countries and avoid tragedy of anti-commons, international cooperation could provide a feasible way to strike a balance among different players. By international cooperation, States would conduct space resources activities with due regard to the corresponding interests of other States and the international community. Developing countries could be enabled to bridge the still widening gap of technology and knowledge, which is beneficiary for their future development and considerably contributes to worldwide stability and peace. And there are various options we could use to carefully balance the interests of developed and developing countries, such as joint ventures, compensation and other measures of economic assistance for affected developing countries.

### *4.4. Promoting environment preservation and protection*

Recently there is also a growing tendency not only towards tackling the problem of environment preservation and protection in the outer space, but also towards trying to avoid their harmful contamination to the maximum possible extent. Space sustainability has become an important concern at modern times. Regarding forward contamination, the fragility of space environment itself and our general lack of much understanding of that environment are behind the reasoning of the precautionary principle. And it is always easier to avoid contamination than deal with it after it emerges. As for backward contamination, we should also take prudent precautions to protect Earth's environment. For instance, Committee on Space Research (COSPAR) promulgated *Planetary Protection Policy* as an international standard on procedures to avoid organic constituent and biological contamination in space exploration. Category V pertains to all Earth-return missions. For the subcategory defined as 'Restricted Earth Return', destructive impact upon return is absolutely prohibited and strict containment and timely analyses are needed after the mission. In addition to maintain the *Planetary Protection Policy*, the proposed international legal regime on space mining could also refer to corresponding measures in the *Draft Building Blocks (DBB) for the Development of An International Framework on Space Resource Activities* of the Hague Space Resources Governance Working Group (HSRGWG). For example, it could require States to review prior to the authorization of space resources activities, develop technical standards, assess conformity to avoid harmful impacts, and conduct response measures if a harmful impact occurs.

## 5. Conclusions

As 'money time' has come, the exploitation and use of space resources and energy require proper governance now. Although a multifaceted approach is being adopted to promote space mining, the current legal framework suffers various inadequacies in addressing this issue effectively. When the Outer Space Treaty was adopted, the issue of exploitation and use of natural resources and energy did not emerge, thus not be addressed *expressis verbis*. But no explicit prohibition, does not amount to lawfulness. Without further regulation, a number of issues can be anticipated, such as over-consumption, disorder, intra-generational inequality even monopoly, and environment contamination. Given that space sustainability is becoming more popular at modern times, sustainable

IOP Conf. Series: Materials Science and Engineering **738** (2020) 012014 doi:10.1088/1757-899X/738/1/012014

development concept would be very useful for the international community to handle these problems. As a result, four approaches should be highlighted for promoting sustainable space mining, including: 1) guaranteeing legal certainty; 2) ensuring sufficient investment; 3) strengthening international cooperation; 4) promoting environment preservation and protection.

**Acknowledgments**
The author is most grateful to the Law School of Xi'an Jiaotong University. And the author also wishes to thank prof. Jinyuan Su and Jun Ou for their valuable comments and suggestions.

**References**
[1] Education for the Information Age. (1999) What is Helium-3 and why is it so important? https://www.edinformatics.com/math_science/what-is-helium-3.html.
[2] Sanders, G.B. (2018) Overview of Past Lunar In Situ Resource Utilization (ISRU) Development by NASA. https://ntrs.nasa.gov/archive/nasa/casi.ntrs.nasa.gov/20190001035.pdf.
[3] Butler, D.A. (2017) Who Owns the Moon, Mars, and Other Celestial Bodies: Lunar Jurisprudence in Corpus Juris Spatialis. J. Air L. & Com., 82: 505-518.
[4] Garner, B.A. (2009) Black's Law Dictionary. West Publishing Co., New York.
[5] Su, J. (2017) Legality of unilateral exploitation of space resources under international law. Int'l & Comp. L.Q., 66: 991-1008.

IOP Conf. Series: Materials Science and Engineering **738** (2020) 012015 doi:10.1088/1757-899X/738/1/012015

# Fault Diagnosis of Power Transformer in Mine

**Lu Chen[1], Shudong Wang[1], Shiyong Fan[1], Haiyan Shi[1], Liang Zhang[1]**

[1]College of Electrical and Information Engineering, Lanzhou University of Technology, Lanzhou, Gansu, 730050, China

*Corresponding author's e-mail: 601038980@qq.com

**Abstract.** In view of the ambiguity of power system transformer faults in mines, the traditional gas analysis method cannot directly judge the possibility of fault occurrence based on gas concentration. A fault diagnosis method for mine transformers based on intuitionistic fuzzy Petri nets is proposed. The relationship between characteristic gas and fault is described by intuitionistic fuzzy set, and a new type of intuitive fault diagnosis model is established. The membership degree and non-affiliation degree are introduced into the model. The intuitionistic fuzzy inference algorithm is designed. By obtaining and processing the specific parameters such as the weight of the connecting arc and the threshold of the transition in the fault diagnosis model, the fault diagnosis process is transformed into the intuitionistic fuzzy inference process using the intuitionistic fuzzy Petri net. Finally, the membership degree and non-affiliation degree of each fault are obtained to judge the possibility of fault occurrence. It is verified by the fact that the above-mentioned mine transformer fault reasoning method can quickly judge the possibility of fault occurrence according to the concentration of gas.

## 1. Introduction

The Mine-used transformer is an important part of the mine power system, and its reliability directly affects the production safety of the mine. Oil-immersed transformers account for a high proportion of mine transformers because of their unique advantages. How to better diagnose them is a subject worth studying. In literature [1-3], the traditional fuzzy Petri net method is used to study the reliability of the transformer, but the diagnostic accuracy is not ideal. The literature [4] starts from the dissolved gas in the transformer oil, and it is of certain reference significance to judge the source of the fault by quantitative analysis of the gas.

With the development of artificial intelligence, various artificial intelligence algorithms such as artificial neural network [5-9], expert system, fuzzy support vector machine have been widely used in transformer fault diagnosis. The application has achieved certain results. In [7], the bat algorithm is used to optimize the parameters and weights of BP neural network, and the optimized parameters are applied to the BP neural network diagnosis model. The iterative convergence is fast, the efficiency is high, but the steps are more complicated. In [8], the Elman network algorithm is used to optimize the initial parameters of the Petri net model, which improves the accuracy of fault diagnosis, but it is impossible to determine which fault is most likely to occur. Literature [10] combines the traditional three-ratio method with statistical methods to construct an expert system for diagnosis, which improves the accuracy of fault diagnosis, but the statistics are time-consuming and labor-intensive.

None of the above methods can directly judge the possibility of fault occurrence based on the concentration of gas. Therefore, this paper proposes a fault diagnosis method for mine transformer based on intuitionistic fuzzy Petri net. The relationship between the gas content in the transformer oil and the fault is represented by the intuitionistic fuzzy set, and the mine transformer intuitive fault diagnosis model is established. Then the membership degree and non-membership degree are introduced into the model, and the fault diagnosis process is transformed into the use of intuition. The intuitionistic fuzzy inference process of fuzzy Petri nets is presented by means of matrix operations. The simulation results show that the mine transformer fault reasoning method can quickly judge the probability of fault occurrence according to the gas concentration.

## 2. Mine-used transformer diagnostic system principle

Transformer oil is a mixture of hydrocarbon molecules of many different molecular weights. The pyrolysis gas production of transformer oil depends on the different stability of hydrocarbon molecules with different chemical bond structures at high temperatures.

When a transformer occurs inside and there is a latent fault, hydrocarbons will oxidize/crack, some C-H bonds and C-C bonds are broken to form unstable H, $CH_3$, $CH_2$, CH, C and other free radicals. These radicals are rapidly recombined by complex chemical reactions, eventually producing hydrogen and low molecular hydrocarbon gases. Such as methane, Ethane, ethylene, acetylene, etc., may also form solid particles of carbon and a hydrocarbon polymer (X wax). Carbon solid particles and hydrocarbon polymers can be deposited inside the equipment. In the early stage of the failure, the formed gas is dissolved in the oil; when the fault energy is large, it is also possible to aggregate the free gas.

The principle of internal fault judgment of mining transformers is based on the above characteristics of gas production. Different faults, due to different energy at the fault point, different temperatures and different insulation materials involved, their gas production is also different:

1) The paper is generally overheated, the main characteristic gases are: $CH_4$, $C_2H_4$, and the secondary characteristic gas: $C_2H_2$.

2) Oil and paper are severely overheated, the main characteristic gases are: $CH_4$, $C_2H_4$, and secondary characteristic gases: $C_2H_2$, CO, $CO_2$.

3) Spark discharge in oil, main characteristic gas: $H_2$, $C_2H_2$.

4) Arc discharge in oil, main characteristic gases: $H_2$, $C_2H_2$, secondary characteristic gases: $CH_4$, $C_2H_4$.

5) Partial discharge in oil, main characteristic gases: $H_2$, $CH_4$, CO, secondary characteristic gases: $C_2H_2$, $CO_2$.

Due to the complex gas content in the transformer oil, the traditional method of fault diagnosis method cannot determine the nature and state of the fault well. It is impossible to quantitatively judge the faults at a certain moment according to the content of various gases, and which faults occur. The possibility is small. To this end, a fault diagnosis modeling method for mining transformers based on IFPN is proposed.

## 3. IFPN-BASED MINE-USED TRANSFORMER FAULT DIAGNOSIS MODELING METHOD

Atanassov K T adds membership and non-affiliation in the traditional Zadeh fuzzy set, and extends the fuzzy set theory to better describe some uncertain information. If X is a given universe A is an intuitionistic fuzzy set on X, then the intuitionistic fuzzy set of X is defined as follows

$$A = \{ < x , \mu_A(x) , \gamma_A(x) > \mid x \in X \}$$

Where $\mu_A(x)$ is the membership function of set A, $X \to [0,1]$, $\gamma_A(x)$ is the non-membership function of set A ,$X \to [0,1]$. For any element $x \in X$ on set A, there is $0 \leq \mu_A(x) + \gamma_A(x) \leq 1$.

The internal mechanism of the transformer is complex, and the mapping between the fault feature quantity and the fault type is fuzzy [14], which is very suitable for fault diagnosis using Intuitionistic Fuzzy Petri Net (IFPN).

### *3.1. Intuitionistic Fuzzy Petri Net Definition*

Intuitionistic fuzzy Petri nets can be expressed as:

IFPN = $(P, T, I, O, W, \tau, S)$, where:

1)$P = \{p_1, p_2, \ldots, p_n\}$, representing a finite set of libraries.

2)$T = \{t_1, t_2, \ldots, t_n\}$, which represents a finite set of transitions, each of which has a corresponding intuitionistic fuzzy rule.

3) $I = [w_{ij}]$is the weighted input matrix $(i = 1,2, \ldots, n; j = 1,2, \ldots, m)$, which represents the input weight of the transition to the library. When the input library of the transition $t_j$ is $p_i$ , $w_{ij}$ is equal to the weight of $p_i$ to $t_j$. When the input library of the transition $t_j$ is not $p_i$, $[w_{ij}] = 0$.

4) $O = [b_{ij}]$is the output matrix, indicating the reliability of the output arc $(I = 1, 2, \ldots n;\ j = 1, 2, \ldots m)$. When the output library of the transition $t_j$ is the library $p_i$, $b_{ij}$ is equal to $c_j$, and $c_j$ is the intuitionistic fuzzy number, which represents the credibility of the transition $t_j$, which can be expressed as $c_j = (\mu_{cj}, \gamma_{cj})$, $(j = 1, 2, \ldots m)$, where $\mu_{cj}$ is the membership degree of the credibility of the transition $t_j$, and $\gamma_{cj}$ is the non-affiliation degree of the credibility of the transition $t_j$. If the library $p_i$ is not the output library of the transition $t_j$, then $b_{ij} = (0,1)$.

5) W is the connected arc weight function, which is divided into the input arc weight function $W_I$ and the output arc weight function $W_O$. $W_I$ is equal to the weight set on the transition of the input library; $W_O$ is the intuitionistic fuzzy number $(\mu_O, \gamma_O)$, indicating The credibility of the arc connected to the output library, $\mu_O$ is the membership degree of the transition credibility, and $\gamma_O$ is the non-affiliation degree of the credibility of the transition.

6) $\tau = \{\lambda_1, \lambda_2, \ldots \lambda_n\}^T$ is the upper threshold of transition, where $\lambda_j$ is the intuitionistic fuzzy number, which can be expressed as $\lambda_j = (\alpha_j, \beta_j)$, $(j = 1,2, \ldots, m; 0 < \alpha_j + \beta_j \leq 1), \alpha_j > 0$ is the threshold of the change credibility, and $\beta_j \geq 0$ is the threshold of the non-confidence of the transition.

7)$S = [S(p_1), S(p_2), \ldots, S(p_n)]^T$, is an n-dimensional column vector representing the token value (identification) of the library node, and $S(p_i)$ is the intuitionistic fuzzy number $(\mu_i, \gamma_i)$, $\mu_i$ is the membership degree of the $p_i$ confidence of the library, and $\gamma_i$ is the non-affiliation degree of the $p_i$ confidence of the library $(i = 1, 2 \ldots n)$. $S_0 = [S_0(p_1), S_0(p_2), \ldots, S_0(p_n)]^T$, which represents the initial token value of the library node.

### *3.2. Intuitionistic fuzzy production rules*

The intuitionistic fuzzy production rule is a tool for intuitionistic fuzzy Petri net knowledge representation. Each change of the intuitionistic fuzzy Petri net can be used as an intuitionistic fuzzy production rule. The input library of the transition is the antecedent proposition of the intuitionistic fuzzy rule, and the output library of the transition is the conclusion proposition of the intuitionistic fuzzy rule. The fuzzy production rule used in this paper is the intuitionistic fuzzy production rule

IF $d_1$ AND $d_2$...AND $d_n$ THEN $d_k$

$$(\lambda, W_{I1}, W_{I2}, \ldots W_{In}, W_O)$$

Among them, $d_n$ is the premise proposition, and $d_k$ is the result proposition. In actual application, it can be replaced by the library. $\lambda$ is the transition threshold; $W_{I1}, W_{I2}, \ldots W_{In}$ are the weights of the input arc, respectively, and the weight is equivalent to the influence factor. Different weights represent the influence of the gas content in the mine transformer oil on the fault result;

$W_O$ is the weight set on the output arc, indicating the credibility of the rule. The corresponding structure diagram is shown in Figure 1.

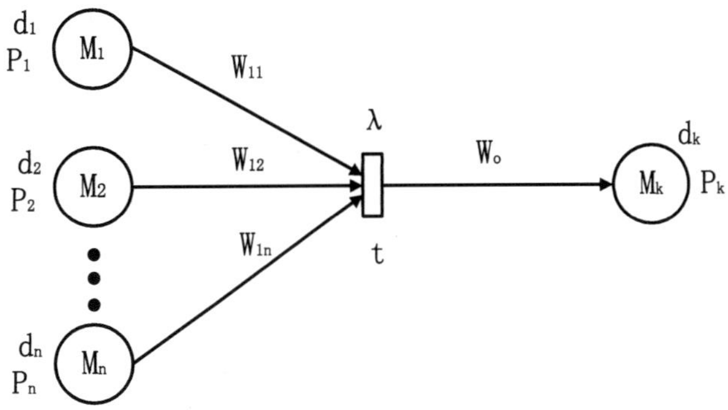

Figure 1. Combined fuzzy production.

### *3.3. IFPN based reasoning method*

Fuzzy reasoning based on intuitionistic fuzzy Petri nets is an uncertainty reasoning, which is mainly used to describe the fuzzy concept of "non-this" and has parallel dynamic reasoning ability. In the process of reasoning, the weight of the connecting arc and the threshold of the transition are determined by using the content of all gases at a certain moment of the mining transformer, and the weighted input matrix $I_{5\times11}$, the output matrix $O_{5\times11}$, the transition threshold vector $\tau$, and the library are obtained. The initial identification vector $S_0$. The transformer fault diagnosis model has 11 libraries and 5 transitions. The reasoning algorithm is as follows:

Input: weighted input matrix $I_{5\times11}$; output matrix $O_{5\times11}$; transition threshold vector $\tau$; library initial identification vector $S_0$;

Output: The identification vector of the library $S = [S(p_1), S(p_2), ..., S(p_{11})]$;

Step 1: Initialize each matrix, determine the input matrix and output matrix values, let $k = 0$, initialize the identification vector value, $S_k = S_0$;

Step 2: Calculate the reserve potential energy H of the change of the library, the intuitionistic fuzzy value $H = I^T \cdot S_k$;

Step 3: Compare H with the threshold $\tau$ and let $\varepsilon = h\infty\tau$. It can be seen that $\varepsilon$ should be an m-dimensional intuitionistic fuzzy vector of $(0,1)$ or $(1,0)$, and then let $\varphi = H \odot \varepsilon$, the premise fuzzy value of the transition is successfully excited, where $\varphi$ is an m-dimensional intuitionistic fuzzy column vector;

Step 4: Calculate the fuzzy value of the output library after one reasoning: $\theta = O \otimes \varphi$;

Step 5: Calculate $S_{k+1}$: $S_{k+1} = S_k \oplus \theta$;

Step 6: Determine the size of $S_{k+1}$ and $S_k$. If $S_{k+1} = S_k$, the reasoning ends, otherwise let $k = k + 1$, return to step 2, continue the reasoning, until the end.

### *3.4. Matrix operators used by the algorithm*

Based on the IFPN inference process as a matrix operation, several matrix operations are as follows:

1) $\otimes$: $C = A \otimes B \Leftrightarrow \max(a_{ik} \cdot b_{kj}) = c_{ij}, 1 \le k \le s$, A, B, C are $n \times s, s \times m, n \times m$ dimensional intuitionistic fuzzy matrix;

2) $\oplus$: $C = A \oplus B \Leftrightarrow \max(a_{ij} \cdot b_{ij})$, A, B, C are $n \times m$ dimensional intuitionistic fuzzy matrices, according to intuitionistic fuzzy logic, $(\max(\mu(a_{ij}), \mu(b_{ij})), \min(\gamma(a_{ij}), \gamma(b_{ij})))$

3) $\odot$: $C = A \odot B \Leftrightarrow a_{ij} \cdot b_{ij} = c_{ij}$, A, B, C are $n \times m$ dimensional intuitionistic fuzzy matrices, according to intuitionistic fuzzy logic, $a_{ij} \cdot b_{ij} = (\mu(a_{ij}) \cdot \mu(b_{ij}), \gamma(a_{ij}) + \gamma(b_{ij}) - \gamma(a_{ij}) \cdot \gamma(b_{ij}))$;

4) $\infty$: $C = A\infty B \Leftrightarrow \begin{cases} 1, a_{ij} > b_{ij} \\ 0, a_{ij} \le b_{ij} \end{cases}$

Where $a_{ij} \ge b_{ij} \Leftrightarrow \mu(a_{ij}) \ge \mu(b_{ij})$ and $\gamma(a_{ij}) \le \gamma(b_{ij})$.

### *3.5. Process based on IFPN mine-used transformer fault diagnosis method*

Based on the IFPN mine transformer fault diagnosis method, it is divided into the following steps:

1) Determine the input library, output library, and transition of the transformer fault.

2) According to the content of characteristic gas in the mining transformer oil at a certain time, set the threshold value of the weight and transition of the IFPN connecting arc, and establish the IFPN-based mining transformer fault fuzzy production rule.

3) Establish a fault diagnosis reasoning model for mine transformer based on IFPN based on fuzzy production rules.

4) Determine the input and output matrix, and the initial identification vector of the library, and use the fuzzy inference algorithm to diagnose the fault diagnosis model.

5) Find the reasoning results and analyze the results.

## 4. Experimental and simulation proof

### *4.1. Model Construction*

Based on the data provided in [8], the fault diagnosis model of mine transformer based on intuitionistic fuzzy Petri net is constructed. The weights obtained by Elman network algorithm are optimized as shown in Table 1.

Table 1. Elman optimized weight.

| Arc | Weight | Arc | Weight | Arc | Weight |
|---|---|---|---|---|---|
| $W_{11}$ | 0.4998 | $W_{25}$ | 0.3001 | $W_{43}$ | 0.0839 |
| $W_{12}$ | 01967 | $W_{26}$ | 0.0121 | $W_{45}$ | 0.1758 |
| $W_{15}$ | 0.3008 | $W_{33}$ | 0.5896 | $W_{46}$ | 0.4961 |
| $W_{23}$ | 0.5116 | $W_{36}$ | 0.4702 | $W_{51}$ | 0.6182 |
| $W_{24}$ | 0.1993 | $W_{42}$ | 0.3596 | $W_{53}$ | 0.3886 |

According to the main gases $CH_4$, $C_2H_4$, $H_2$, $C_2H_2$, CO, $CO_2$ generated in the transformer oil as input reservoirs, corresponding to $P_1$, $P_2$, $P_3$, $P_4$, $P_5$, $P_6$, respectively, five faults of the transformer: oil and paper are generally overheated, Oil and paper are severely overheated, spark discharge in oil, arc discharge in oil, and partial discharge in oil as output reservoirs, corresponding to $P_7$, $P_8$, $P_9$, $P_{10}$, and $P_{11}$, respectively. The fault fuzzy production rule is as follows:

$R_1$: if $CH_4$ is slightly higher ($W_{11}(0.4998)$) and $C_2H_4$ is lower ($W_{12}(0.1967)$) and $CO_2$ is lower ($W_{15}(0.3008)$) then oil and paper are generally overheated ($W_0(0.7,0.28)$);

$R_2$: if $H_2$ high ($W_{23}(0.5116)$) and CO high ($W_{24}(0.1993)$) and $CO_2$ high ($W_{25}(0.300)$) $C_2H_2$ is not high ($W_{26}(0.0121)$) then oil and paper Severe overheating ($W_0(0.7,0.28)$);

$R_3$: if $H_2$ is higher ($W_{33}(0.5896)$) $C_2H_2$ is slightly higher ($W_{36}(0.4702)$) then spark discharge in oil ($W_0(0.7,0.28)$);

$R_4$: if $C_2H_4$ is slightly higher ($W_{42}(0.3596)$) and $H_2$ is higher ($W_{43}(0.0839)$) and $CO_2$ is higher ($W_{45}(0.1758)$) and $C_2H_2$ is higher ($W_{46}(0.4961)$) then oil Medium arc discharge ($W_0(0.7,0.28)$);

$R_5$: if $CH_4$ is slightly higher ($W_{51}(0.6182)$) and $H_2$ is higher ($W_{53}(0.3886)$) then partial discharge in oil ($W_0(0.7,0.28)$);

Transform the above fuzzy production rule into the IFPN reasoning model, as shown in Figure 2.

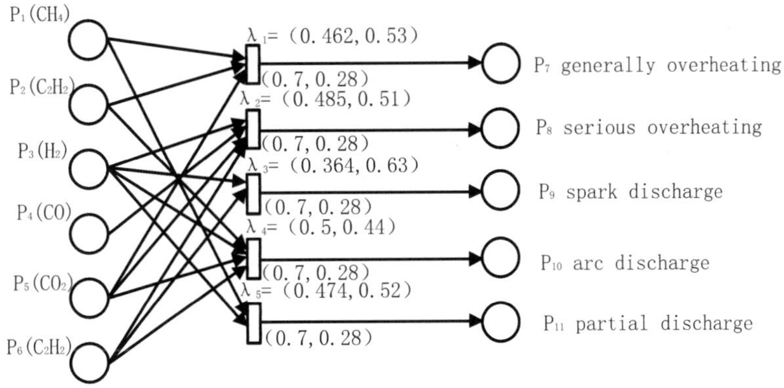


Figure 2. IFPN fault diagnosis reasoning model.

### *4.2. Example verification*

The algorithm is applied to the IFPN fault diagnosis reasoning model in Figure 2. The diagnostic reasoning process is as follows:

First, let $k = 0$, according to the data in the fault diagnosis reasoning model, the input matrix I, the output matrix $O$, the initial identification matrix $S_0$ and the transition threshold $\tau$ are determined according to the definition of the intuitionistic fuzzy Petri net and the intuitionistic fuzzy production rule.

$$I = \begin{bmatrix} 0.4998 & 0.1976 & 0 & 0 & 0.3008 & 0 & 0 & 0 & 0 & 0 & 0 \\ 0 & 0 & 0.5116 & 0.1993 & 0.3001 & 0.0121 & 0 & 0 & 0 & 0 & 0 \\ 0 & 0 & 0.5896 & 0 & 0 & 0.4702 & 0 & 0 & 0 & 0 & 0 \\ 0 & 0.3569 & 0.0839 & 0 & 0.1758 & 0.4961 & 0 & 0 & 0 & 0 & 0 \\ 0.6182 & 0 & 0.3886 & 0 & 0 & 0 & 0 & 0 & 0 & 0 & 0 \end{bmatrix}$$

$$S_0 = [(1,0)(0.56,0.4)(0.8,0.15)(0.3,0.65)(0.49,0.45)(0.7,0.24)(0,1)(0,1)(0,1)(0,1)(0,1)]^T$$

$$\tau = [(0.462,0.53)(0.485,0.51)(0.364,0.63)(0.5,0.44)(0.474,0.52)]^T$$

Let $k = 0$, initialize $S_k = S_0$

$$H = I^T \cdot S_k = [(0.757,0.214)(0.625,0.344)(0.80,0.196)(0.7,0.274)(0.929,0.058)]^T \quad (1)$$

$$O = \begin{bmatrix} (0,1)(0,1)(0,1)(0,1)(0,1)(0,1) & (0.7,0.28) & (0,1) & (0,1) & (0,1) & (0,1) \\ (0,1)(0,1)(0,1)(0,1)(0,1)(0,1) & (0,1) & (0.7,0.28) & (0,1) & (0,1) & (0,1) \\ (0,1)(0,1)(0,1)(0,1)(0,1)(0,1) & (0,1) & (0,1) & (0.7,0.28) & (0,1) & (0,1) \\ (0,1)(0,1)(0,1)(0,1)(0,1)(0,1) & (0,1) & (0,1) & (0,1) & (0.7,0.28) & (0,1) \\ (0,1)(0,1)(0,1)(0,1)(0,1)(0,1) & (0,1) & (0,1) & (0,1) & (0,1) & (0.7,0.28) \end{bmatrix}$$

$$\varepsilon = H \infty \tau = [(1,0)(1,0)(1,0)(1,0)(1,0)]^T \quad (2)$$

$$\varphi = H \odot \varepsilon = [(0.757,0.214)(0.625,0.344)(0.80,0.196)(0.7,0.274)(0.929,0.058)]^T \quad (3)$$

$$\theta = O \otimes \varphi = [(0,0.344)(0,0.344)(0,0.344)(0,0.344)(0,0.344)(0,0.344)(0.53,0.344)(0.438,0.344)(0.56,0.344)(0.49,0.344)(0.65,0.344)]^T \quad (4)$$

$$S_1 = S_0 \oplus \theta = [(1,0)(0.56,0.344)(0.8,0.15)(0.3,0.344)(0.49,0.344)(0.7,0.24)(0.53,0.344)(0.438,0.344)(0.56,0.344)(0.49,0.344)(0.65,0.344)]^T \quad (5)$$

According to the above method, we can get:

$$S_2 = [(1,0)(0.56,0.3)(0.8,0.15)(0.3,0.344)(0.49,0.3)(0.7,0.24)(0.53,0.3)(0.438,0.3)(0.56,0.3)(0.49,0.251)(0.65,0.3)]^T$$

$$S_3 = [(1,0)(0.56,0.3)(0.8,0.15)(0.3,0.3)(0.49,0.3)(0.7,0.24)(0.53,0.3)(0.438,0.3)(0.56,0.3)(0.49,0.23)(0.65,0.3)]^T$$

$S_3 = S_2$, the reasoning ends.

It can be seen from the results that the library with the highest degree of membership in the possible failure is $P_{11}$, and its identifier is $P.P_{11} = (0.65, 0.3)$, which indicates that the fault "partial discharge" corresponding to the $P_{11}$ of the library is most likely to occur. The library with the least degree of membership in the possible failures is $P_8$, which is identified as $P_8$= (0.438,0.3); This indicates that the fault of the library $P_8$ corresponding to the "oil and paper severe overheating" is the least likely.

## 5. Conclusion

For oil-immersed transformers working in mines with high coal dust content for a long time, this paper proposes a fault diagnosis method for mine transformers based on intuitionistic fuzzy Petri nets. The conclusions are as follows:

1) Mine-used transformer fault diagnosis method based on intuitionistic fuzzy Petri nets introduce information such as membership degree and non-affiliation degree into the traditional Petri net model, and realize the intuitionistic fuzzy inference algorithm through matrix operation, which is more conducive to computer programming, and the reasoning process is more efficient, reasoning results can get more information.

2) The diagnostic model can directly determine the fault of the transformer by analyzing the gas in the transformer oil, and obtain the membership degree and non-affiliation degree of each fault cause, which provides a new fault for oil-immersed transformer fault diagnosis idea.

## References

[1] Li Jianglin, etc. A FPN-based substation fault reasoning mechanism [J]. Power System Protection and Control, 2012, 40(17): 13-18.

[2] Dai Chenxi, etc. High-speed rail traction transformer based on model and fuzzy Petri net fusion disability diagnosis [J]. Power System Protection and Control, 2016, 44(11): 26-32.

[3] Wang Jianyuan, Ji Yanchao. Fuzzy Petri network knowledge representation method and its application in obstacle diagnosis [J]. Journal of China Electrical Engineering, 2003, 23(1): 122-126.

[4] S Okabe, G Ueta. Partial discharge criterion in AC test of oil immersed transformer and gas filled transformer in terms of harmful partial discharge level and signal transmission rate [J]. IEEE Trans on Dielectrics and Electrical Insulation, 2012, 19(4): 1431-1439.

[5] Zhang Yifei, etc. Oil-immersed transformer model based on chemical reaction optimization neural network and fusion DGA algorithm [J]. High Voltage Technology, 2016, 42(4): 1275-1281.

[6] Gongmao Law, etc. Power Transformer Fault Diagnosis Based on BP Network Algorithm Optimized Fuzzy Petri Net [J]. Power System Protection and Control, 2015, 43(3): 113-117.

[7] Gong Maofa, etc. Power transformer fault diagnosis based on improved bat algorithm optimization BP neural network [J]. Journal of Shandong University of Science and Technology, 2017, 3(3): 70-74.

[8] Li Shiguang, etc. Fault diagnosis of mine transformer based on optimized fuzzy Petri net [J]. Industrial and mining automation, 2017, 43(5): 54-57.

[9] Gao Lihui, etc. Transformer partial discharge monitoring and improvement BP neural network prediction model research [J]. Sensors and Microsystems, 2017, 36(3): 40-42.

[10] Shi Ruifeng, etc. Fault diagnosis expert system for dissolved gas power transformer in oil [J]. Journal of Power Systems and Automation, 2014, 26(12): 49-54.

IOP Conf. Series: Materials Science and Engineering **738** (2020) 012016 doi:10.1088/1757-899X/738/1/012016

# Enhanced Thermoelectric Performance in Li Doped SnS *via* Carrier Concentration Optimization

**Yi Niu, Yide Chen, Jing Jiang, Yan Pan, Chengcheng Yang, Chao Wang***

Clean Energy Materials and Engineering Center, State Key Laboratory of Electronic Thin Films and Integrated Devices, School of Electronic Science and Engineering, University of Electronic Science and Technology of China, 611731, China

cwang@uestc.edu.cn

**Abstract**. Tin sulfide (SnS) has attracted increasing attentions due to its similar band structure and crystal structure to tin selenide (SnSe). The layered structure and strong anharmonicity of these materials result in an ultralow lattice thermal conductivity, which is advantageous for the thermoelectric performance. Further enhancement in thermoelectric properties is expected to improve the carrier concentration and power factor by band structure optimization. In this work, lithium (Li) is proved to be an efficient dopant for SnS and the carrier concentration is increased to $1.3 \times 10^{18}$ $cm^{-3}$ at room temperature for $Sn_{0.98}Li_{0.02}S$. Calculations show that Li doping flattens the edge of the valence band and increases the number of carrier pockets in SnS, which collectively enhance the electronic transport properties. Moreover, the phonon scattering was enhanced by the nano-precipitates and increased boundaries in $Sn_{1-x}Li_xS$, resulting in 40% reduction in thermal conductivity at room temperature. Combined with the enhanced power factor (~3 μW $cm^{-1}$ $K^{-2}$ at 848 K) and low thermal conductivity (0.36 W $m^{-1}$ $K^{-1}$ at 848 K), the maximum ZT of 0.66 is achieved in $Sn_{1-x}Li_xS$ bulk samples at 848 K.

## 1. Introduction

Thermoelectric materials provide a simple and eco-friendly solution for direct conversion between heat and electrical energy without any moving parts or greenhouse gasses [1-3]. The efficiency of thermoelectric materials is generally evaluated by the dimensionless figure of merit ZT ($ZT = \alpha^2\sigma T/\kappa$), where $\alpha$ is the Seebeck coefficient, $\sigma$ is the electrical conductivity, T is absolute temperature, and $\kappa$ is the thermal conductivity consisting of electronic thermal conductivity ($\kappa_e$) and lattice thermal conductivity ($\kappa_L$) [4]. High power factor $PF = \alpha^2\sigma$ and low thermal conductivity are required to achieve high ZT values from the perspective of formulas. Nevertheless, $\alpha$, $\sigma$ and $\kappa_e$ are strongly coupled and conflict with each other through the transport properties of electrons, which leads to poor ZT values and difficult optimization [5, 6]. For example, increasing the electrical conductivity, usually realized by increasing the carrier concentration, but leads to a reduced Seebeck coefficient and increased electronic thermal conductivity [7, 8]. In a word, it's difficult to improve the thermoelectric performance by optimizing one thermoelectric parameter individually. Except for the lattice thermal conductivity [9-11], the only parameter that can be independently regulated, is an effective breakthrough to optimize the thermoelectric performance. The reduction in the lattice thermal conductivity is usually achieved by two approaches: (1) nanostructuring [12, 13], introducing defects [14] and all-scale hierarchical architecturing, [15] (2) looking for intrinsic low thermal conductivity materials with complex crystal conductivity, weak chemical bonds, heavy mass [2, 9, 16] and so on.

Tin selenide (SnSe), one of the most potential thermoelectric materials, owns the intrinsic ultralow thermal conductivity in terms of density functional theory (DFT) and calculated large Grüneisen parameters due to its strong anharmonic bonding [16, 17]. Recently, a record high ZT of ~2.8 ± 0.5 out of plane at 773 K was achieved in SnSe single crystal [18]. However, due to the long duration of synthesis, expensive preparation cost and poor mechanical properties for single crystalline [19], polycrystalline materials with similar structure have been widely studied. As an analogue of SnSe, tin sulfide (SnS) shows the similar crystal and band structure as SnSe, which inspire more research in SnS based materials. The crystal of SnS adopts an orthorhombic structure with double layers of tightly bound Sn-S atoms and the layers pile up mainly with the weak and long Sn-S interactions, [20] resulting in ultralow lattice thermal conductivity close to the predicted amorphous limit (0.26 W $m^{-1}$ $K^{-1}$), which has been reported by experiments and calculations [21-23]. However, pure SnS still possesses a poor ZT because of the extremely low carrier concentration ($2.47\times10^{15}$ $cm^{-3}$ in this work) [22, 24], therefore targeting enhancement of ZT is achieved successfully by increasing carrier concentration via p-type doping like Ag or Na, [25, 26] and a maximal ZT of 0.8 at 850 K when the carrier concentration is further increased to ~$1 \times 10^{20}$ $cm^{-3}$ has been predicted according to the SPB model in Pei's work [25].

In this work, lithium (Li) is selected as a p-type dopant that can lower the Fermi level into valence bands and flatten the top of valence band edge, thus an increased carrier concentration of $1.3 \times 10^{18}$ $cm^{-3}$ is achieved in Li doped SnS bulk samples. Even though Li doping increased the electrical conductivity by 1~3 orders of magnitude, the Seebeck coefficient maintains impressively high level of 450 ~ 650 μV $K^{-1}$ at the entire temperature range. Besides, a remarkably low thermal conductivity of 0.36 W/m/K is obtained at 848 K for SnS-1% Li sample. The resultant optimized carrier concentration and increased nanoinclusions believed to be the key factors to the enhancement of thermoelectric performance.

## 2. Materials and methods

### 2.1. *Synthesis*

Commercial flakes of 99.95% Li, powders of 99.99% Sn and 99.5% S were used as raw materials without further purification. The elements were stoichiometrically weighted of $Sn_{1-x}Li_xS$ (x = 0, 0.005, 0.0075, 0.01, 0.02 and 0.03) and then loaded in stainless steel jars in a purified argon (>99.999%) atmosphere for mechanical alloying (MA) by a high-energy ball milling (SPEX 8000 D) for 15 h. The MA-derived powders were consolidated by hot-pressing at 853 K in vacuum for 6 min under an axial pressure of 75 MPa, results in a highly dense disk-shaped pellets sample with dimension of 13 mm and 2.5 mm thickness.

### 2.2. *Phase structures and microstructures*

The phase structures were examined by X-ray diffraction (XRD) using Cu Kα radiation (DX-2700B, HAOYUAN China). The structural and morphological characterizations were obtained with scanning electron microscopy (SEM) using FEI Inspect F and transmission electron microscopy (TEM) at room temperature.

### 2.3. *Thermoelectric properties*

The Seebeck coefficient and electrical resistivity were measured with CTA-3. The total thermal conductivity was calculated by relationship $\kappa=\alpha\rho Cp$, where ρ is the density of bulk sample, α is the thermal diffusivity, and Cp is the specific heat capacity. Both Cp and α were measured by Netzsch LFA 457, in which the reference sample is Pyroceram 9606. The density of bulk sample was measured by the Archimedes method. The lattice thermal conductivity ($\kappa_L$) was obtained by subtracting the electronic thermal conductivity ($\kappa_e$) from the total thermal conductivity (($\kappa_L = \kappa - \kappa_e$). The electronic conductivity contribution is calculated according to Wiedemann-Franz law $\kappa_e = L\sigma T$. Here, L is the Lorenz number denoted by SPB model, ranging from $1.48\times10^{-8}$ to $1.49\times10^{-8}$ $V^2$ $K^{-2}$. Hall resistances

($R_H$) of bulk samples were measured with Ecopia AMP55. Hall coefficient ($R_H$) was measured under a reversible magnetic field (0.8 T) by the Van der Pauw method by using a Hall measurement system (Lake Shore 8400 Series, Model 8404, USA) in a wide temperature range. The hole mobility ($\mu_H$) and carrier concentration (n) were calculated from $\mu_H = \sigma R_H$ and $n = 1/eR_H$, where e is the elementary charge.

### *2.4. Electronic calculation methods*

The electronic structure was calculated by the density-functional theory (DFT) using the projector augmented wave (PAW) method that was implemented in the Vienna ab initio simulation package (VASP). The exchange-correlation function was taken from the generalized gradient approximation (GGA) from the parameterization of Perdew–Burke–Ernzerhof (PBE). A plane-wave basis was adopted for the expansion of the valence electron wave functions and charge densities with kinetic-energy cutoffs of 400 eV. We use a well-converged Monhorst–Pack k-point grid (9 × 9 × 9) for relaxation and self-consistent calculation in Brillouin zone. Experimental lattice constants and internal atomic coordinates were used throughout the calculations.

## 3. Results

### *3.1. Crystal structure characterization*

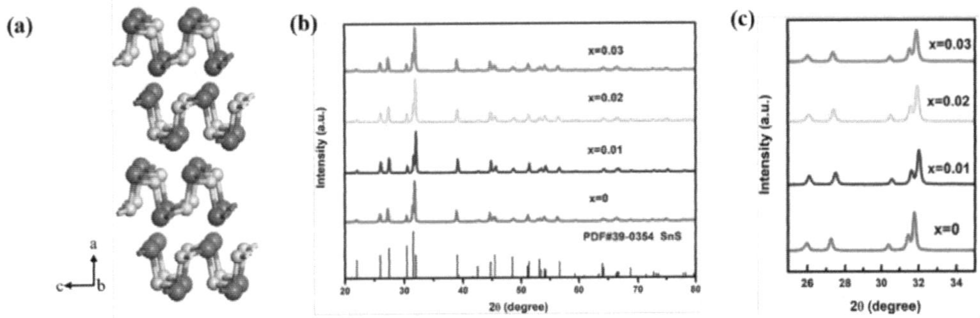


Figure 1. Crystal structure of SnS, with gray and yellow balls for Sn and S atoms, respectively (a). Bulk X-ray diffraction (XRD) patterns (b) and enlarged XRD patterns within the angular range of 25~35° (c) of $Sn_{1-x}Li_xS$ (x = 0, 0.01, 0.02 and 0.03)

Figure 1(a) shows the crystal structure of orthorhombic (α-phase) SnS with a Pbmn space group (a = 11.19 ± 0.05 Å, b = 3.98 ± 0.02 Å and c = 4.33 ± 0.02 Å). The double Sn-S planes are linked by van der Waals force and the atoms in each layer are joined with the three nearest neighboring atoms by covalent bonds, forming accordion shaped corrugated slabs. The crystal structures of the as-synthesized $Sn_{1-x}Li_xS$ (x = 0, 0.005, 0.0075, 0.01, 0.02 and 0.03) bulk samples are determined by X-ray diffraction (XRD), and all of the peaks are well indexed to the orthorhombic structure of SnS (PDF#39-0354) as shown in figure 1(b). However, the peak positions for Li doped samples shift slightly to higher angles relative to the pure SnS due to the substitution of a larger atom Sn (112 pm) by a smaller atom Li (76 pm), as shown in figure 1(c), an enlarged view of figure 1(b) within the angular range of 25~35°. This seems to indicate that Li goes into the SnS lattice and the stress field is generated due to the different atomic radius of Sn and Li, which is characterized by Raman spectra. The pure SnS displays the peaks at 163, 193, 219 and 285 $cm^{-1}$ [27], while the Raman peaks shift to lower wave number can be observed in $Sn_{0.98}Li_{0.02}S$ as shown in figure 2(a). Further, the Li-S bond and Sn-S bond have been detected by X-ray photoelectron spectroscopy (XPS), in which the spectrum of S 2p shows two peaks at binding energies of 161.97 eV [28] and 160.87 eV [29] as shown in figure 2(b).

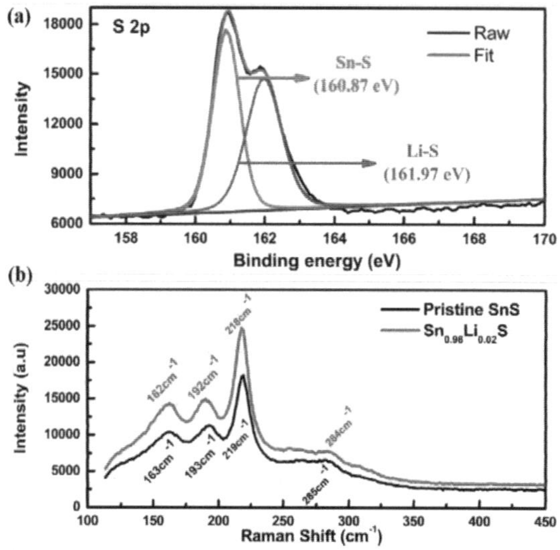


Figure 2. (a) Raman spectra of SnS and $Sn_{0.98}Li_{0.02}S$, (b) X-ray photoelectron spectrum (XPS) of the 2p orbitals for the S element of $Sn_{0.98}Li_{0.02}S$.

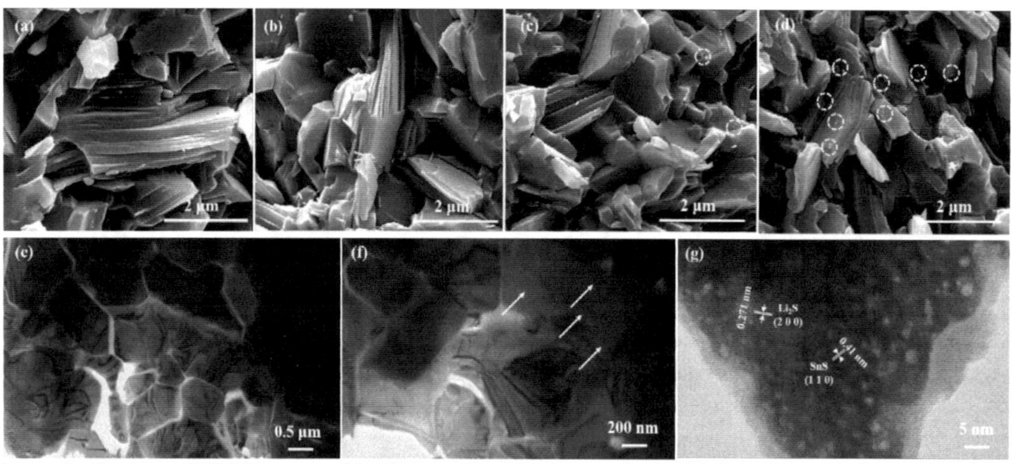


Figure 3. Fracture SEM images of bulk samples of (a) pure SnS, (b) $Sn_{0.99}Li_{0.01}S$, $Sn_{0.98}Li_{0.02}S$, (d) $Sn_{0.97}Li_{0.03}S$ and (e-g) TEM for $Sn_{0.98}Li_{0.02}S$

Figure 3(a)-(d) exhibit the scanning electron microscopy (SEM) for bulk $Sn_{1-x}Li_xS$. Pure SnS exhibits a typical lamellar morphology as shown in figure 3(a) and the strong anisotropy of pure SnS has been studied in our pre-work [30]. As the Li content increases, the grain size gradually decreases and the layered structure gradually disappears, meanwhile some nano-precipitates occur on the surface of SnS matrix, as marked in figure 3. To further clarify the composition and structure of doped samples, the microstructures for $Sn_{0.98}Li_{0.02}S$ are investigated by transmission electron microscope (TEM) and shown in figure 2(e)-(g). The regular parallel fringes spacing of about 0.271 nm is obtained for the nano-precipitates, corresponding to (2 0 0) lattice plane of $Li_2S$. The increased grain boundaries and the nanoinclusions of $Li_2S$ in Li-doped samples play an important role in the enhancement of thermoelectric performance, which will be described in details below.

### *3.2. Electronic structure calculations*

Figure 4 shows the electronic band structure and density of states (DOS) of pure and doped SnS, which indicates that the DOS near the Fermi level is mainly derived from Sn and S, and the contribution of Li is negligible (inset Figure 4(d)). But the band structure of SnS is modified significantly by Li doping in two aspects like alkali elements doped SnSe samples [31, 32]. Firstly, the number of carrier pockets are increased because the Fermi level ($E_f$) of Li doped SnS is shifted into the two main valence bands VB1 and VB2, and approached VB3. Secondly, flatting the top of the valence band, resulting in an enhanced effective mass. Therefore, an enhanced Seebeck coefficient by the increment of the local DOS near the Fermi level via Mott expression as shown in the following.

$$\frac{1}{\hbar^2}\left(\frac{\partial^2 E}{\partial \kappa^2}\right)_{k=k_i} = \frac{1}{m^*}$$

$$\alpha = \frac{\pi^2 {k_B}^2 T}{3q}\left\{\frac{dn(E)}{ndE} + \frac{d\mu(E)}{\mu dE}\right\}_{E=E_F}$$

With

$$n[E] = g[E]f[E]$$

where ħ is the Planck constant and ki is the wave vector of the i-th carrier pocket, $k_B$ is the Boltzmann constant, f(E) is the Fermi function, q is the carrier charge, n(E) is the carrier mobility, and g (E) is the density of states (DOS).

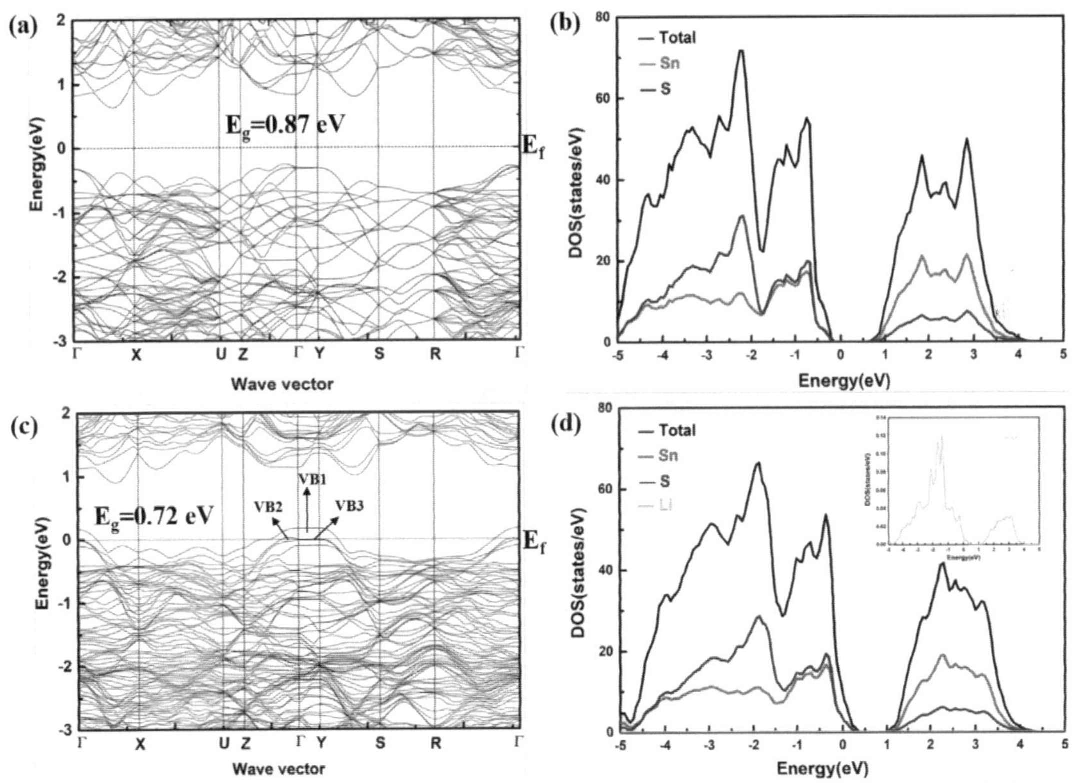


Figure 4. Electronic band structure (a) and density of states (DOS) (b) of pure SnS, electronic band structure (c) DOS (d) of Li-doped SnS. The Fermi energy was set to 0.

*3.3. Thermoelectric properties*

Thermoelectric properties for all samples are measured below the phase transition temperature (around 878 K), shown in figure 5. The pure SnS exhibits typical semiconducting thermally activated transport behavior and owns an extreme low electrical conductivity due to the low

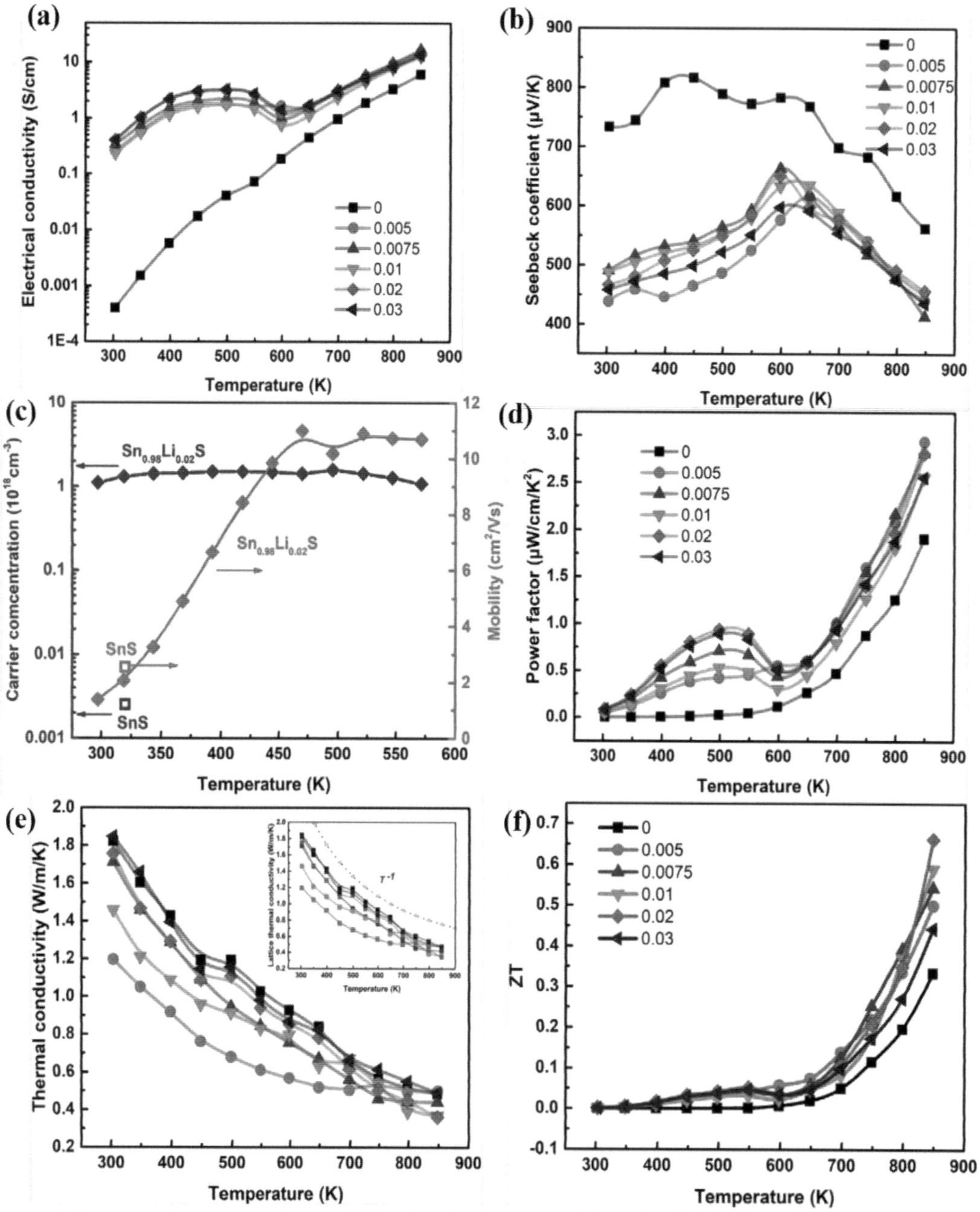


Figure 5. Temperature dependence of (a) electrical conductivity, (b) Seebeck coefficient, (c) carrier concentration and mobility, (d) power factor, (e) thermal conductivity and (f) ZT values of $Sn_{1-x}Li_xS$ (x = 0, 0.005, 0,0075, 0.01, 0.02 and 0.03).

carrier concentration. After introducing Li, the electrical conductivity dramatically enhanced from $3.98\times10^{-4}$ S cm$^{-1}$ to 0.4 S cm$^{-1}$ at room temperature as the carrier concentration increased from almost $2.47\times10^{15}$ cm$^{-3}$ to $1.32\times10^{18}$ cm$^{-3}$, as shown in figure 5(a) and (c). The electrical conductivities of all doped samples show the same trend, and the upturn when T > 598 K can be attributed to the thermal excitation of minority carriers like Ag doped SnS [26] and Pb, Zn co-doped SnSe [33] samples. Figure 5(b) shows the temperature dependent Seebeck coefficient of $Sn_{1-x}Li_xS$. Pure SnS has a high Seebeck coefficient between 562 μV K$^{-1}$ and 816 μV $K^{-1}$ at the whole testing range, consistent with a p-type semiconductor. Even though Li doping increased the electrical conductivity by 1~3 orders of magnitude, the Seebeck coefficient maintains impressively high level (above 400 μV $K^{-1}$) due to the increased effective mass, as described in figure 4. The Seebeck coefficient of doped samples increase as a function of temperature, showing the peak of Seebeck coefficient appears at 598 ~ 648 K, and then decrease due to the intrinsic excitation of carriers, which is consistent with electrical conductivity. According to the Goldsmid method [34, 35], the effective bandgap could be determined by the Seebeck coefficient measurement for a simple band structure such as SnS via $E_g = 2eS_{max}T_{max}$, where $S_{max}$ and $T_{max}$ are the maxium Seebeck coefficient and the temperature at which $S_{max}$ occurs. This yields band gap of 0.7 ~ 0.8 eV for Li doped SnS samples, which is highly consistent with the calculated band gap of pure SnS and $Sn_{0.98}Li_{0.02}S$ as shown in figure 4. Although the doped samples show a lower Seebeck coefficient than that of the undoped SnS, the much higher electrical conductivity of the former yields larger PF as shown in figure 5(d). The highest power factor of doped samples reaches ~3 μW $cm^{-1}$ $K^{-2}$ at 848 K for $Sn_{0.995}Li_{0.005}S$, which is significantly higher than pure SnS (1.9 μW $cm^{-1}$ $K^{-2}$ at 848 K).

Thermal conductivities for all samples are below 2 W $m^{-1}$ $K^{-1}$ and decreases as temperature increases as shown in figure 5(e). Obviously, the κ comes mainly from the lattice thermal conductivity (inset, figure 5(e)) because of the negligible electronic contribution. It is found that Umklapp processes dominate the scattering of phonons in all of the samples because $\kappa_L$ decreases with increase in temperature via a $T^{-1}$. Moreover, the doped samples own a thermal conductivity as low as 0.36 W $m^{-1}$ $K^{-1}$ for $Sn_{0.98}Li_{0.02}S$ at 848 K, even lower than that of Ag and Na doped SnS samples, which is attributed to the larger atomic size difference between Li and Sn and therefore a stronger effect on the scattering of phonons. Moreover, the extremely low thermal conductivity benefits from the grain boundaries and nanoinclusions introduced by Li doping, as shown in Fig. 3.

ZT values of $Sn_{1-x}Li_xS$ samples are calculated and compared with Na and Ag doped SnS samples, which is shown in figure 5(f). All samples show poor thermoelectric performance at low temperature because of the poor electrical conductivity and increase with increasing testing temperature. The highest ZT value of 0.66 is obtained for $Sn_{0.98}Li_{0.02}S$ at 848 K, which is about 100% higher than that of pure SnS. The maximum value is comparable to that of Na doped SnS [25] and slightly higher than that of Ag doped SnS [26] samples. According to the present work, lithium is an efficient dopant for SnS, leading to an increased carrier concentration, strongly phonon scattering and optimized band structure. Therefore, the enhanced ZT values are obtained due to the increased PF and decreased thermal conductivity through decoupling and simultaneous optimizing each thermoelectric parameter.

## 4. Conclusions

SnS is one of the promising thermoelectric materials like SnSe because of the strong lattice anharmonicity. However, the ZT values of SnS are suppressed by the poor electrical conductivity caused by the low carrier concentration. In this work, Li doped SnS bulk samples are obtained successfully by mechanical alloying method. Calculations show that the Fermi level enters into valence band and the valence band edge is flattened after Li doping, both the two effects have a positive influence on the thermoelectric performance enhancement. The highest ZT value of 0.66 is obtained for $Sn_{0.98}Li_{0.02}S$ at 848 K contributed by the carrier concentration optimization and reduced thermal conductivity. The thermoelectric performance could be further improved by continuously increasing carrier concentration, but the solid solubility may be a stumbling block. Besides, microstructure and band structure regulating, compositional optimization, solid solution are also efficient approaches for the further improvement of SnS.

## References

[1] Yang L, Chen Z G, Dargusch M S and Zou J 2018 *Adv. Energy Mater.* **8** 1701797.
[2] Wang C, Niu Y, Jiang J, Chen Y D, Tian H Q, Zhang R, Zhou T, Xia J F, Pan Y and Wang S Y 2018 *Nano Energy* **45** 432-38.
[3] Bell L E 2008 *Science* **321** 1457-61.
[4] He J and Tritt T M 2017 *Science* **357** eaak9997.
[5] Snyder G F and Toberer E S 2008 *Nat. Mater.* **7** 105-14.
[6] Tan G J, Zhao L D and Kanatzidis M G 2016 *Chem Rev.* **116** 12123-49.
[7] Wu R N, Li Z L, Li Y B, You L, Luo P F, Yang J and Luo J 2019 *Journal of Materiomics* **5** 489-95.
[8] Zhu T J, Liu Y T, Fu C G, Heremans J P, Snyder J G and Zhao X B 2017 *Adv. Mater.* **29** 1605884.
[9] Chen Z W, Zhang X Y and Pei Y Z 2018 *Adv. Mater.***30** e1705617.
[10] Hanus R, Agne M T, Rettie A J E, Chen Z W, Tan G J, Chung D Y, Kanatzidis M G, Pei Y Z, Voorhees P W and Snyder G F 2019 *Adv. Mater.* 1900108.
[11] Zhou T, Mao J, Jiang J, Song S W, Zhu H T, Zhu Q, Zhang Q Y, Ren W Y, Wang Z M, Wang C and Ren Z F 2019 *J. Mater. Chem.* C, **7** 434.
[12] Hsu K F, Loo S, Guo F, Chen W, Dyck J S, Uher C, Tim H, Polychroniadis E K, Kanatzidis M G 2004 *Science* **303** 818-21.
[13] Li J F, Liu W S, Zhao L D and Zhou M 2010 *NPG Asia Mater.* **2** 152-58.
[14] Zhou C J, Lee Y K, Cha J, Yoo B, Cho S P, Hyeon T and Chung I 2018 *J Am. Chem. Soc.* **140** 9282-90.
[15] Biswas K, He J Q, Blum I D, Wu C I, Hogan T P, Seidman D N, Dravid V P, Kanatzidi M G 2012 *Nature* **489** 414-18.
[16] Zhao L D, Lo S H, Zhang Y S, Sun H, Tan G J, Uher C, Wolverton C, Dravid V P and Kanatzidis M G 2014 *Nature* **508** 373-77.
[17] Chen Y X, Ge Z H, Yin M J, Feng D, Huang X Q, Zhao W Y and He J Q 2016 *Adv. Funct. Mater.* **26** 6836-45.
[18] Chang C, Wu M H, He D S, Pei Y L, Wu C F, Wu X F, Yu H L, Zhu F Y, Wang K D, Chen Y, Huang L, Li J F, He J Q and Zhao L D 2018 *Science* **360** 778-83.
[19] Lee Y K, Luo Z Z, Cho S P, Kanatzidis M G, Chung I 2019 *Joule* **3** 1-13.
[20] Li S, Li X F, Ren Z F and Zhang Q 2018 *J. Mater. Chem.* A **6** 2432-48.
[21] Parker D and Singh D J 2010 *J. Appl. Phys.* **108** 083712.
[22] Tan Q and Li J F 2014 *J. Electron. Mater.* **43** 2435-39.
[23] Sun B Z, Ma Z J, He C and Wu K C 2015 *RSC Adv.* **5** 56382-90.
[24] Tan Q, Wu C F, Sun W and Li J F 2016 *RSC Adv.* **6** 43985-88.
[25] Zhou B Q, Li S, Li W, Li J, Zhang X Y, Lin S Q, Chen Z W and Pei Y Z 2017 *ACS Appl. Mater. Interfaces* **9** 34033-41.
[26] Tan Q, Zhao L D, Li J F, Wu C F, Wei T R, Xing Z B and Kanatzidis M G 2014 *J. Mater. Chem.* A **2** 17302-06.
[27] Kumar M, Patel M, Kim, Kim J, Kim J and Kim B S 2018 *Mater. Lett.* **213** 122-25.
[28] Abraham K M and Chaudhri S M 1986 *J. Electrochem. Soc.* **133** 1307-11.
[29] Shalvoy R B, Fisher G B, Stiles P J 1977 *Phys. Rev.* B **15** 1680-97.
[30] Wang C, Chen Y D, Jiang J, Zhang R, Niu Y, Zhou T, Xia J F, Tian H Q, Hu J and Yang P 2017 *RSC Adv.* **7** 16795-800.
[31] Ge Z H, Song D S, Chong X Y, Zheng F S, Jin L, Qian X, Zheng L, Dunin-Borkowski R E, Qin P, Feng J and Zhao L D 2017 J. *Am. Chem. Soc.* **139** 9714-20.
[32] Peng K L, Lu X, Zhan H, Hui S, Tang X D, Wang G W, Dai J Y, Uher C, Wang G Y and Zhou X Y *Energy Environ. Sci.* **9** 454-60.
[33] Liu J, Wang P, Wang M Y, Xu R, Zhang J, Liu J Z, Li D, Liang N N, Du Y W, Chen G and Tang G D 2018 *Nano Energy* **53** 683-89.
[34] Pei Y Z, Heimz N A and Snyder G J 2011 *J. Mater. Chem.* **21** 18256.

IOP Conf. Series: Materials Science and Engineering **738** (2020) 012017 doi:10.1088/1757-899X/738/1/012017

# Research on Ship Electronic Power System Fault Diagnosis Based on Expert System

**Chao Liang[1], Shudong Wang[1], Rui Chen[1], Shuo Zhao[1] and Yongyong Wu[1]**

[1]College of Electrical and Information Engineering, Lanzhou University of Technology, Lanzhou, Gansu, 730050, China

*Corresponding author's e-mail: 543294659@qq.com

**Abstract.** Aiming at the difficulty in locating the fault point quickly and accurately after the ship electronic power system fault. In this paper, fault diagnosis expert system is used to diagnose the fault occurred to ensure stability of power supply of ship power system. The possible faults are divided into four categories on the base of analysis of ship electrical system fault characteristics. Frame structure is used to describe the connections between the various parts of the ship power system in knowledge presentation. Expert system uses production rules to describe the specific fault and combine the ideal of fuzzy inference to dealing with the uncertainty in fault diagnosis at the same time. The expert system designed in this paper can find fault points and provide solutions quickly and accurately and improves the reliability of the ship power system by test.

## 1. Introduction

With the continuous development of China's shipbuilding industry, marine power system is more and more tend to be more and more complicated, the hair of the ship power grid power distribution module and the complexity of the load also constantly improve, once the failure occurs somewhere, largely caused the secondary faults, which makes the engineer's work has greatly increased, it is often difficult to rapidly and accurately locating the fault point and troubleshooting, seriously affected the stability of the vessel operation[1-3]. And ship power equipment along with the development of the shipbuilding industry are also constantly upgrading, and the original utility engineer is very difficult to cope with the new equipment failure, therefore, it is necessary to construct a set of can integrate the expert knowledge and can constantly update the knowledge base in the field of fault diagnosis expert system[4-5], improving the work efficiency and the accuracy of the engineer, ensure the continuity of vessel operation. Traditional power system fault diagnosis is based on the research of the characteristics of the power system network topology model is established, and the expansion of system correlation matrix, vary according to the trend of the event of a failure to fault location, the method based on mathematical calculation due to the complexity of the electric power system and increasingly expanded scale, at the same time under the influence of various aspects of uncertainty is difficult to achieve an ideal effect [6]. At present, intelligent methods are mainly adopted for fault diagnosis of power system, including those based on signal processing, analytical model and knowledge. Among them, literature [7] adopted an analytical model-based approach, first analyzing the characteristics of the diagnosed object, and then Establish the corresponding mathematical model, an observer is constructed to estimate the output of the system, and then each sensor measurement value compared with the estimate, so as to extract the fault information, this method because of the difficulty of modeling and the event of a failure may cause the change of the model structure or parameters, affect the accuracy of the conclusion. Literature [8] adopts the method based on signal processing. The advantage of this method is that it can avoid modeling complex models and directly adopt models such as wavelet transform, correlation function, autoregressive sliding average and frequency spectrum. This method is relatively simple to implement, but it is not sensitive to the detection of minor faults. Only when the faults develop to a certain degree can it have a better detection effect. Literature [9] used the method based on knowledge, the method not only inherits the advantages of the signal processing does not need to build mathematical model, while avoiding the defect insufficient sensitivity of the signal processing method, the method introduces expert diagnosis knowledge, can give full play to human experts in the process of diagnosis and subjective experience, improve The efficiency and accuracy of fault diagnosis, which is based on knowledge, has become the focus of many researchers. On the basis of above research, combining with the characteristics of ship power system, through the analysis of power system, the possibility of failure and its causes, to produce the fault classification, frame type knowledge are used to indicate the structure of the whole power system, at the same time use the production rule knowledge of specific faults are described, the fuzzy reasoning to solve the uncertainty in the process of reasoning, designed a fault diagnosis expert system. This system combines the reasoning knowledge and diagnosis experience of several experts, and the diagnosis ability of some faults can even surpass that of a single expert. Through tests, this expert system can make rapid and accurate diagnosis of various faults.

## 2. The overall structure of expert system and fault classification of Marine power system

### *2.1. overall design of fault diagnosis expert system*

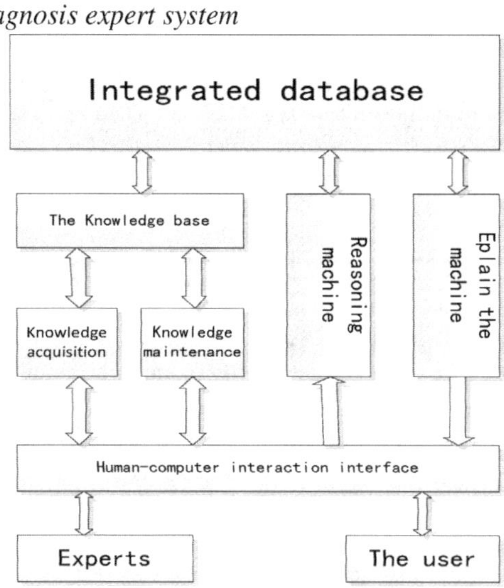


Picture 1 The overall structure of the expert system

Fault diagnosis expert system of Marine power system can be divided into five parts, including comprehensive database, knowledge base, inference machine, explanation machine and human-computer interaction interface, as shown in figure 1. Its working principle is as follows: the user inputs fault features into the expert system by human-computer interaction interface, the inference machine loads rules and matches them with the knowledge in the knowledge base, sends the matched information into the comprehensive database, and feeds back to the user through the interpretation machine [10]. Experts can also upgrade and maintain the information in the knowledge base through human-computer interaction interface, so that the expert system can constantly adapt to the updating of equipment. Through the above operations can achieve the output. Enter the fault to judge, so as to get the diagnosis results.

IOP Conf. Series: Materials Science and Engineering **738** (2020) 012017 doi:10.1088/1757-899X/738/1/012017

### *2.2. Classification of Marine power system faults*
Ship power system due to its special working environment and the layout of the intensive, under the influence of environmental factors is more, fault phenomenon and the cause of the problem has a certain randomness and disruptive, based on the relay protection knowledge and ship electric experts experience, common fault of Marine power system are classified and summarized. In this paper, the faults of Marine power system studied can be divided into four categories, namely generator fault, distribution device fault, power grid fault and power load fault. The cable wiring of ship power system is very close and the working environment is bad, and the external insulation layer of cable is vulnerable to corrosion. Therefore, the power grid fault is the most frequent place for the fault of ship power system, and the power grid fault often leads to the damage of other parts and causes secondary fault[11-12]. Therefore, power grid faults should be given priority in fault diagnosis.

## 3. Knowledge base construction of expert system
Knowledge base and reasoning machine separate is marked characteristics from the traditional application of expert system, knowledge base is used to store all kinds of diagnosis expert system rules, fault reasoning machine is used for the input and the knowledge base of knowledge is compared to diagnosis conclusion, the separation of knowledge base and reasoning machine makes it to expert system knowledge to upgrade the need to modify the contents of the knowledge base without having to modify the reasoning machine, which greatly reduce the cost of the expert system upgrade, this expert system USES Microsoft Access 2008 for the construction of knowledge base, To realize the characteristics of the knowledge base that is easy to modify and input [13].

### *3.1. Representation of knowledge*
The expert system knowledge representation of adopts frame type and the method of combination of production, first set up the research of power system network topology structure, framework representation is used to describe the topology structure, again to play to the frame type is easy to describe attributes of objects and the advantage of internal relations, the same type of power components with similar groove include: inheritance framework name, type, description, and fault description, certainty and default values, as shown in figure 2. Through the failure phenomenon to diagnose the cause of the problem in the process, this system USES production rules of fault diagnosis knowledge, production rule is usually can be expressed as an IF P THENQ < CF >, in which P and Q respectively corresponding rules and conclusions, the premise of CF said confidence, a rule generally includes rules number, premise, rules, conclusion and confidence, for example: rule 11 IF the generator start-up failure, and the excitation current is close to zero THEN Reason for failure: loss of excitation power source confidence: 0.7, where confidence is used for inexact reasoning.

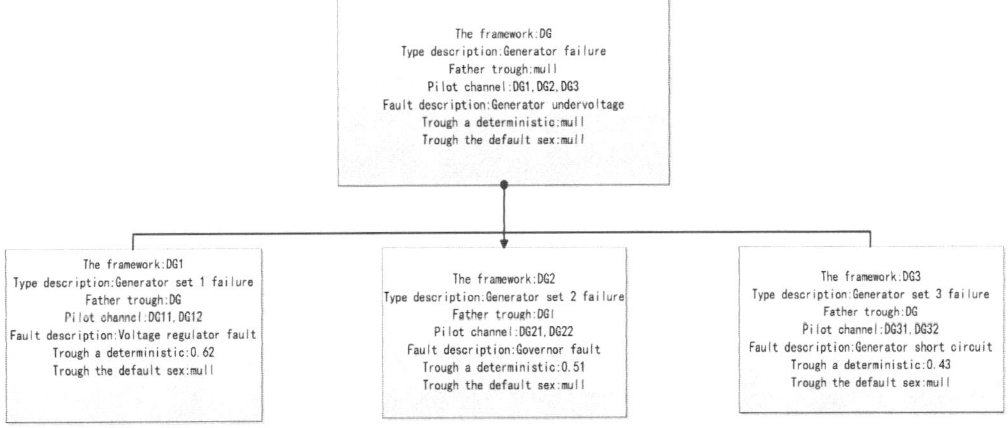


Picture 2 Frame rule

### *3.2. Fuzzy rules*

Based on past experience, the ship power system fault occurred most frequently in grid part, so need to focus on the grid fault analysis, usually when transmission line grounding or short circuit fault occur, tend to appear increase with the decrease of or current voltage, voltage and current value and the load characteristic and fault points are closely associated with power distance, and often use a certain threshold to determine whether the increase of current and voltage reduce method does not apply here, here, using the fuzzy rules are used to diagnose the faults of transmission lines [14-16]. Since there is generally no condition that voltage increases while current decreases in the case of transmission line fault, the linear membership function in FIG. 3 is used to conduct the paste treatment on the voltage and current values

$$u(x)=\begin{cases} 1.0 & x<10 \\ 1.25-0.025x & 10\le x\le 30 \end{cases} \tag{1}$$

When the current is high, the membership function is

$$u(x)=\begin{cases} 0.05x-1 & 30<x<40 \\ 1.0 & 40\le x \end{cases} \tag{2}$$

For example, when the current value is 20A, u(x)=0.75

When Ix=20A, the reliability of normal current is 0.75. When the voltage is low, the membership function is

$$u(x)=\begin{cases} 1.0 & x<320 \\ (380-x)/60 & 320\le x\le 350 \end{cases} \tag{3}$$

That is V x=360, when the reliability of the normal voltage is 0.75, the voltage and current reliability obtained is the reliability of the fuzzy rule, and the preprocessed voltage and current value can be matched with the fuzzy rule shown in table 1 to get the diagnosis result.

## 4. Reasoning mechanism of expert system

### *4.1. reasoning mechanism of power grid module*

The inference machine is responsible for how the expert system invokes rules in the process of fault diagnosis. The perfection of the inference machine design will directly affect the inference accuracy and efficiency of the expert system. According to the characteristics of the ship power system, compared with other types of fault, fault module [17], the highest frequency of failure, therefore, preferred fault diagnosis module of power grids will greatly improve the working efficiency of the expert system, in order to improve the reasoning speed, module for power grid fault reasoning using the forward reasoning method, expert system will be known the fault feature information of loaded into comprehensive database, and then one by one with the diagnosis rules knowledge base pairing [18-20], will successfully matching rule conclusion as a new fact is added to the comprehensive database, using the updated database to match again, Until a conclusion is reached or no new knowledge is added.

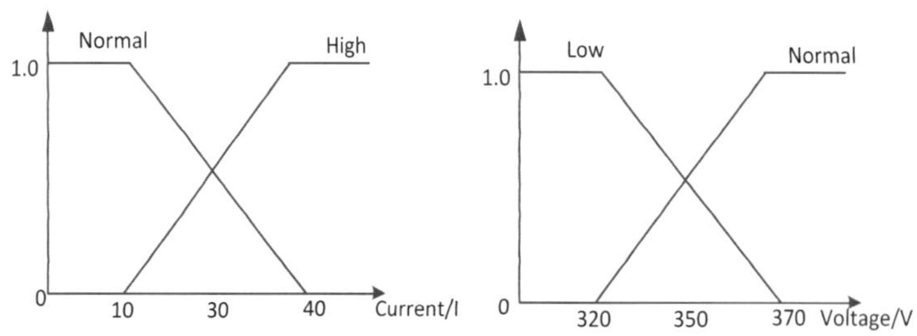

Picture 3 Membership function of voltage and current

Table 1 Fuzzy rules

| Rule number | Ia | Ib | Ic | Ua | Ub | Uc | The fault types |
|---|---|---|---|---|---|---|---|
| F1 | Normal | Normal | Normal | Normal | Normal | Normal | Normal |
| F2 | High | Normal | Normal | Low | Normal | Normal | A phase short circuit |
| F3 | Normal | High | Normal | Normal | Low | Normal | B phase short circuit |
| F4 | Normal | Normal | High | Normal | Normal | Low | C phase short circuit |
| F5 | High | High | Normal | Low | Low | Normal | AB phase short circuit |
| F6 | Normal | High | High | Normal | Low | Low | BC phase short circuit |
| F7 | High | Normal | High | Low | Normal | Low | AC phase short circuit |
| F8 | High | High | High | Low | Low | Low | ABC phase short circuit |

### *4.2. Reasoning mechanism of other modules*

When the expert system of fault diagnosis module is not found after finishing point, fault happened in other modules, because the level of the rest of the equipment failure point is less, can be relatively easy to find fault point, at this time can use the forward reasoning and reverse reasoning methods at the same time, to further improve the accuracy of reasoning, the first use of forward reasoning to put forward the hypothesis of fault point, and then prove the existence of the hypothesis, through a reverse reasoning to find fault point [21].

## 5. Conclusion

This paper USES the expert system method to diagnose the fault of the Marine electric power system. According to the characteristics of the Marine electric power system, the fault is divided into four modules. The frame structure is used to describe the overall structure of the ship's power grid and the relationship between the modules. The specific diagnostic knowledge is described by production rules. At the same time, fuzzy rules are adopted to solve the uncertainty problems in fault diagnosis. After power system simulation, the fault diagnosis expert system of Marine power system designed in this paper can accurately diagnose the fault and put forward the solution.

## References

[1] Jialin Wang, Li Xia, Zhengguo Wu, Xuanjie Yang. Fault diagnosis method of ship power grid using quantum genetic algorithm [J]. Journal of power systems and automation, 2013,25 (03): 35-39.

[2] Wensheng Zheng. Application of fault diagnosis expert system in fault diagnosis of Marine power system [J]. China water transport (next month), 2010,10 (04): 88-89.

[3] Haiyan Zhang, Fei Xia. Fault diagnosis of Marine power system based on Petri net [J]. Journal of Harbin engineering university, 2008 (09): 933-937.

[4] Cong Wang, Xiaohui Ji, Jinbo Liu. Design of fault diagnosis expert system for Marine electric power system [J]. Automation technology and application, 2002 (01): 24-26

[5] Xiaohui Zhu, Renlong Qi. Fault diagnosis and optimization method for ship power grid based on genetic algorithm [J]. Ship science and technology, 2016,38 (12): 88-90.

[6] Xiaohui Ji, Cong Wang, Yong Liu. Intelligent research on fault diagnosis expert system of ship electric power system [J]. Ship electric technology, 2001 (06): 32-35.

[7] Dazhi Wang, Xuecchen Jiang,Yi Ning, Linlin Zhao. An improved analytical model for power grid fault diagnosis [J]. Journal of Northeastern University (natural science edition), 2016,37 (08): 1065-1069.

[8] Lei Wang, Qing Chen, Zhanjun Gao. Knowledge representation method of fault diagnosis of transmission network and its application [J]. China electrical engineering bulletin, 2012,32 (04): 85-92.

[9] Ziguan Zhou, Xiaomin Bai, Zaihua Li, Jing Xu, Xiaojun Li, Huiling Li. Fault diagnosis method of intelligent transmission network based on knowledge grid technology [J]. Chinese journal of

IOP Conf. Series: Materials Science and Engineering **738** (2020) 012017 doi:10.1088/1757-899X/738/1/012017

electrical engineering, 2010,30 (04): 8-15.
[10] Shizhen Dong. Research method for fault diagnosis of power grid [J]. Guangdong electric power, 2009,22 (05): 32-36.
[11] Peng Miao, Yang Xu, Ping Jiang. Review and development of fault diagnosis of power transmission network [J]. Relay, 2005 (22): 84-91.
[12] Zhixiang Li. Design of Marine electric power support device using fuzzy comprehensive evaluation system [J]. Ship science and technology, 2016,
[13] Wei He, Tao Ren. Research on power grid fault diagnosis expert system based on fuzzy inference [J]. Power system communications, 2008 (09) : 47-50.
[14] Yi Du, Peichao Zhang, Weiyong Yu, Yuben Nie. Fault diagnosis system for distributed power grid based on fault recording data [J]. Relay, 2003 (01): 26-29, 38.
[15] Dongmei Zhao, Dongying Zhang, Kaili Xu, Jun Liu, Shu Gao, Yihan Yang. Design of fault diagnosis expert system for hierarchical distributed power grid [J]. Modern electric power, 2001 (03): 41-46.
[16] Ping Xie, Cuiru Wang, Junlin Chen, Fengzhan Tian. Object-oriented representation of power grid fault diagnosis knowledge [J]. China electric power, 1997 (02): 41-43.
[17] Xin Ping, Yunhui Yin. Design of a new power grid fault diagnosis expert system [J]. Jiangsu electrical engineering, 2007 (03): 15-17.
[18] Xinguo Hou, Xin Pan. Ship power system simulation based on Simulink [J]. Computing technology and automation, 2016,35 (01): 18-21.
[19] Han Zhou, Weifeng Shi. Ship power system fault condition simulation [J]. Ship power technology, 2008 (04): 193-196.

IOP Conf. Series: Materials Science and Engineering **738** (2020) 012018 doi:10.1088/1757-899X/738/1/012018

# Research on compressed air energy storage technology based on photovoltaic power generation system

**Yongyong Wu[1], Shudong Wang[1], Rui Chen[1], Shuo Zhao[1]and Chao Liang[1]**

[1]College of Electrical and Information Engineering, Lanzhou University of Technology, Lanzhou, Gansu, 730050, China

*Corresponding author's e-mail: wyy2117@163.com

**Abstract:** At the present time of rapid development of social economy, the upsurge of "energy revolution" has come. Clean, low-carbon, safe and efficient modern energy system is undoubtedly the current research hotspot. Micro-grid with photovoltaic power supply has become an important way for renewable energy utilization. However, there are still some problems in the operation of micro-grid with photovoltaic power supply in terms of power supply reliability and power quality. By adopting energy storage control technology to control the operation of the micro-grid with photovoltaic power supply, the reliability of the micro-grid operation can be effectively improved, so that the renewable energy utilization form can be effectively applied.

## 1. Introduction

Electric energy is an indispensable application energy in today's society. Its influence range is very wide. At the same time of technology development, the field of electric energy production has also made innovative adjustments to its process projects. Only by fundamentally optimizing the structure of electric energy production can we meet increasingly. The increased demand for electrical energy, and the power industry has gradually developed into a basic industry for the operation of the national economy. In the past, the grid operation system has certain defects, which has a lag effect on the supply of electric energy to a certain extent. The emergence and application of energy storage technology can effectively improve the flexibility of power production and provide economic construction. It is of great significance to promote the application and development of energy storage technology.

## 2. The role and mode of energy storage technology in the power system

### *2.1. The role of energy storage technology in the power system*

Energy storage technology can help increase the penetration rate of renewable energy, promote the development of distributed generation; improve the stability of power systems and achieve adequate peak shaving, reduce peak load and corresponding power system investment and power investment; Promote the liberalization of the electricity market.

### *2.2. Pumped energy storage technology*

The pumped storage power station is constructed according to the relevant capacity. Under normal circumstances, the stored energy can be released for a few hours, several days, and the efficiency can be maintained at about 75%. This approach is commonly used in the areas of energy management and frequency control. One of the biggest characteristics of pumped energy storage is that the stored energy is very large, and because of this feature, it is very compatible with the development of the power system.

### *2.3. Battery energy storage technology*
The NaS battery used for energy storage can output pulse power, and because of the high power, the power output is usually six times the rated power, but the whole work usually does not exceed 30s. This feature allows us to better adjust and improve the electrical energy, and effectively improve and develop the overall equipment economy [1].

### *2.4. Super capacitor energy storage technology*
In general, super capacitors have more permittivity, pressure resistance and larger surface area than conventional capacitors. For example, the insulation ability of ceramic super capacitor is relatively strong, and the withstand voltage level of ceramic super capacitor is relatively high, so this technology also provides a good foundation for the application and development of energy storage in the future.

### *2.5. Compressed air energy storage*
Compressed air storage energy is often used for peaking gas turbine power plants. For the same power output, CAES units consume 40 less gas than conventional gas turbines. This is because conventional gas turbines need to consume about two-thirds of the input fuel for air compression when generating electricity, while CAES can pre-compress air using cheap electricity at the bottom of the grid load, and then release stored energy as needed. Add some gas to generate electricity.

## 3. Brief introduction and development of compressed air energy
Compressed air energy storage is the use of high-pressure air for power storage. The basic compressed air energy storage system shall include compressor, air storage room, generator and expander, etc. It takes the air storage room as the energy pool, and the compressor, expander and generator as the equipment for energy conversion. In the process of energy storage, electric compressor (or directly driven by mechanical energy) is used to compress air and store it in the gas storage chamber; in the process of energy release, the work is done by the expansion of released high-pressure air, which drives the generator to output electric power. In this system, the maximum storage capacity is determined by the maximum storage capacity of the gas chamber and the circulating pressure difference, and the response time is also related to the start-up characteristics of the power equipment. As a large-scale energy storage technology, compressed air energy storage has unique system advantages. Its use of air as energy storage medium determines that it is extremely insensitive to climate and temperature, and has stronger geographical adaptability, even in areas with poor conditions, it still has great potential [3].

## 4. Operating principle of CAES
CAES is a kind of energy storage method that compresses the air by using low valley power, wind power, photo electricity and so on, and seals the high-pressure air in the underground salt cavern, underground mine hole, expired oil and gas well or new gas storage room, and releases the compressed air to promote the turbine power generation during the peak load period of power grid. According to the operation principle, it can be divided into two types: supplementary combustion type and non supplementary combustion type [3]. In order to realize the cycle operation of the system, the supplementary combustion of fuel is needed for the supplementary combustion of CAES. During energy storage, the motor drives the compressor to compress the air to a high pressure and store it in the gas storage chamber; during energy release, the high pressure air in the gas storage chamber enters the gas turbine and burns with the fuel in the combustion chamber to drive the gas turbine to do work, so as to drive the generator to output electric energy. Due to the use of fuel for supplementary combustion, there is a problem of pollution emission, and there is a dependence on natural gas and

other fuels, which to some extent limits its application. Based on the development of conventional supplementary combustion CAES, non supplementary combustion CAES collects and stores the compressed heat generated in the compression process during energy storage by adopting regenerative technology, and heats the high-pressure air entering the turbine when the system releases energy. Non supplementary combustion CAES can not only eliminate the dependence on fuel and achieve zero emission of harmful gases, but also use the compressed heat and low-temperature exhaust of the turbine for external supply. Heating and cooling, and then realize the combined cooling, heat and power supply, realize the comprehensive utilization of energy, and improve the system efficiency.

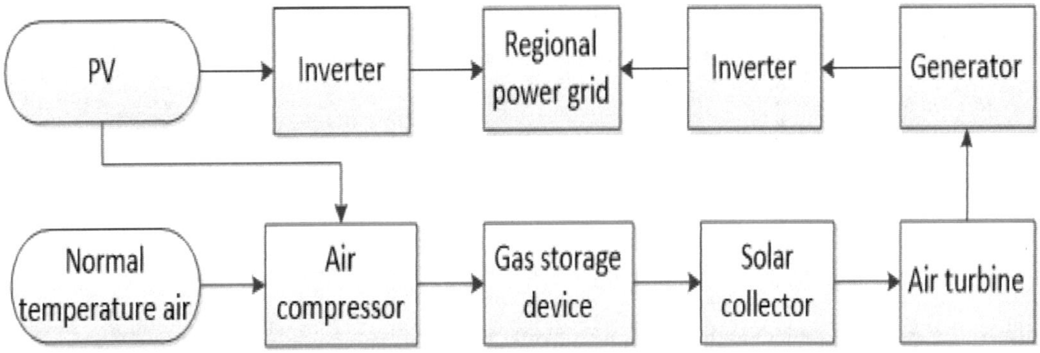


Fig. 1 compressed air energy storage schematic diagram of photovoltaic power generation system

In the compression process, the compressor absorbs the electric energy of the power grid, converts the outside air into high temperature and pressure air, and then stores part of the compressed energy in the heat storage system in the form of heat energy (such as high temperature hot water), while the other part mainly stores the energy in the gas storage tank in the form of molecular potential energy (high pressure air). The process of energy conversion is very complex, but when the parameters of the system are given, the relationship between the electric energy absorbed by the system from the grid and the technical functions of the outside world to the compressed air can be constructed. To this end, this section first gives the definition of compression efficiency as follows:

Compression link efficiency: in the process of compression and energy storage, the ratio between the technical function amount of compressed air by the outside world and the electric energy provided by the grid to the system, expressed in $\eta_c$. The efficiency of the compression link is determined by the conversion efficiency of electrical energy to mechanical energy and the adiabatic efficiency of the compressor. In engineering, the conversion efficiency of electric and mechanical energy of compressor can be approximated as a constant, expressed in $\eta_{\mathrm{m}}^{c}$; the adiabatic efficiency of compressor can also be regarded as a constant, expressed in $\eta_{\mathrm{a,\,i}}^{c}(\mathrm{i}=1,\cdots,N_c)$, Where $N_c$ represents teetotal number of stages of the compressor, then $\eta_{c,i}=\eta_m^c\eta_{a,i}^c(i=1,\cdots,N_c)$. Therefore, if we know the technical functions of the outside world on the compressed air at all levels of the compressor, we can calculate the electric energy absorbed by the system from the grid.

## 5. High speed air turbine expander model

The working process of high-speed air turbine expander includes three parts: gas suction, expansion and discharge. The simulation model of turbine expansion is shown in the figure below.

IOP Conf. Series: Materials Science and Engineering **738** (2020) 012018 doi:10.1088/1757-899X/738/1/012018

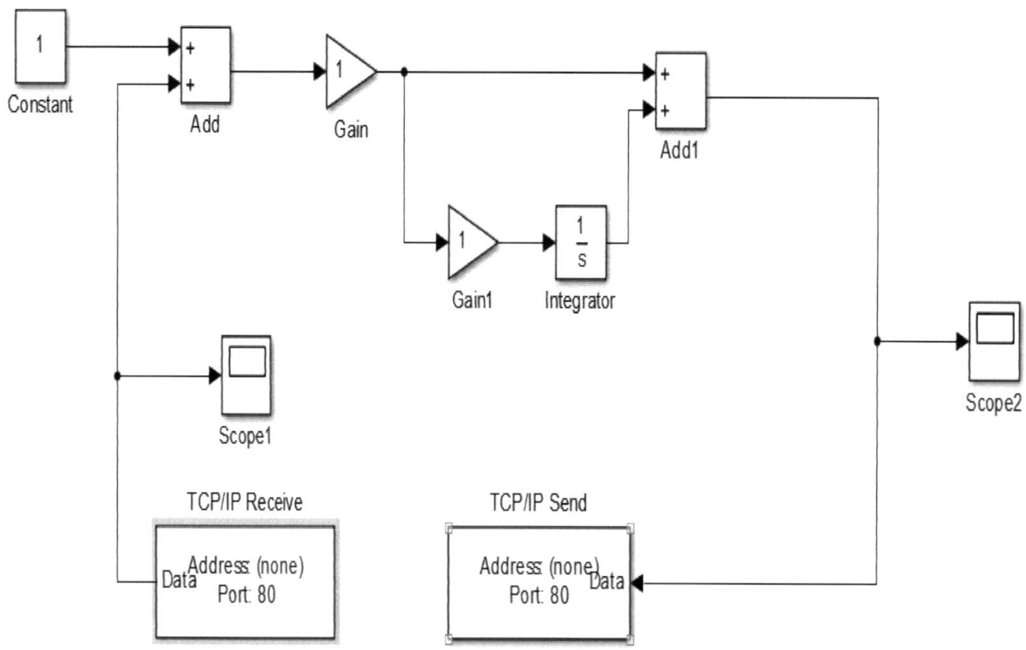


Fig. 2 model simulation of high speed air turbine expander

## 6. Simulation model of gas storage device

The simulation model of gas storage device is shown in the figure below.

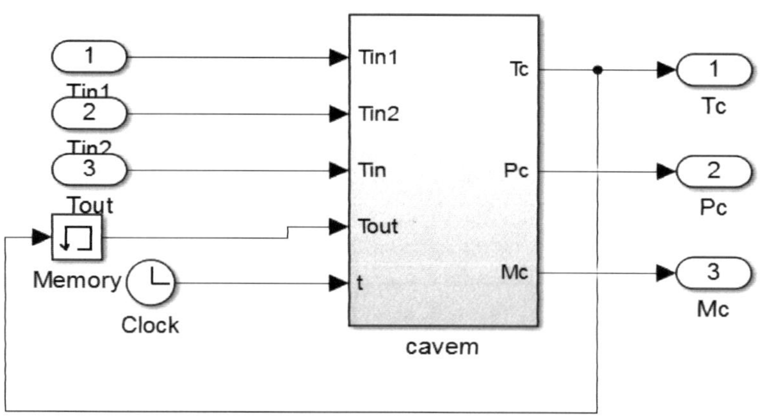


Fig. 3 simulation model of gas storage device

## 7. Multi-stage compression process simulation model

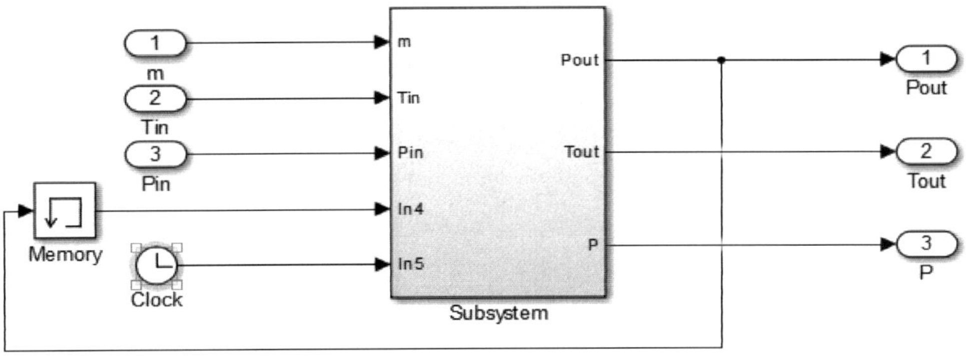


Fig. 4 Compression process simulation

## 8. Simulation results analysis

Simulation of compressed air energy storage expansion power generation control system

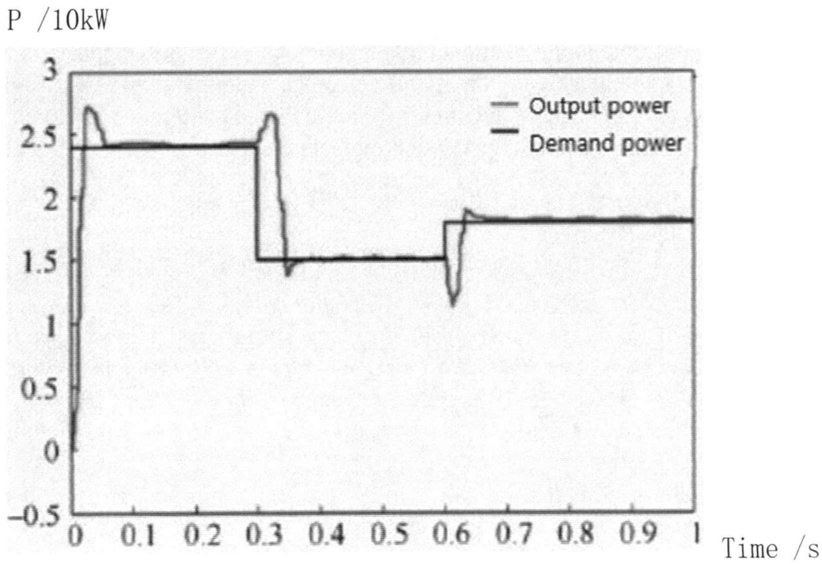


Fig. 5 Time-output power curve

Figure 5 shows the output power simulation results. When the demand changes, the output power can quickly track the demand set value and meet the load side power command requirements.

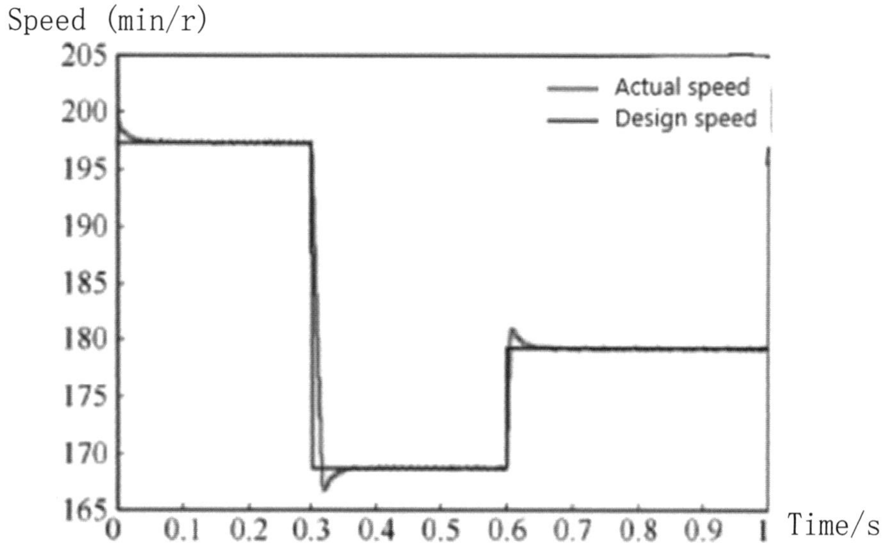


Fig. 6 time-speed curve

Fig. 6 is the simulation result of the speed. The blue line is the change of the speed reference value. The value changes at 0.3 S and 0.6 S. This is related to the flow change. The goal is to make the expansion machine operate with the highest efficiency. The real line is the actual speed value. It can be seen that the system speed coincides with the speed reference value, and when the grid demand changes, the system speed can quickly and accurately track the speed reference value change within 0.1 S, and can achieve the highest efficiency point tracking of the expansion power generation system. Make the system work best.

## 9. Conclusion

As one of the most promising large-scale power storage technologies, compressed air has a very broad development prospect. It is considered to be the most promising large-scale energy storage technology because of its advantages such as large storage capacity, long storage period and little influence by geographical conditions. However, compressed air energy storage still faces many problems. For example, how to effectively improve the efficiency of energy storage system, capacity efficiency, environmental protection efficiency and energy conversion efficiency and so on. Therefore, it is of great significance to study the mathematical model of compressed air energy storage system to provide guidance on the control level for the industrialization of compressed air energy storage technology. At present, as a new energy storage technology, compressed air energy storage has significant comparative advantages and market prospects. And its key subsystem also has a good prospect.

## Reference

[1] Zhiyuan Tan. The specific application and future development of energy storage technology in the field of electrical engineering [J] Construction • Building Materials • Decoration, 2016(15): 81-82.

[2] Hao Fu, Yuying Zhang, Yan Cui, Lulu Zhang, Tong Jiang. Progress in research on compressed air energy storage technology [J] Introduction to Science and Technology, 2016, 34(23): 81-87 .1575.

[3] Zhaokui Qin, Yan Yuan. Current status and application prospects of compressed air energy storage technology [J] Shanghai Gas, 2017(06): 35-41 .901.

[4] Jianjun Zhang, Shengni Zhou, Shuaiqi Li, Wenji Song, Ziping Feng. Current status and development trend of compressed air energy storage technology [J] New Energy Progress, 2018, 6(02): 140-150 .744.

[5] Ning Luo, Heqing Li, Wenyi Liu. Research status and analysis of gas storage device of compressed air energy storage system [J] Energy Storage Science and Technology, 2018, 7(03) :489 -494.448.
[6] Zhenjia Zhao. Research on Dynamic Modelling Simulation and Control Strategy of Non-Compressed Air Energy Storage System [D] Hebei University, 2016.3748.
[7] Afang Feng, Lei Zhang. Talked about the status quo of energy storage technology and its application prospects in the field of electrical engineering [J] Modern Manufacturing, 2013(36): 96-97.

IOP Conf. Series: Materials Science and Engineering **738** (2020) 012019 doi:10.1088/1757-899X/738/1/012019

# Design and Development of Electric Locomotive Converter Technology Integrated Experiment Platform

**Rui Chen[1], Shudong Wang[1], Chao Liang[1], Shuo Zhao[1] and Yongyong Wu[1]**

College of Electrical and Information Engineering, Lanzhou University of Technology, Lanzhou, Gansu, 730050, China

*Corresponding author's e-mail: 1090118371@qq.com

**Abstract:** Devices for power electronic technology experiment are generally backward and fail to satisfy the need of teaching and technology development nowadays. The design and manufacture of comprehensive experimental platform combined with traditional experiment and advanced electric locomotive converter technology. This platform realized the operation of converter system of electric locomotive. The platform is used for experimental operation of undergraduate and research and innovation of graduate.

## 1. Introduction

Power electronic technology as a multi-discipline cross discipline, its application field is from traditional roles to more extension exhibition sunrise industry, especially in the field of locomotive traction and rail transit successfully applied as a result, some complex control algorithm is applied to the field of power electronics control, such as nonlinear adaptive fuzzy logic control the use of these complex control algorithm, makes the work performance of power electronic technology to change, is suitable for the development of more fields [1] therefore, for the power electronic technology basic knowledge teaching, more can't be ignored However, at present, the experimental teaching of power and electronic technology course is still in the stage of the teacher's verification of a single knowledge point, which seriously ignores the cultivation of students' hands-on ability and innovation ability and carries out necessary experimental reform to meet the needs of The Times.

## 2. Problems existing in traditional experiments

The implementation of the experiment needs the support of experimental equipment and experimental platform. The design and development of the experimental platform need to be combined with two aspects. The second is the combination of experiment and practical engineering [2] only in this way can the effects of the experiment be truly brought into play:

1) Traditional power electronic experiment combined with experimental platform for the problem of the traditional power electronic rectifier inverter circuit, not only on books have image display and itself has specific image, easy to remember, easy to be accepted by the students' cognitive, but in practical application process has been basically realize modular encapsulation [3] so the design of the experimental platform, to give attention to two or more things traditional components of the form, and to highlight the application and practice of related modules.

2) How to carry out the power electronic technology experiment, lets the student to master the knowledge of the power electronic technology from the hard and software from the hardware platform, the current experiment is to introduce students to the working process of the circuit, and not able to

IOP Conf. Series: Materials Science and Engineering **738** (2020) 012019 doi:10.1088/1757-899X/738/1/012019

make students participate in the circuit device performance analysis of the working process of the main circuit analysis and dynamic analysis, the simulation and the simulation experiment is blank according to the research idea, students should choose their interested in in-depth analysis, thus forming his unique research method, it needs to have a good experimental platform for support.

3) How to integrate the new technology into the experiment the application field of power electronics technology is expanding continuously and the electromagnetic compatibility and energy saving of smart grid are developing rapidly. At the same time, with the development of power devices, power electronics technology has played a role in promoting the development of electric locomotives [4] in view of this situation, the design of the experimental platform should be combined with the current popular practical projects while leaving an interface for upgrading and further expansion.

## 3. Experimental platform design

In light of the deficiency of the power electronics experiment platform, the research of power electronic technology experiment principle and development trend, on the basis of considering the actual needs of experimental teaching, and with the rapid development of electric locomotive converter technology, the combination of integrative innovation has been designed electric locomotive converter technology experimental platform for electric locomotive converter technology design and development of comprehensive experimental platform mainly includes alternating rectangular transmission system circuit design, platform design and remote monitoring platform design three parts experiment platform diagram is shown in figure 1.

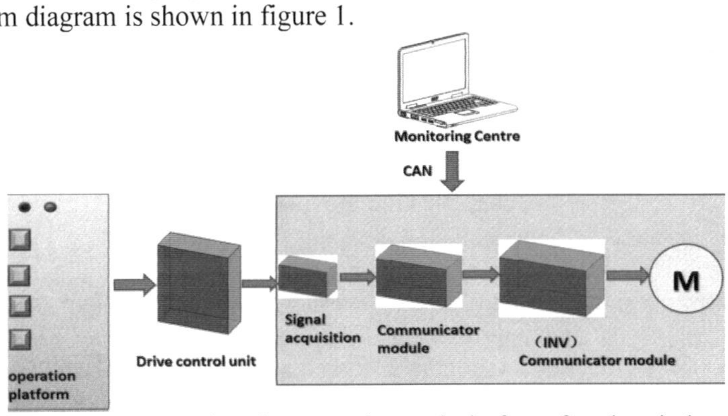


Fig.1 schematic diagram of comprehensive experimental platform for electric locomotive converter technology

### *3.1. Main circuit design of ac and dc transmission system*

The main circuit of the ac/dc transmission system mainly includes four quadrant rectifier intermediate link four quadrant inverter induction motor and five modules of signal acquisition circuit, as shown in figure 2.

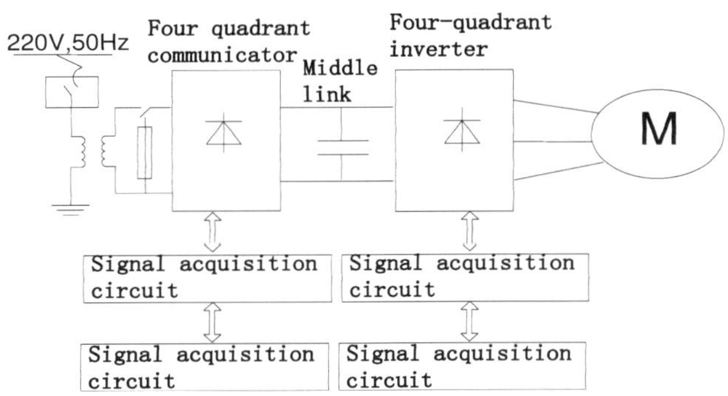


Fig.2 main circuit structure drawing

1) Four quadrant four quadrant rectifier inverter hardware design with intelligent power module SPM45 series, in order to improve experiment platform should use sex, therefore adopts the modular design method, the four quadrant sorting machine and four quadrant inverter alone made into two independent sandwich plate, and is equipped with convenient connection interface connected with other parts.

2) Acquisition circuit and control circuit design in order to ensure the system can work in two kinds of working condition, traction brake needs of four quadrant the whole flow of the intermediate links, and four quadrant inverter to collect the information of the voltage and current sampling circuit design is very important to the system design process, hall-effect sensor module USES the voltage type and current type of hall sensor module for testing the circuit principle diagram as shown in figure 3 is shown in figure 4.

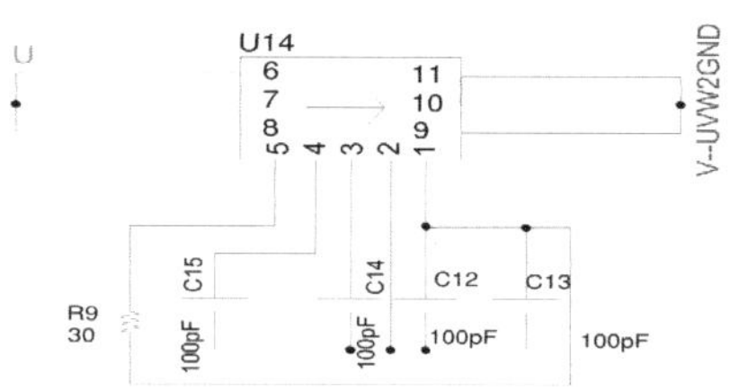


Fig.3 voltage acquisition circuit schematic diagram

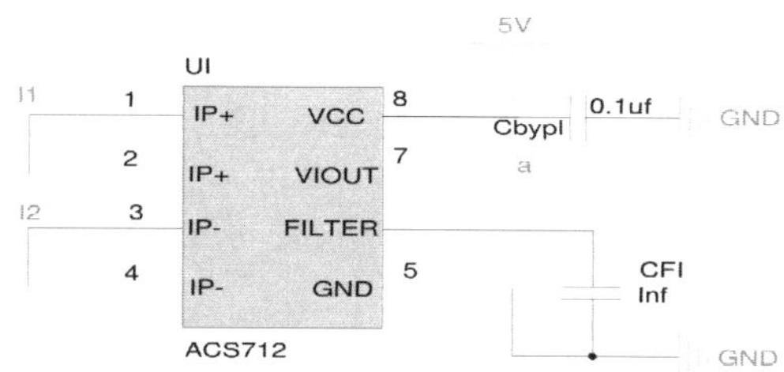


Fig.4 schematic diagram of current acquisition circuit

Figure 4 is the schematic diagram of the current acquisition circuit. The linear current sensor ACS712 based on hall effect is adopted to measure the current. ACS712both ac and dc can be measured. The output voltage is proportional to the measured ac or dc, and proportional to the chip's power supply voltage, which is convenient for software calculation.

The power module is designed to supply power to each module of the experimental platform. The dc voltage required by the experimental platform is 3.3v, 5v and 15v, among which the dc 5v voltage provides voltage for the intelligent power chip current sensor and the operational amplifier, which is provided by the external switching power module.3.3v power supply for main chip. The intelligent power chip is provided with electric voltage. The voltage is PTN78000W, which converts 5v voltage into 15v dc power module, as shown in figure 5.

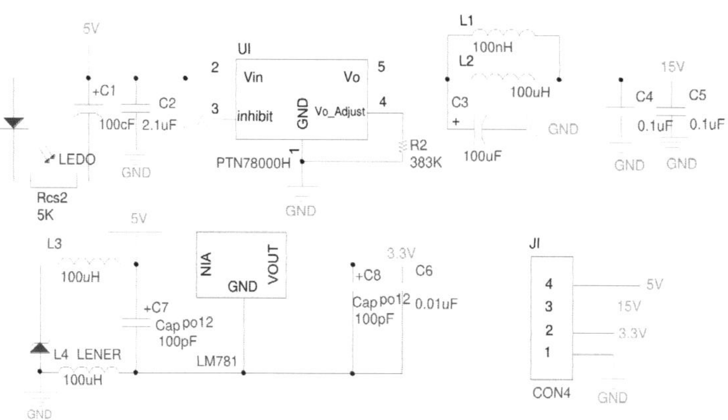


Fig.5 schematic diagram of power module circuit

4) In the whole process of circuit design, the selection of ac side inductance and dc side capacitance is of vital importance. For ac side inductance, its size will affect the power factor of the whole system,

dc side voltage and other aspects of the output power of the system:

$$\frac{(2U_{dc}-3e_m)e_mT_s}{2U_{dc}\Delta i_{max}} \leq L \leq \frac{2U_{dc}}{3i_m\omega} \tag{1}$$

Where, $e_m$ is the peak voltage of the power grid; $T_s$ is the control switching cycle of the rectifier;$i_m$ is the base wave peak value of the grid side phase current. Maximum allowed $\Delta i_{max}$to harmonic current.

Through a large number of experiments and theoretical formula calculation, 2 m h was selected as the experimental platform.

For the dc side capacitance, the choice of its model directly affects the dc side voltage tracking to a given voltage and the stability of the output voltage:

$$\frac{1}{2\Delta U_{dc}^{*}R_L} \leq C \leq \frac{T_r}{0.74R_L} \tag{2}$$

Type,$\Delta U_{dc}^{*}$ to dc voltage maximum dynamic change; $T_r$is the time required for the dc side voltage value to determine the steady-state voltage.

Through a large number of experiments, combined with the theoretical formula calculation, the experimental platform to choose 2800μF.

### *3.2. Control circuit design of ac and dc drive system*

The dynamic control single mining with mesh before into STM32F4 as the core control chip, adopts IFCO indirect field-oriented control method and SVPWM modulation type, output PWM wave, complete control of the hand in rectangular transmission main electrical path and using CAN bus to complete the data transmission control circuit of the hardware system block diagram as shown in figure 6 considering the experiment system of gradation and the characteristics of the secondary development, the modular design of main circuit module, then on System main circuit different parts of the line control.

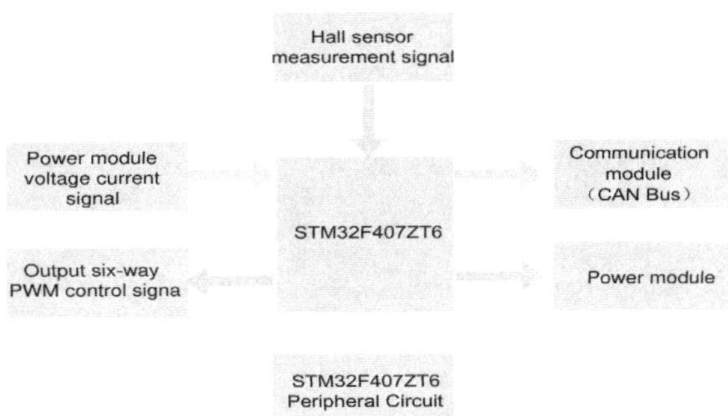


Fig.6 control circuit system diagram

### *3.3. Remote console design*

In order to realize the control of the ac and dc system, it is necessary to carry out real-time monitoring of key parameters in the operation process. If fault is found, the indicator light of fault will turn on in time to facilitate the timely detection and treatment of fault. Meanwhile, the monitoring system has the function of saving data output data. The data in the experimental process are exported to the excel table for further research. The design of this part mainly adopts the Lab VIEW developed by NI company to compile the upper computer interface [5], as shown in figure 7.

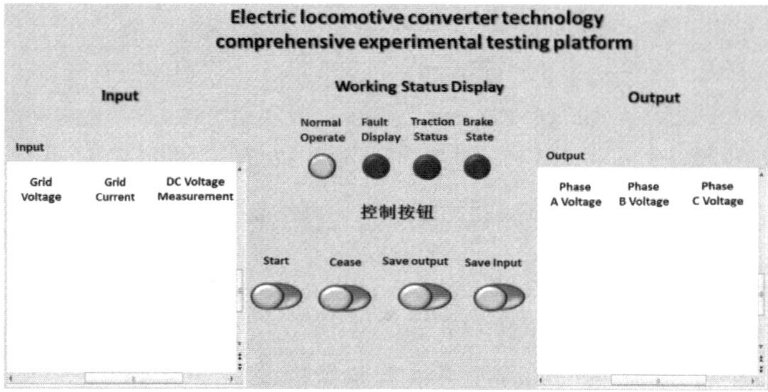

Fig.7 system detection interface

## 4. Advantages of experimental platform

### *4.1. Application level extensiveness*

This experimental platform is not only applicable to major majors of power electronics courses, such as electrical engineering and automation, but also applicable to other non-major majors, such as electronic information engineering. At the same time, it can not only satisfy the outstanding undergraduate students to expand the experiment, but also satisfy the graduate students to innovate in scientific research.

### *4.2. Extensibility*

This experimental platform in the process of design, considering the different levels of the experimenter, in the process of design, will pay the rectangular main circuit part of the drive system has carried on the modular design, it can be divided into four quadrant rectifier part of the inverter and motor control part, and each part has its own core control module, this can be either a separate experiment, and various modules can be connected to a comprehensive experiments at the same time, using programmable devices to design, students can according to the experimental purposes and requirements, to write programs for undergraduate students and graduate students to carry out scientific research innovation.

### *4.3. Innovativeness*

Innovative fills the experiment project and electric locomotive converter technology of combining the blank of this experimental platform, with reference to the locomotive train with typical circuit of current transformer, producing electric locomotive converter circuit model of the power electronic technology curriculum knowledge combined with the actual electric locomotive converter technology, effectively make up for the inadequacy of field practice teaching link, fully arouse the enthusiasm of students learning in the process of experiment, begin to do, want to method was verified by thinking innovation to verify the experimental analysis results of the four process, greatly improve the quality of teaching.

## 5. Conclusion

Power electronic technology as a multi-discipline cross discipline, is also the foundation knowledge of a variety of new project, so for the electric power electronic technology knowledge teaching quality must be improved, keep up with the forefront of The Times is the only way to improve students' practical ability, solve the students knowledge and the actual project phase out of the question for the design of the electric locomotive comprehensive experiment platform, can greatly improve the quality of teach, effectively will be the mainstream of the electric locomotive converter technology combined with power electronic technology knowledge to students majoring in electrical have a more intuitive understanding of power electronics technology knowledge, more profound understanding and more practical application.

**Reference**

[1]. Ying Yang. Student-oriented innovation of electronic technology experimental course teaching [J]. Lab research and exploration, 2010, 29 (6) : 80-83.
[2]. Zhidong Wang, Zhuoyong Liang , Gang Wang. Experimental support platform for microcomputer and network communication in electrical engineering [J]. Journal of power systems and automation, 2010, 22 (4) : 151-155.
[3]. Mei Liang. Research and construction of integrated experimental platform for power electronics technology [J]. China modern educational equipment, 2012, 15 (13) : 16-17.
[4]. Jiangtao Zhou. Development of ac-dc integrated transmission test platform [D]. Chengdu: southwest jiaotong university, 2011.
[5]. Rujin Dong. Experimental study on online motor monitoring system based on Lab VIEW [D]. Shanghai: east China university of science and technology, 2011.

IOP Conf. Series: Materials Science and Engineering **738** (2020) 012020 doi:10.1088/1757-899X/738/1/012020

# Design of Resonant Wireless Charging Device for Public Electric Bicycle

**Shuo Zhao[1], Shudong Wang[1], RuiChen[1], Chao Liang[1] and Yongyong Wu[1]**

[1]College of Electrical and Information Engineering, Lanzhou University of Technology, Lanzhou, Gansu, 730050, China

*Corresponding author's e-mail: special5376@163.com

**Abstract:** As a starting point for the difficulty of charging public electric bicycles, this paper proposed a convenient and efficient charging device design concept, namely the design of resonant wireless charging device applied to public electric bicycles. Firstly, the main wireless charging technology on the market is introduced. Secondly, the design of the resonant wireless charging device is analysed including the charging principle, main circuit design and its advantages. The device supply power in centimeters or meters, even if there are obstacles in the charging range, energy can be efficiently transmitted, and it can be applied to the public electric bicycle parking hall, which can effectively solve the problem of difficult charging.

## 1. Introduction

In the age of household appliances, smart human beings to cable transmission as the main way, so that electrical energy is widely used in every aspect of life. In recent years, as a public transportation tool with high efficiency, convenient use and low price, public e-bike has been increasingly popular with people. Figure 1 shows an exhibition of public e-bike. However, the new convenience will also bring new problems, the emergence of public electric bicycles has brought their own charging difficult problem. Such as inadequate charging facilities,

FIG. 1 Public electric bicycle

The charging device is not easy to carry and other problems, but more in the past through the cable transmission of electrical energy charging equipment because of the disorderly pull wire caused accidents. Based on these problems, this paper proposes a wireless and efficient resonant wireless charging device for public e-bike. Based on the difference of power transmission principle and transmission path, there are three main transmission modes of radio energy in the research circle of power transmission up to now [1]: one is resonant coupling mode, and the other two are electromagnetic induction coupling mode and microwave radiation mode. In addition, there are other emerging power transmission methods such as electric field coupling, ultrasonic wave and laser, but they are still in the initial stage of research.

### *1.1. Electromagnetic induction coupled wireless charging technology*

Electromagnetic induction coupled wireless charging technology [2] is proposed based on the theory of induced current generated by changing magnetic field. The principle of electromagnetic induction coupled power transmission is shown in figure 2. It works like a transformer. In the radio energy transmission system based on the electromagnetic induction coupling theory, the coupling between the coil at the transmitting end and the coil at the receiving end requires a high degree of directivity, and is almost not coupled when it is relatively vertical. What's more important is that it restricts the distance and can only supply power within a short distance.

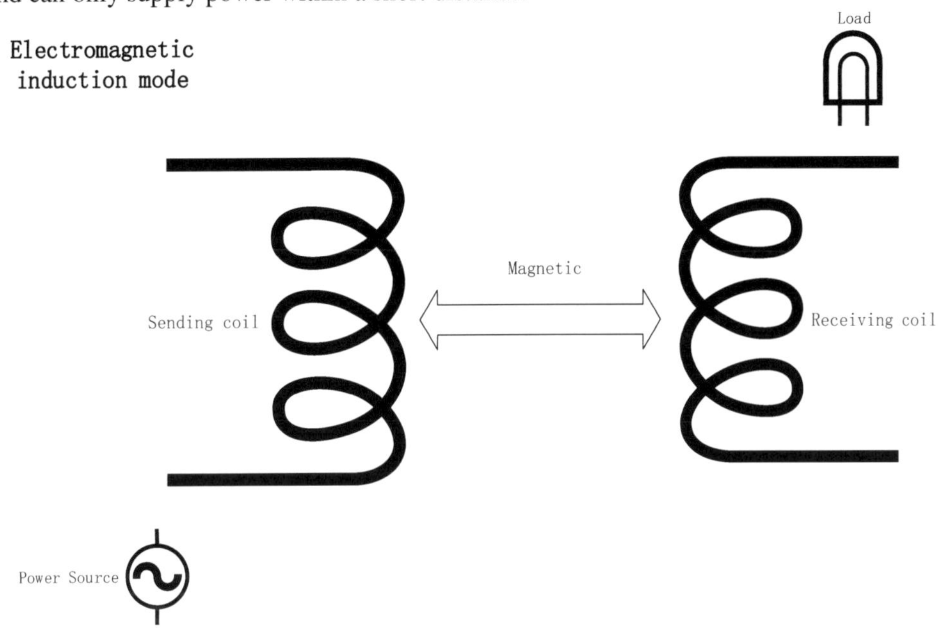


FIG. 2 Schematic diagram of electromagnetic induction coupled power transmission

### *1.2. Microwave radiation wireless charging technology*

As shown in figure 3, the system first converts the electric energy into Microwave form, transmits it through the antenna at a long distance, and then transmits it through free space. Finally, it receives the signal from the receiving end and converts it into direct current through rectifying circuit to supply Power to the load. However, although the use of microwave power transmission can achieve a relatively long distance power transmission, but the transmission power is small, and the microwave energy transmission system is generally used in special occasions, such as microwave aircraft, satellite solar power station.

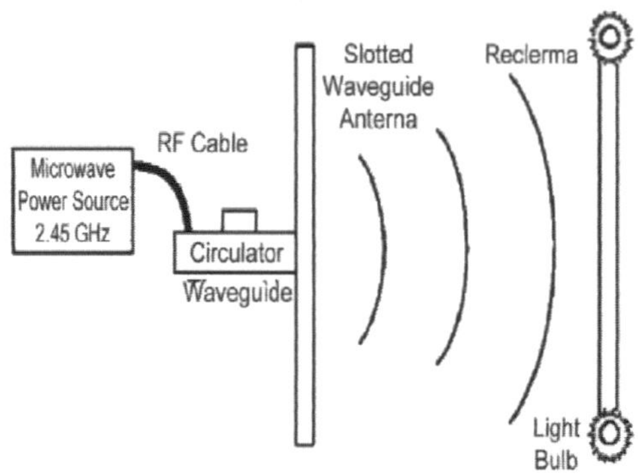


FIG. 3 Schematic diagram of microwave radiation power transmission

*1.3. Resonant wireless charging technology*

Resonant wireless charging technology [4] is established mainly on the basis of the non-radiation near-field resonance strong coupling theory, which is also the wireless charging technology of public electric bicycles to be introduced in this paper. Its basic principle is to use two objects with the same frequency as the medium to achieve efficient energy exchange. Compared with inductive coupling, it can realize more efficient power transmission over longer distance. In addition, the resonant radio energy transmission system can provide wireless power to two or more devices simultaneously.

## 2. Principle of resonant wireless charging for public electric bicycles

The so-called Wireless charging, as its name implies, is a kind of application of Wireless Power Transfer (WPT), also known as Contactless Power Transmission (CP).It refers to the technology that can store electric energy without the help of metal wires and other physical connections, so as to use the electric field, magnetic field and other space invisible medium as the transmission way, at the beginning of the electric energy output, the end of the electrical energy receiving and storage. Resonant wireless charging technology is a kind of technology based on electromagnetic wave induction theory and other related ac induction theory. As shown in figure 4, the charging device is mainly composed of the power supply, inverter circuit,, resonant circuit, rectifying and filtering circuit, booster circuit and load that provide direct current to the whole system.

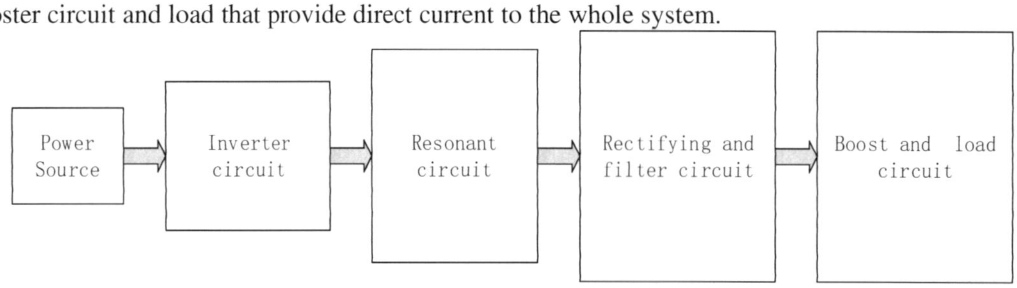


FIG. 4 Theoretical composition of resonant wireless charging device

The power supply section is used to provide direct current to the entire public electric bicycle charging system. Inverter circuit adopts high frequency inverter principle to convert direct current into alternating current. The resonant circuit is composed of primary side resonance and secondary side resonance. The primary side resonance circuit receives the alternating current converted by the inverter, and then transmits the electric energy to the secondary side resonance circuit through the magnetic coupling of the resonant circuit. The rectifying and filtering circuit converts the alternating current from the resonant circuit into direct current after rectifying and filtering. Finally, the booster circuit is used to convert it into a voltage that can meet the needs of the load. The booster circuit used in this system is Boost chopper. Boost circuit converts the input direct current into another kind of direct

IOP Conf. Series: Materials Science and Engineering **738** (2020) 012020 doi:10.1088/1757-899X/738/1/012020

current by controlling the switching device. In this design, as the impedance matching and power regulator of load, its conversion efficiency is high, and the output is controllable, which can fully meet the requirements of wireless charging system. The load of this design system is the public electric bicycle.

## 3. Core design of resonant wireless charging device

In the system design of radio energy transmission device, its composition is mainly based on three problems.

### *3.1. Design of inverter circuit*

The inverter circuit is the basis to ensure the smooth transmission of electric energy. The higher the inverter frequency, the higher the power transmission efficiency. As for this system, it is inevitable that the interference signal should be as small as possible. The harmonic of bridge inverter circuit is very small, and the inverter frequency is high, which meets the requirements of the design system .If the load is resonant, the current can be kept in the same phase with the voltage to achieve zero voltage and zero current switching, which can reduce the switching loss to the minimum. As is shown in figure 5, is the schematic diagram of bridge inverter circuit.

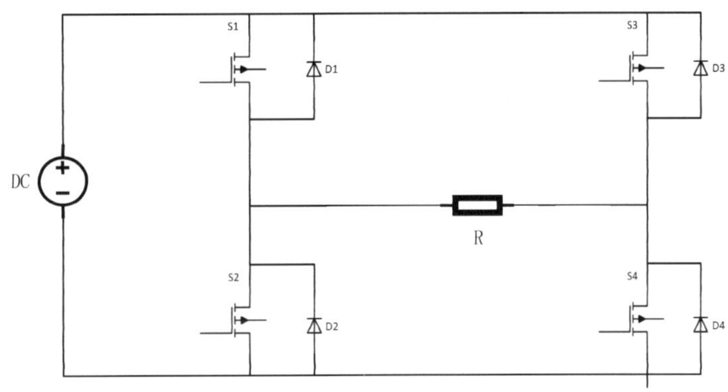


FIG. 5 Schematic diagram of bridge inverter circuit

### *3.2. Resonant circuit design*

Resonant circuit is the core of resonant wireless charging device for public electric bicycle. Wireless charging system can effectively complete the transmission of electric energy within a certain distance. The key lies in whether the resonant topology circuit can be truly close to full resonance. In this system, LC series resonance is adopted to realize the coupling of primary and secondary energy. The series resonance of the primary side circuit can bear the electromotive force generated by the coil, reduce the damage to the switching devices in the inverter circuit, and improve the power factor of the primary side resonance circuit. The auxiliary side circuit adopts series resonance, and the voltage of the load will not change with the change of the load, which will lay a foundation for the stability of the system output.

### *3.3. The design of rectifying filter circuit*

The function of rectifying filter circuit is to convert alternating current into pulsating direct current. Considering the rectifying efficiency and effect of the system, as well as the maximum reverse peak voltage assumed by the diode in this circuit, the principle of bridge rectifying circuit is shown in figure 6.

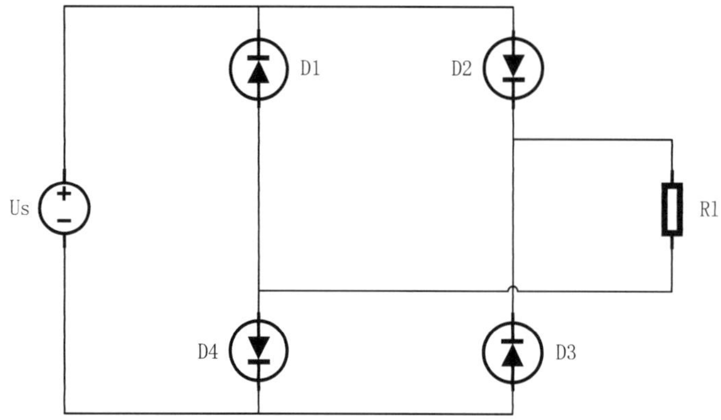


FIG. 6 Schematic diagram of bridge rectifier circuit

There is still a small amount of ac ripple in the dc obtained through rectification, so it cannot be directly supplied to the load. Therefore, this system adopts the function of filter capacitor to filter out the useless ac ripple through charging and discharging.

## 4. Advantages of resonant wireless charging

In terms of the current development situation of public electric bicycles, wireless charging technology is at the forefront of power supply technology and is worth studying. In this paper, resonant wireless charging device has many advantages: firstly, in terms of its energy transmission form, there is no physical connection between the power supply end and the load, which is very convenient compared with the use of bulky cables in traditional charging technology .In terms of its safety, resonant wireless charging device can be made into a completely closed structure, even in a humid environment, do not have to worry about the leakage of electricity often occurred in the wired charging equipment. Since the principle of resonant wireless charging is based on the strong coupling between two objects with the same frequency, the same coupling device can be used for the equipment to be supplied with similar power, which can save space to a great extent.

## 5. Conclusion

Nowadays, with the popularization of electronic products in every aspect of human life, with the continuous development of science and technology, people's requirements are getting higher and higher. In terms of efficiency and convenience, resonant wireless charging devices of public electric bicycles are at the forefront of wireless energy transmission. Resonant wireless charging technology not only has the advantage of providing power to multiple devices at the same time, but also can meet the requirement of long distance energy transmission. In terms of its development prospects and application prospects, the public e-bike is very popular today. Because the wired charging equipment itself has a lot of defects, such as cable equipment occupies a lot of space, low safety performance. For resonant wireless charging device, it has good sealing performance and high safety performance. In such a comparison, the development prospects and application prospects of wireless charging devices are indeed very broad. Of course, in today's rapid development, users will have higher and higher requirements on wireless charging equipment, so the research on resonant wireless charging device of public e-bike is still on the way.

## References

[1] Mixiaoyan, song yuqing, et al. Analysis and design of electric vehicle wireless charging system [J]. Modern manufacturing technology and equipment, 2016(10):65-67
[2] Qu linan.Research on inductive wireless energy transmission mechanism [D]. Harbin Institute of Technology, 2016
[3] Wu jiaxin, wujunyong, zhangning, et al. Experimental study on wireless energy transmission based

on induction [J]. Modern electric power, 2012,29(1):24-28
[4] Gaoyuqing. Characteristics research and system design of magnetic resonance radio energy transmission system [D]. Zhejiang university, 2017
[5] Development of inductive and resonant radio energy transmission technology [J]. Zhang bo, shu xujian, huang runhong. Journal of electrical technology. 2017(18)
[6] Research review on electromagnetic induction non-contact power transmission technology [J]. Gao jianxin, wu xusheng, gao wei, li guangyi. Journal of power supply. 2017(02)
[7] Analysis on characteristics of inductive system of radio energy transmission [J]. Zhang jiantao, zhu chunbo, lei Yang, song kai, lu rengui, wei guo, Chen qingquan. Journal of electrical technology. 2015(S1)
[8] Research on different compensation topologies of inductively coupled power transmission system [J]. Zhou wenqi, ma hao, he xiangning. Journal of electrical technology. 2009(01)

IOP Conf. Series: Materials Science and Engineering **738** (2020) 012021 doi:10.1088/1757-899X/738/1/012021

# Study of the influence of heat treatment process on the mechanical performance of wedge shaped components

**Guangfeng Shi, Change Huang, Weiqi Liao, Jiangchun Li, Xie Minxiong**

National Experimental Teaching Demonstration Center for Electromechanical Integrated Engineering Training, Guilin University of Electronic Technology, Guilin, Guang Xi, China

Corresponding author's e-mail: 515908042@qq.com/hche668@163.com

**Abstract:** Mine Splitter is widely used in mining machinery, and wedge assembly components are very important parts of it. In this work, we studied wedge assembly's element content by EDS, and through different heat treatments to improve the performance of wedge assembly components of Mine splitter. The heat treatment process of wedge assembly components is divided into quenching, quenching and tempering to find the changes on microstructure and mechanical properties. It is found that after quenching at 840°C and tempering at 450°C, the structure is tempered Sorbite with the highest strength and wear resistance and the best comprehensive mechanical properties.

## 1. Introduction

Mine Splitter refers to the equipment for mechanical expansion and splitting of stone. Its working process is as follows: firstly, a hole with a specific diameter and depth is drilled on the split stone, and then the working part of the Minesplitter (an intermediate wedge and two reverse wedges) is put into the hole. Through the action of external thrust, the intermediate wedge moves forward, and the two wedges are propped up and turned into transverse motion. The rock to be split will split in a predetermined direction in a few seconds [1]. Therefore, wedge components are the main parts of the Mine splitter. Wedge components, as the executive part of the Mine splitter, are subjected to tremendous impact. Therefore, wedge components are the first failure parts of the whole Mine splitter. Therefore, it is of great significance for the research of wedge components. In this paper, wedge-shaped components will be studied, and the influence of heat treatment process on the performance of wedge-shaped components will be studied. In this paper, a wedge-shaped component produced by a company is used. Its structure is shown in Figure 1.

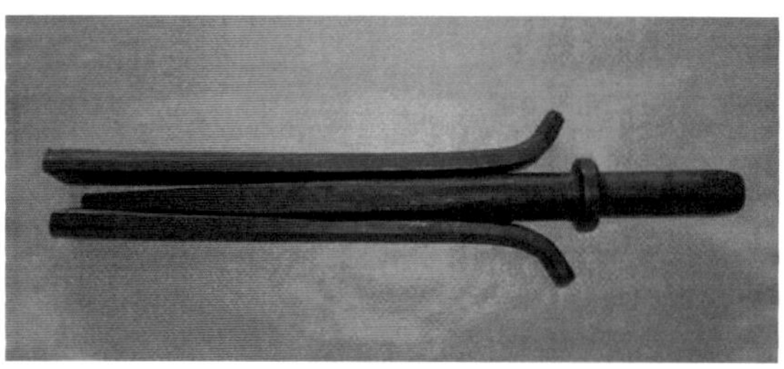

Figure 1. Structural diagram of wedge-shaped components

## 2. Test materials and methods

In order to determine the material element composition of the wedge assembly, EDS was used to detect the structure as shown in Figure 2.

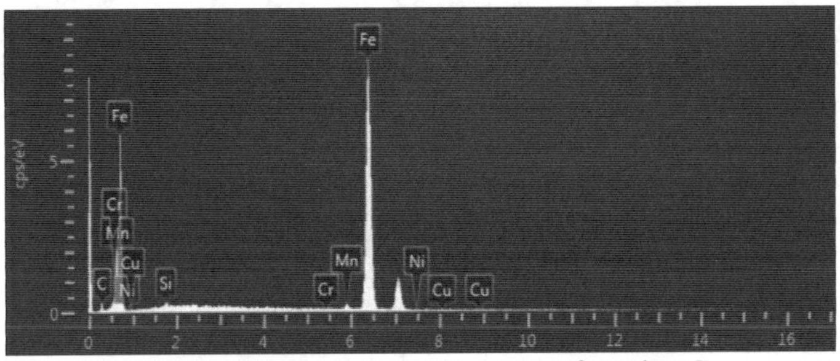


Figure 2. Material Composition Diagram of Wedge Component

From the data in Fig. 2, we can know the element content of wedge-shaped component material. According to GB/T 2975-1998 "Sampling position and sample preparation for mechanical properties test of steel and steel products", the hardness and metallographic specimens are cut from test rods by wire cutting machine [2]. The heat treatment process and experimental scheme of wedge assembly components of hydraulic splitter are shown in Table 1. The cooling method adopted is Oil cooling.

Table 1. Heat treatment process parameters for wedge-shaped components of splitter

| | Heat treatment | Process parameter |
|---|---|---|
| 1 | Quenched | 820°CQuenched X30min Oil cooling |
| 2 | Quenched + Tempered | 840°CQuenched X30min Oil cooling+350°CX 1h Tempered (Air cooling) |
| 3 | Quenched + Tempered | 840°CQuenched X30min Oil cooling +400°CX 1h Tempered (Air cooling) |
| 4 | Quenched + Tempered | 840°CQuenched X30min Oil cooling +450°CX 1h Tempered (Air cooling) |
| 5 | Quenched + Tempered | 840°CQuenched X30min Oil cooling +500°CX 1h Tempered (Air cooling) |
| 6 | Quenched | 860°CQuenched X30min Oil cooling |

In order to study the mechanical properties of the wedge-shaped component, heat treatment experiments were carried out. Firstly, the wedge-shaped component was intercepted by wire cutting, and the surface of the wedge-shaped component was cleaned. The heat treatment experiments were carried out using high temperature box resistance furnace and programmable box resistance furnace. After heat treatment according to different experimental groups, the cross sections of different test blocks were taken, after grinding, polishing and corrosion with 4% nitric acid and alcohol solution[3].

## 3. Test results and analysis

### *3.1. Effect of Quenching Heating Temperature on Microstructure and Hardness of Wedge Component Materials*

The mechanical properties of materials with different heat treatment results were compared by heat treatment. Firstly, the samples were quenched, and heated to 820°C, 840°C and 860°C, respectively. After 60 minutes of heat preservation, the samples were discharged from the furnace. Oil-cooled to room temperature. After quenching, the samples were observed by body microscope, and the microstructure was obtained as shown in Figure 3.

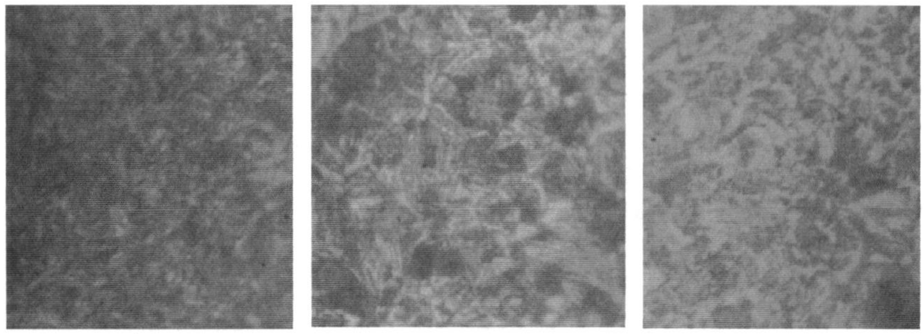

(a) (b) (c)

Figure 3. Microstructure of Wedge Component Materials at Different Quenching Temperatures(a)820°C(b)840°C(c)860°C

In Figure 3(a) (b) (c) are the microstructures of the samples after quenching at 820°C, 840°C, and 860 °C, respectively. The quenched lath martensite, flake martensite and retained austenite structure of the wedge-shaped assembly sample. From (a) (b) (c) in Fig. 3, it can be seen that the lath martensite bundles increase gradually with the increase of quenching temperature. The reason is that the ferrite in the wedge-shaped assembly material dissolves gradually into austenite, which gradually transforms the material into martensite. At 820 °C, the grain growth of wedge-shaped components is small. With the increase of quenching temperature, the grain size of wedge-shaped components grows obviously. This is because the austenite grain grows continuously and forms coarse martensite, which reduces the strength and stiffness of materials. After three groups of experiments, it was found that when the heating temperature was 840 °C, the martensite bundle had larger grain size, and the hardness of the wedge component sample was the highest among the three quenching temperatures [4]. Rockwell hardness values of three groups of samples were shown in table 2.

Table 2. Rockwell hardness values of wedge component samples at different quenching temperatures

| | Initial（HRC） | 820°C（HRC） | 840°C（HRC） | 860°C（HRC） |
|---|---|---|---|---|
| 1 | 53 | 55 | 60 | 53 |
| 2 | 54 | 51 | 65 | 47 |
| 3 | 50 | 59 | 59 | 55 |

After three times of testing, the average value of the data is processed and analyzed. The point-line diagram is shown in Figure 4. It can be seen from the diagram that the Rockwell hardness of the wedge-shaped component sample is the maximum value in the three groups of tests after quenching at 840 C, and the hardness of the wedge-shaped component sample is 20% higher than that of the original material. From Figure 4, it can be seen that the Rockwell hardness of the wedge-shaped component sample is the maximum of the three groups of tests at 840 °C.

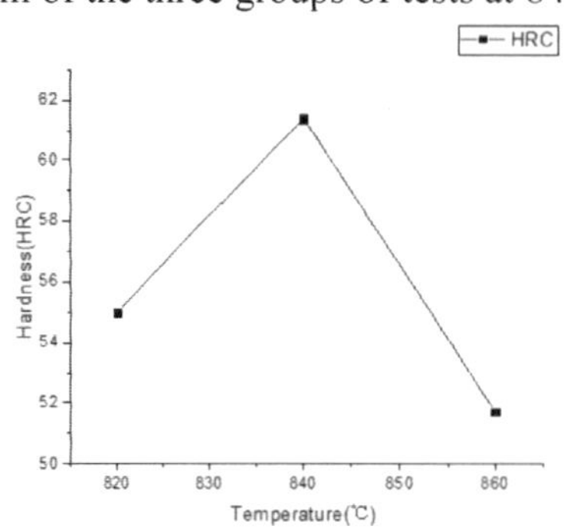


Figure 4. The effect of different quenching temperatures on the hardness of wedge-shaped components

### *3.2. Effect of Tempering Temperature on the Microstructure of Wedge Component Materials*

In order to study the effect of heat treatment process on the properties of wedge-shaped assembly materials, the quenched samples at 840 were divided into four groups, and then air-cooled tempered at 350°C 400°C 450°C 500°C for 60 minutes. The changes of structure and properties of samples with tempering temperature were analyzed. The microstructures at four tempering temperatures are shown in Figure 5 below.

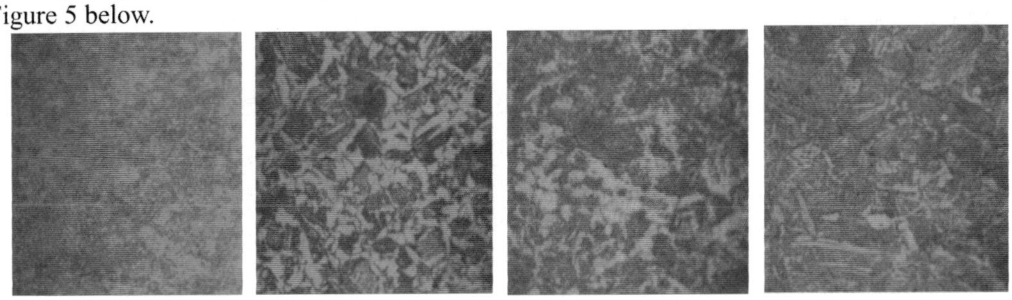

(a) (b) (c) (d)

Figure 5. Microstructure of Wedge Component Materials at Different Tempering Temperatures(a)350°C、(b)400°C、(c)450°C、(d)500°C

Twelve samples treated by oil quenching at 840°C for 1 h were divided into four groups. Three samples were tempered at 350°C 400 °C450°C 500 °Cand then their metallographic structure was observed by body microscope. It was found that tempered sorbite was obtained by tempering at 350°C and 500 °C. The carbides were mainly flaky and partially granular, and the carbides were small and uniform [5]. The carbides of the structure treated at tempering temperatures of 400°C and 450 °C are mainly flake-like particles, but when tempering temperatures exceed 450 °C, the carbides are flake-needle-like and fine-grained [6].

### *3.3. Effect of Tempering Temperature on Hardness and Wear Resistance of Wedge Component Materials*

The hardness of the wedge-shaped component sample after tempering is measured. In the process of measuring the hardness, the hardness value is measured by means of the average value of several times. The hardness value at different tempering temperatures is shown in Table 3.

Table 3. Rockwell hardness values of wedge component samples at different tempering temperatures

| | 350°C（HRC） | 400°C（HRC） | 450°C（HRC） | 500°C（HRC） |
|---|---|---|---|---|
| 1 | 53 | 57 | 55 | 48 |
| 2 | 50 | 56 | 59 | 46 |
| 3 | 49 | 54 | 57 | 52 |

According to the hardness value obtained from the experiment, the point-line diagram of the data is shown in Figure 6. By measuring the hardness of the wedge-shaped component sample, it is found that the hardness value of the sample increases with the increase of tempering temperature, but it does not continue to increase. The hardness value of the sample decreases gradually when the tempering temperature exceeds 450 °C degrees. Because wedge-shaped components of Mine splitter not only have certain requirements for hardness in practical work, but also need good wear resistance, so after the experiments of various heat treatment schemes, in order to get the best combination of hardness and impact toughness of wedge-shaped components [7]. the wear resistance experiments of wedge-shaped components materials under various heat treatment processes are carried out finally, the results are shown in Figure 7.

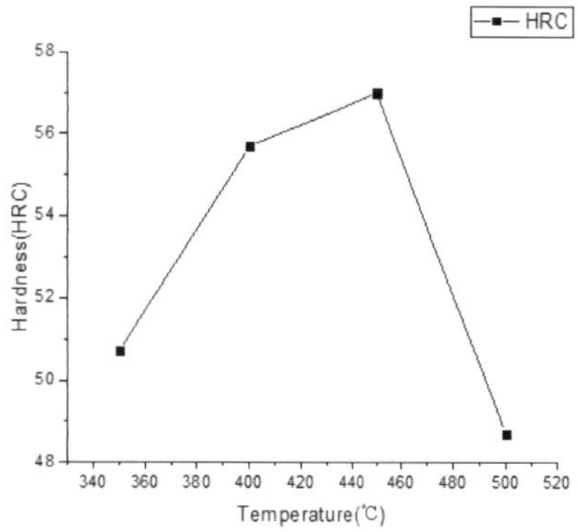


Figure 6. The effect of different tempering temperature on material hardness of wedge-shaped components

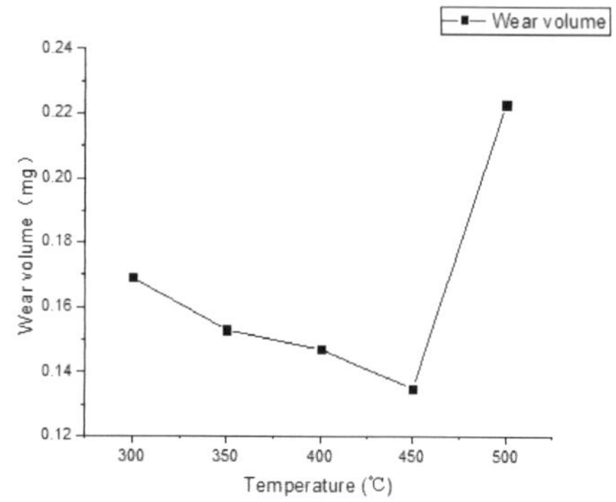


Figure 7. The effect of different tempering temperature on wear resistance of wedge-shaped components

It is verified by experiments that the best combination of hardness and impact toughness of wedge-shaped component material can be obtained by the heat treatment process of quenching at 840°C and tempering at 450°C.

## 4. conclusion

(1) It is found that the heat treatment process of wedge component material has certain influence on its performance. The performance of wedge-shaped component material after quenching at 840°C and tempering at 450°C is the best. The hardness of wedge-shaped component material is increased to 57 (HRC), the hardness value is increased by 8% as a whole, and the wear resistance is also improved correspondingly.

(2) By comparing the quenching process of three groups of experiments at 820°C, 840°C and 860°C, it is found that the wedge-shaped component material after quenching at 840°C has larger grain size and uniform distribution, so higher hardness can be obtained.

(3) After 840°C quenching and 450°C tempering, the wedge assembly components has the best mechanical properties combination. the structure of the wedge-shaped component material is flake like particles, the distribution of carbide is uniform, and it shows strong strength, and the wear resistance is also improved correspondingly.

## Acknowledgments

This paper is the 2018 Guangxi university student innovation and entrepreneurship training project "structural design and optimization of mine splitting machine" (201810595159); it is the phased research result of Guangxi university student innovation and entrepreneurship training project "design and optimization of wedge components of mine splitting machine" (201810595160).

## Reference

[1] Chen, H. Zhou, T.P. Chen,Z.J. (2019) Effect of tempering temperature on Microstructure and mechanical properties of 30cr15mo1n. Steel, 5: 60-67.
[2] Yang, M. Cui, M. Li, Z.Y. (2019) Effect of tempering temperature on Microstructure and mechanical properties of 42CrMo steel. Modern Metallurgy, 1:4-5.
[3] Fan,Y.Q. Ma,C.W. Li,S.P. (2019) Effect of heat treatment process on Microstructure and mechanical properties of 550MPa steel.heat treatment process, 2:189-196.
[4] Wu, Z.K. Chen,W. Zhou,H.M. Xiang,J.B. (2019) Effect of tempering temperature on mechanical properties of 30CrMnSiA steel. metal heat treatment, 2: 163-166.
[5] Yang, P.Y. Ai, Y.L. Chen,W.H. He,W. (2018) Effect of quenching temperature on Microstructure and properties of B3 steel for grinding balls. hot working process, 22:174-177.
[6] Guo,W. Zhang.L.P. Chai,R.G. Zhang,C.W. (2018) Effect of tempering process on wear resistance

IOP Conf. Series: Materials Science and Engineering **738** (2020) 012021 doi:10.1088/1757-899X/738/1/012021

of Fe based alloy cladding layer. hot working process, 6:153-157.
[7] Huang,X.L. He,Y.H. Zhang,Q.K. (2017) Effect of heat treatment process on the structure and mechanical properties of 18% Cr Martensitic Stainless Steel.Powder metallurgy materials science and engineering,4:503-509.

IOP Conf. Series: Materials Science and Engineering **738** (2020) 012022 doi:10.1088/1757-899X/738/1/012022

# Study of Hydrothermally Lignite and Rice Husk blend: Thermogravimetric Analysis

**Miqdad Mehdi[1*], Muhammad Hamid Siddiqi[2]**

[1]School of Law, Xi'an Jiaotong University, China.

[2]School of Energy and Power Engineering, Xi'an Jiaotong University, China.

*Corresponding author e-mail: miqdadmehdi93@gmail.com

**Abstract**. The increased realization in greenhouse emission has encouraged to grow new technologies to accommodate the confinement and seizure of $CO_2$. Keeping this situation, the thermal behaviour of lignite, Rice husk, besides their mergers throughout pyrolysis and combustion procedures, was examined in the study for their possible usage as a sustainable and substitute energy resource. The experiments were done in a thermo-gravimetric analyser at a heating rate of 40 °C/min and range from standard room-temp up to 950°C. These blends presented diverse ignition as well as combustion actions dependent on the proportion of rice husk and atmosphere (Oxy-fuel or Nitrogen). These investigational outcomes support to describe as well as expect the performance of lignite and rice husk blends in real applications.

## 1. Introduction

Consistent, useful, as well as the clean energy resource, remains one of the elementary requirements of humanity. Air pollution is a major global environmental problem, with various adverse effects on health and the environment. International and regional agreements bound nations to reduce emissions of air pollutants. Several conventions regulate atmospheric and related issues, yet there is still no coherent legal framework addressing the protection of the atmosphere. Durable confirmation proposes that the mutually average global temperature and $CO_2$ absorption have expressively improved the beginning of industrial advancement. Also, they remain sound connected [1]. Anxieties above environmental variation have directed to growing energies, going on emerging technologies near to condense carbon dioxide productions as of humanoid actions [2-3].

Coal has been and will endure some of the key energy resources in the long-term as of its abundant reserves in addition to comparably low-price, particularly intended for the usage of base-load power generation. Intended for illustration, the coal percentage in the world energy feeding remained 29.4% in 2009, as compared to the oil as 34.8% as well as 23.8% natural gas [3]. In the tenure of power generation, in the USA, coal endures the leading fuel, causal around 43.3% of the entire power generation in 2011 [4], and approximately 79% in China.

As shown in Fig 1, the largest producer of power from coal is in China, as compared to the other countries, which has coal and biomass reserves. The new principle intended to condense the GHG emissions by emerging new technologies [4]. The combustion of fossil fuels is the donor to the emissions of carbon dioxide and Green House consequence [5]. The left-over to energy perception is one of the most suitable solutions to handle by the agro-waste as well as to diminish the GHG emissions by cultivating the adeptness of fossil fuels [6].

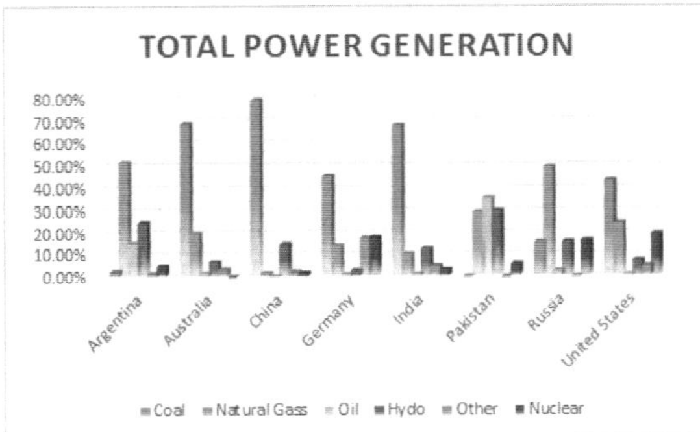


Figure 1 Total generation of Power from different fuels [5]

Pakistani coal has 97% lignite, and residual are sub-bituminous and bituminous coal [7]. Nevertheless, the direct use of lignite has the numeral of Limitations related to the environment adulteration, maximum cost, and minimum energy [8].

Numerous biomass resources, for instance, agricultural, forestry, municipal as well as industrial wastes be able to use as fuel for the generation [9–11]. As described by authors, [12] huge distinctions of biomass features are typically owing to alterations in the methods, heating rate, atmosphere, in addition to the alterations in the chemical configurations.

This work is based on the study of thermal properties of blends using the TG apparatus under iso-thermal and non-isothermal situations and, similarly, discover out the conduct of the blend linked to the emission-related to the environment as well as energy production. The combustion and pyrolysis study of numerous blends were conceded out to classify the finest configuration of the blend to accomplish the objective.

## 2. Material and Methods

TG study of given samples was conceded out by the rate of heating 40 °C/min in the Oxy-Fuel and nitrogen environment. The temperature range was ambient to 950 °C; the sample quantity was 10 mg max in gross weight. Oxygen for combustion and nitrogen for pyrolysis in the thermo-gravimetric analyser, the rate of flow was fixed on 20 ml/min, with the rate of heating 40 °C/min, in addition to staying additional 60 mins at 950° C in pyrolysis. The experiments were completed through SDT q600 America Thermo-Gravimetric Analyzer, according to the ISO 11358-1 standard.

The calorific values of blends were taken through subsequent the ASTM D5865-13 by LECO AC 500 bomb calorimeter (Germany).

Proximate analysis was conceded out in the muffle furnace to measure the moisture, volatile matter, and ash content, according to ASTM standards D3302/D3173, D3175-11, D3174-12, respectively. LECO628 CHN-X analyzer (Germany) remained to amount the sulfur and carbon substances in the samples (ASTM D4239-14, D3172-13, respectively).

### *2.1. Sample Formation*

Pakistani lignite was found from Baluchistan, and the rice husk was collected from the local village to form blends. Three number of diverse arrangements were prepared by variable the attentiveness of the local rice husk ($L_{RH}$) and lignite of Pakistan ($P_L$). The first composition 30% by weight, lignite, and 50% by weight rice husk was mixed to make $P_L$30-$L_{RH}$50 blend. Similarly, $P_L$40-$L_{RH}$60 and $P_L$50-$L_{RH}$50 found through resolving the $P_L$ and $L_{RH}$ in given relations. The ultimate analysis stood accomplished with CHNOS elemental analyzer to analyze carbon, fixed carbon as well as sulfur contents. By using the bomb calorimeter, the calorific value can be found. The Calorific values and ultimate & proximately analysis of all blends are shown in Table 1.

## 3. Result and Discussion

All samples are predominantly confined Carbon & Oxygen. In $P_L50/L_{RH}50$, the number of carbon contents was in large quantity as compared to the others, which is relatively virtuous in terms of thermal perspective. In $P_L50/L_{RH}50$, low Sulphur contents found as compared to the parent fuels and environment-friendly and no special treatment required for flue gases.

Table 1: Ultimate & Proximate Analysis of parent's fuels and their blends

| | | | **Local Rice Husk** | **Pakistani Lignite** | **$P_L$30%-$L_{RH}$70%** | **$P_L$40%-$L_{RH}$60%** | **$P_L$50%-$L_{RH}$50%** |
|---|---|---|---|---|---|---|---|
| 1 | Gross Calorific Value | K.cal/Kg | 4082 | 5998 | 4458 | 4980 | 4990 |
| 2 | Fixed Carbon | % | 22.35 | 43.9 | 28.12 | 32.57 | 33.29 |
| 3 | Volatile matters | % | 63.25 | 40.09 | 56.07 | 50.05 | 49.56 |
| 4 | Ash Contents | % | 13.65 | 15.56 | 15.34 | 16.35 | 16.65 |
| 5 | Total Moisture | % | 7.75 | 6.65 | 5.42 | 7.04 | 5.11 |
| 6 | Inherent Moisture | % | 0.78 | 0.65 | 0.38 | 1.02 | 0.42 |
| 7 | Sulphur Contents | % | 0.19 | 4.05 | 1.272 | 1.696 | 2.12 |

Significances of the ultimate and proximate analysis presented that the $P_L50/L_{RH}50$ is the finest sample combination because of its supreme calorific value (4990 Kcal/Kg), which means $P_L50/L_{RH}50$ is the appropriate combination of combustion and energy production at the level of industrial production.

### *3.1. Pyrolysis*

The lingo-cellulosic compounds study is related to pyrolysis because the components decay happens on multiple temperatures between the ranges. Cellulose, as well as hemicellulose, remain polymeric carbohydrate configurations, named polysaccharides. Fig 2 (A, B, C), are combined TG, as well as DTG outlines, show the thermal degradation features of different blends at a 40 °C/min heating rate. Coal-biomass blends can be separated into three main regions in the process of pyrolysis: Firstly, moisture as well as slightly removal of volatile matters (<100 °C); secondly, hemicellulose the degradation of $L_{RH}$ (180-320 °C) as well as lignin and cellulose decomposition of lignite (320-420 °C) and thirdly, lignin degradation of lignite (>420 °C) [13, 14]. From the TG curve, it was detected that the ultimate strength of the second zone was considerably greater than the first zone. From 330 °C toward 950 °C, the weight degradation rate minimized because of the slow decay of lignite in blends. The loss of carbon atoms characteristically happens at a slower rate over a considerable temperature variation of 180-950 °C. $CO_2$ production is detected because of the oxygen contents of lignite and rice husk [15]. The creation of $CO_2$ in the nitrogen environment was found chief-contributor to un-restricted gases through the high absorbance-intensity. Correspondingly, methane is produced as a result of de-volatilization. As shown in Fig 2, the moisture degradation and very small amount of volatile constituent's production is monitored in the first region of weight degradation. Abundant, the de-volatilization happened in the first region of weight degradation, the breakage of thermal bounds of fragile in the polymeric structure of biomass, and the creation of stronger as well as additional steady bonds to take their place [16].

DTG curves show the temperature range upon which the maximum amount of weight degradation happened stay labelled through the peak location in the curve. DTG peaks can be allocated as follow: The first peak of DTG at 80°C, 60°C and 90°C for $P_L50\text{-}L_{RH}50$, $P_L40\text{-}L_{RH}60$, and $P_L30\text{-}L_{RH}70$ respectively, is possibly due to removal of moisture and slight volatile matters and the second because of maximum degradation of weight, 315°C, 320°C and 335°C respectively. The peaks of DTG are considerable nearer of all blends. The third region saw a considerable lower-rate of weight degradation as compared to the second stage.

An evaluation of DTG curves of lignite and rice husk blends established that the peak position moved to low-temperature through the increase of rice husk in the blend [15]. Sulphur dioxide production may be accredited to sulphur in the lignite, whereas methane creation might be produced because of de-volatilization [16].

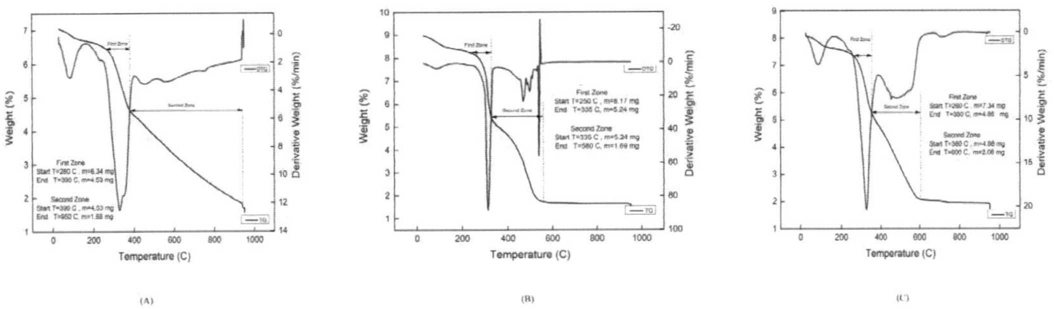


Figure 2 Pyrolysis TG and DTG profile (A) $P_L50$-$L_{RH}50$ (B) $P_L40$-$L_{RH}60$ (C) $P_L30$-$L_{RH}70$

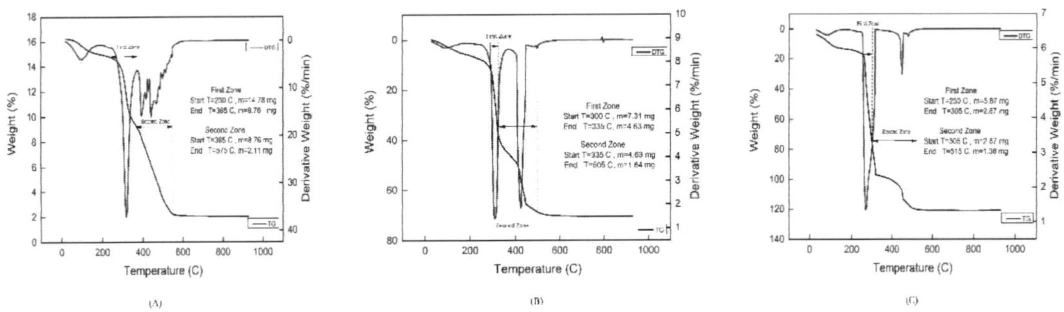


Figure 3 Combustion TG and DTG profile (A) $P_L50$-$L_{RH}50$ (B) $P_L40$-$L_{RH}60$ (C) $P_L30$-$L_{RH}70$

*3.2. Combustion*

Thermal stability and thermal decomposition of blends are resolute with thermogravimetric analysis to observe the alteration in mass. Combustion of lignite and rice husk blends $P_L50/L_{RH}50$, $P_L40/L_{RH}60$, and $P_L30/L_{RH}70$ are shown in Fig 3. The first zone can be weight degradation as an outcome exclusion of the moisture contents. The thermal degradation of local rice husks can be done in two different steps. First, the de-volatilization of cellulose and hemicellulose components and second, degradation can be done because of char combustion.

Fig 3 (A, B, C), demonstrate the weight degradation because of moisture contents between 25-100 °C and show a low peak at 100°C, 80°C and 60°C of $P_L50/L_{RH}50$, $P_L40/L_{RH}60$, and $P_L30/L_{RH}70$ respectively, which determine the material stability [17]. About 100°C of all samples, volatile production is on the go. In the oxy-fuel situation, water production happened for all samples in the range of 100-200 °C. The crucial influence to develop carbon dioxide gas is due to its greater intensity of absorbance. The DTG curve of blends reveals, a high peak at 320°C, 305°C and 275°C for $P_L50/L_{RH}50$, $P_L40/L_{RH}60$ and $P_L30/L_{RH}70$ respectively and the extreme weight loss at 250 °C to 340°C was perceived in TG curves because of weak internal bonding in rice husk. From 320-375°C for $P_L50/L_{RH}50$, 305-350°C for $P_L40/L_{RH}60$, and 275-305°C for $P_L30/L_{RH}70$, decomposition of lignite is occurring. The degradation of lignite was moderately sluggish than rice husk due to strong bonding between carbon-carbon atoms [17, 18].

## 4. Conclusion

Co-firing of coal and biomass might deliver an attractive choice for the dumping of rice husk, permitting for energy retrieval, along with economical and eco-friendly supports. Combustion analysis indicated that the thermal stability of the lignite was found to be reduced with the increase of rice husk. The increase in the decay rate through the increase of rice husk amount caused the reactivity of rice husk. $P_L50\%$-$L_{RH}50\%$ was originated to be practicable for the energy creation point of view like power plant

& boilers because of high calorific value and low volatile matters, low ash contents and low ignition temperature as associated for other blends and profitable than the imported coal.

**Acknowledgments**
The author is grateful to the Law School and School of Energy and Power Engineering of Xi'an Jiaotong University. And the author also wishes to thank Prof. Jinyuan Su for their valuable comments and suggestions.

**References**

[1] IPCC. Contribution of working groups I, II, and III to the fourth assessment report of the intergovernmental panel on climate change. Intergovernmental Panel on Climate Change; 2007.
[2] G. AF. Needs, resources, and climate change: clean and efficient conversion technologies. Progress in Energy and Combustion Science 2011; 37:15e51.
[3] BP. Statistical review of world energy 2010; 2010.
[4] Annual energy review 2009. U.S. Energy Information Administration; 2010.
[5] C. Snyder, T. Bruulsema, T. Jensen, and P. Fixen, "Review of greenhouse gas emissions from crop production systems and fertilizer management effects," Agriculture, Ecosystems & Environment, vol. 133, pp. 247-266, 2009.
[6] H. Boerrigter and R. Rauch, "Review of applications of gases from biomass gasification," ECN Biomassa, Kolen en Milieuonderzoek, vol. 20, 2006.
[7] M. S. Malkani, "A review of coal and water resources of Pakistan," Journal of Science, Technology, and Development, vol. 31, pp. 202-218, 2012.
[8] S. Su, Y. Jin, X. Yu, and R. Worrall, "Preliminary experimental studies of waste coal gasification," in Cleaner Combustion and Sustainable World, ed: Springer, pp. 719-723, 2013.
[9] DA. Mortari, I. Ávila, AM. dos Santos, PM. Crnkovic, "Study of thermal decomposition and ignition temperature of bagasse, coal and their blends," Therm Eng; pp. 9:81–6, 2010.
[10] L. Wilson, W. Yang, W. Blasiak, GR. John, CF. Mhilu, "Thermal characterization of tropical biomass feedstock's," Energy Convers Manage; pp. 52:191–8, 2011.
[11] JE. White, WJ. Catallo, BL. Legendre, "Biomass pyrolysis kinetics: a comparative critical review with relevant agricultural residue case studies," J Anal Appl Pyrol; pp. 91:1–33, 2011.
[12] M. Rieth, A. Clements, M. Rabaçal, F. Proch, O. Stein, and A. Kempf, "Flamelet LES modeling of coal combustion with detailed de-volatilization by directly coupled CPD," Proceedings of the Combustion Institute, vol. 36, pp. 2181-2189, 2017.
[13] SS. Idris, NA. Rahman, K. Ismail, AB. Alias, ZA. Rashid, MJ. Aris, "Investigation on thermochemical behavior of low rank Malaysian coal, oil palm biomass and their blends during pyrolysis via thermogravimetric analysis (TGA)," Bioresource technology, vol. 101, pp. 4584-4592, 2010.
[14] NS. Yuzbasi, N. Selçuk, "Air and oxy-fuel combustion characteristics of biomass/lignite blends in TGA-FTIR," Fuel processing technology, vol. 92, pp. 1101-1108, 2011.
[15] M. Muthuraman, T. Namioka, K. Yoshikawa, "Characteristics of co-combustion and kinetic study on hydrothermally treated municipal solid waste with different rank coals: A thermogravimetric analysis" Applied Energy, pp. 89:141–148, 2010.
[16] P. Phasee, C. Areeprasert. "Thermal Decomposition Behavior during Combustion of Hydrothermally Treated MSW by Thermogravimetric Analysis," Energy Procedia; pp. 138:616–621, 2017.
[17] MH. Siddiqi, MS. Hussain, T. Iqbal, AN. Tabish, W. Mehmood, T. Qureshi, "Combustion and Pyrolysis Study of Pakistani Coal, Rice Husk, and Their Blends via TGA Analysis," 2018 2nd International Conference on Energy Conservation and Efficiency (ICECE), 2018.
[18] L. Yanfen, M. Xiaoqian, "Thermogravimetric analysis of the co-combustion of coal and paper mill sludge," Applied Energy; pp. 87:3526–3532, 2010.

# Research Recommendations for Food Safety Testing Standard Sample System

**Dong Tianfei**

Chifeng Industrial Vocational Technical College, Chifeng City, Inner Mongolian Autonomous Region, China, 024000

**Abstract:** The research on the standard sample system for food safety testing is mainly to solve the problems of forward-looking, systematic and scientific in the development of food safety testing standard sample system. This research work combines the necessity of research and the analysis of the current research situation, classifies and constructs the standard samples in the food safety testing standard sample system, establishes the food safety testing standard sample system, and compiles the corresponding system table and system management. Besides we also give some suggestions on how to promote the improvement and improvement of the system, such as factors and technical requirements.

## 1. Introduction

In recent years, food safety issues have remained a hot topic that has attracted wide attention from all walks of life. In the new era of people's living standards, people are paying more attention to food safety and physical health issues. Governments have also joined the food safety research work queue, investing a lot of manpower, material resources and financial resources to study food safety. And the research of detection technology has placed an important position in various research work. This is because food safety is not only related to people's physical health and life safety, but also related to the social stability and economic development of a country. The reason why governments have increased their attention to food safety issues and the study of food safety testing techniques. It is to ensure the safety of food and to make people eat safe and secure food. Therefore, it is necessary to conduct research on the food safety testing standard sample system.

## 2. Significance and Value of Research on Food Safety Testing Standard Sample System

Standard sample refers to an organic component in the entire standardization work. Standardized samples are needed to quantitatively or qualitatively describe the standard content and requirements of the substance to be tested. It is used to analyze the quality assessment and control of the substance to be tested, and to calibrate the analysis. Instrumentation, assessment and analysis, analysis of data comparison and product quality evaluation. It is characterized by uniformity, stability, and the advantages of intuitive image and high precision. In recent years, due to the development of technology industries such as molecular biology and instrumental analysis, standard samples have appeared in the field of food safety testing, but the scope is also limited to the monitoring of food additives and residues, for genetically modified products, biotech crop products, food microorganisms. Standard samples are still in the preliminary development stage, and standard sample studies in the field of phytosanitary and animal quarantine have not yet entered formal research. In order to meet the needs of international trade and the resolution of domestic food safety issues, it is necessary to strengthen the study of standard samples.

## 3. Research Status and Research Scope of Food Safety Testing Standard Samples

### *3.1. Research Status of Food Safety Testing Standard Samples*

The research work on food safety testing standards is mainly for the study of special standard samples for the analysis of pesticide residues. This is because in the 1980s and 1990s, in order to increase agricultural production and income, the use of pesticides began to be used, resulting in a large number of toxic substances. Substances entering the soil, water bodies and crops not only pollute the food, but also cause environmental pollution. These pollutants remain in the soil or water body for a long time, affecting the subsequent plant growth environment, and also cause extreme health. Big damage. In order to protect the environment, ensure that people eat safe agricultural products, and increase the detection and analysis of toxic pesticide residues in agricultural products, there have been food safety testing standards for pesticide residues in crops, so that pesticide pollution has been controlled. It also makes pesticide pollution control a top priority in food safety testing. With the analysis of pesticide residues, the demand for special standard samples for the analysis of pesticide residues has increased, but at the outset these samples are dependent on exports, so the cost is extremely high and the operation is not convenient. Therefore, it is an inevitable trend to strengthen the research on standard samples. In these years, not only standard samples of pesticide residues have been developed, but also standard samples for food additives, genetically modified products and animal and plant quarantine standards have emerged to meet the satisfaction of agricultural ecological environment and biological samples needs. The research on the standard sample has also promoted the research work of the food safety test standard sample system in China, and the research work on the standard sample in the field of food safety testing has made certain progress.

### *3.2. Food Safety Testing Standard Sample Research Scope*

Food safety refers to the safety of foods and raw materials that people eat in their daily lives. These foods do not cause any harm to human health under normal and reasonable eating methods. Food safety testing includes animal and plant product safety testing, food microbial testing, food physicochemical testing and cosmetic testing in the fields of agroforestry, animal husbandry and fisheries. The majors and disciplines involved include agronomy, agroecology, animal husbandry and veterinary medicine, food safety engineering, microbiology, crop breeding, organic chemistry, inorganic chemistry, agrochemistry, physiology, zoology and botany. Food safety testing involves multiple industries and multiple disciplines. These standard samples of different industries and disciplines related to food safety testing are common in some respects, such as phytosanitary standards including weeds, pests, viruses, bacterial fungi, and Plant genetically modified ingredients, etc., and the standard samples for animal quarantine also contain viruses and bacteria. Although these standard samples are not identical in use and content, they are identical in technical requirements and characteristics, so the standards for food safety testing are samples which can also be interdisciplinary and professional, domain and other boundaries, and then can be classified and divided according to the needs of specific quarantine work. Therefore, the main research contents of the standard sample system for food safety testing can also be used in the fields of phytosanitary, animal quarantine, microbial detection and physical and chemical detection.

## 4. Construction of Food Safety Testing Standard Sample System Framework

### *4.1. Food Safety Testing Standard Sample Classification and Grading*

According to international standards, the principles of scientific, systematic, forward-looking and practical should be followed when classifying food safety testing standards. First of all, we must fully follow the laws of science, combine the characteristics of standard samples in different fields to systematically divide the system, and form a scientific and standardized standard sample system. Secondly, to classify the standard samples, it is necessary to fully study the current standard samples of various disciplines in the economic, scientific, technical and management needs, and to list the standard samples in detail according to the specific application and actual needs, in order to make a scientific classification and form a systematic standard sample system. Once again, there must be sufficient room for development, and the standard samples should be appropriately expanded in conjunction with the actual needs of scientific development, so that the standard sample system has unlimited development space. In addition, the user's needs should be fully considered, and a scientific retrieval system should be established in combination with standard sample requirements and daily use to facilitate quick retrieval of relevant data and standard samples. According to these standards and classification principles, standard samples in the field of food safety testing can be divided into three categories: the first category is based on the characteristics of the standard sample itself, divided into biological, chemical and physical properties, engineering characteristics and other standard samples. Classes are the division of the standard sample application field; the third category is the description of the analyte and matrix. On the basis of this, combined with the definition of food safety testing standards and the actual needs of laboratory research work, the standard samples are listed and systematically classified. When classifying, the same characteristics should be divided together, and then the common feature forming categories should be refined. Different types of standard samples are further searched for other common characteristics, and the relevant common characteristics are refined to form another level category through repeated iterations and classifications, different levels can be formed.

According to the above induction and analysis, four main levels can be formed: the fourth layer is a standard sample of the personality, and the third layer is a standard sample of the door, which is extracted from the fourth layer. The second layer is the field standard sample, which is extracted from the third layer. The first layer is a characteristic standard sample that is extracted from the second layer. The first layer of the characteristic standard sample is consistent with the second layer of the field standard sample and classification principle. The third layer of the standard sample is based on the above classification principle combined with the analyte description and the matrix description, and is reorganized and returned. class. In addition, in order to facilitate management and use, analysis, etc., a fourth layer of individual standard samples has been added, which is an independent standard sample for each category of standard samples.

### *4.2. Construction of Food Safety Testing Standard Sample System*

According to the classification and division of the food safety testing standard samples above, the construction of the standard sample system can be carried out. In the food safety test standard sample, the first layer is a characteristic standard sample, including biological characteristics, chemical characteristics, engineering characteristics, and the like. The second layer is the field standard sample, including the field of phytosanitary, animal quarantine, microbial testing, physical and chemical testing. The third layer is a standard sample of foods, including food classification, food additives, pesticide residues, metal elements, pests, weeds, fungi, bacteria, and the like. The fourth layer is a standard sample of personality, including a standard sample of toxic wheat, a standard sample of tobacco ring spot virus, a standard sample of avian influenza, and a standard sample of Staphylococcus aureus. At present, a food safety testing standard sample system framework has been formed, but it is also necessary to formulate a standard sample system table that is in line with China's national conditions and food safety testing work. The system table should include the development, planning, management and other work of food safety testing standard samples, and promote the food safety testing work smoothly through the uniform standard sample regulations. The standard system framework is to be continuously enriched with the advancement of science and technology.

### *4.3. Food Safety Testing Standard Sample System Table and Standard Sample Code*

The food safety testing standard sample system table contains the hierarchical diagram of the system table, the standard sample code, and the standard sample information. Combined with the hierarchical structure diagram, the structure and framework of the standard samples of different levels can be clearly divided, and the standard samples of different levels can be managed according to the actual situation. Through the standard sample system table and standard sample code, it is convenient to carry out the food safety test standard sample retrieval and classification management work, and make the standard sample information more standardized and perfect according to the relevant information in the table. Different levels of standard sample categories and individual standard samples constitute a food safety testing standard sample system table, and can be arranged according to a certain level to form a standard sample system table in the field of food safety testing.

Combined with the classification and grading of standard samples, the code for food safety testing standard samples consists of three letters and 13 Arabic numerals. The letters represent the national standard samples. The first two numbers represent the characteristics of the standard samples, the third and fourth. The numbers represent the application areas of the standard samples, while the fifth and sixth numbers represent the categories of the standard samples. The fifth and eighth ninth numbers represent the sequence numbers of the standard samples, and the last four numbers represent the release year number of the standard sample.

## 5. Implementation and Improvement of the Food Safety Testing Standard Sample System

The food safety testing standard system is a standard sample system developed in a unified manner for food safety testing and unified planning and management. The purpose is to better carry out food safety testing through uniform standard samples. The system includes samples of food safety testing standards that have been and are expected to be developed. The system will also be continuously improved and improved with the advancement of science and technology. At this stage, the framework of China's food safety testing standards system has been formed. The research and development, facility conditions and technical means of food safety testing standards need to be further optimized and improved. A more complete system operation mechanism should be established to ensure food. The safety test standard sample system operates normally and orderly. First, a complete and complete organization that is compatible with the food safety testing standard sample system should be established, and the food safety testing standard samples can be comprehensively planned and managed. Secondly, in order to ensure the orderly development of the system, a set of systematic documents should be established to standardize the normal work of the food safety testing standard sample system, such as further improving the food safety testing standards work guide and working technical requirements, especially to strengthen the food microbial standard samples. The research and improvement of professional technical requirements, the construction documents must meet the requirements of national document standards, and must be uniformly released and implemented. In addition, continuous improvement of the food safety testing standard sample system should be carried out, not only with a sound mechanism, organizational structure and regulatory documents, but also a clear blueprint for future development. Specifically, we should do a unified planning for the construction of the standard system, scientifically and rationally set up the types of sample development, and form a complete long-term food safety testing standard sample system in combination with future development. It is also necessary to arrange various research resources reasonably, do a good job in communication and communication between different departments, and promote the development of research and development. Strengthen the construction of the resource guarantee system for system research and construction, that is, rationally arrange funds, technical equipment, etc., and appropriately increase the input of manpower, material resources and financial resources, strengthen investment in the construction and implementation of key food safety testing standards and sample systems, and build advanced experiments. The room draws on the experience of foreign research, development, planning and management of the food safety testing standard sample system, and explores a standard sample system that conforms to international standards and is consistent with the actual work needs of China's food safety testing. Finally, through the construction of the feedback mechanism, we actively listen to the recommendations of relevant scientific research personnel and institutions to promote the continuous improvement and development of the food safety testing standard sample system.

## 6. Conclusion

In summary, this paper studies the construction of the standard sample system in the field of food safety testing, builds a system framework based on the classification of standard samples, and establishes a food safety testing standard sample system. In the future, in order to meet the needs of actual food safety testing, it is also to promote the integration with the international standard sample system. It should also strengthen the research and development of normative documents and professional technical requirements, and further explore the improvement and continuous improvement measures of the standard sample system. The food safety testing standard sample system is guaranteed to operate efficiently and stably.

**References:**

[1] Qiao Xing. Classification of food safety testing standards samples [J]. Food Safety Guide, 2018 (36): 92-92.
[2] Zhai Qingrun, Yang Lian. Research on the Standard System of National Standards for Food Safety Inspection [J]. Food Safety Guide, 2018(15):43.
[3] Jia Dong, Wang Jinling, Xu Dajun, Jia Jinsheng, Fu Haibin, Cao Zhijun. Study on the Standard Sample System of Food Safety Testing[J]. Standard Science,2013(10):27-31.

# Study on the migration of bisphenols from plastic food contact materials to food simulants

**Yue Qiu[1,2], Yan Ruan[1], Lu Zhang[1], Genrong Li[1], Chaolan Tan[1] and Zhining Xia[2]**

[1]Chongqing Academy of Metrology and Quality Inspection, Chongqing, 401123, China

[2]College of Chemistry and Chemical Engineering, Chongqing University, Chongqing, 400030, China

Corresponding Author: Yue Qiu, E-mail: qiuyuecqu@163.com

**Abstract.** A rapid method for the detection of bisphenols was established by ultra performance convergence chromatography (UPC$^2$). The migration rules of bisphenols in four different simulants including of pure water, acetic acid, ethanol and isooctane were first studied, and the effect of extraction solvent on recovery of bisphenols in aqueous simulants was also investigated. Furthermore, the influence of extraction solvent on recovery of bisphenols in aqueous simulants was discussed, and the relationship between migration amount of bisphenols and food simulants, migration time, migration temperature and other factors were explored. The experimental results showed that toluene exhibited the optimal extraction efficiency for bisphenols in aqueous simulants; bisphenols presented the most obvious migration in fatty food simulants; the migration amount of bisphenols increased with the extension of time; the migration rate and amount of bisphenols also enhanced with the increment of contact temperature.

## 1. Introduction

With the increasing public attention for food safety, higher requirements are put forward for food contact materials which are closely related to food quality. Bisphenols are a new class of hazardous migratory substances. These compounds are important raw materials for the production of polycarbonate, epoxy resin and other polymer materials [1-3]. Bisphenol A (BPA) is currently the largest amount of a kind of bisphenol substances, which is widely used in various of plastic food contact materials [4]. Studies showed that BPA has estrogen-like activity mutagenic, endocrine disrupting and carcinogenic effects [5-10]. Therefore, the European Union, United States, China and other countries have stipulated and restricted the use of BPA in food contact materials.

In recent years, the bisphenols which have similar structure with BPA were regarded as the substitutes for BPA, which have been gradually applied in the production of plastic food contact materials [11-14]. At present, bisphenol F (BPF) and bisphenol S (BPS) are commonly used to replace the use of bisphenol A, but researches have shown that BPF and BPS are also harmful to human health [15-17]. As an environmental hormone, bisphenol A-diglycidyl ether (BADGE) is destructive to endocrine system and immune system, as well as the genotoxicity and mutagenesis [18]. Tetrabromobisphenol A (TBBPA) and hexafluorobisphenol A (BPAF) also have reproductive toxicity which could lead to fetal dysplasia [19,20]. According to the national standard GB 9685-2016, the specific migration amount of BPA in food should not exceed 0.6 mg/kg, and that of bisphenol S should not exceed 0.05 mg/kg. The EU stipulates that the total migration of BADGE and its hydrolyzed derivatives in food shall not exceed 9 mg/kg. However, relevant regulations on the migration of BPF, BPAF and other commonly used bisphenols have not been formulated.

Due to the similar structure and great variety of bisphenols, together with the complexity of food matrix, the detection interference of trace bisphenols is serious. Therefore, it is urgent to establish a detection method that can simultaneously analyze a variety of bisphenols and solve the problems of low extraction efficiency and large impurity interference in the sample pretreatment method. In this study, a detection method for four bisphenols (BPF, BPAF, BADGE and TBBPA) in plastic food contact materials was established by using the solid-phase extraction (SPE) and ultra performance convergence chromatography, and its migration behavior in food simulants was studied, which could provide data reference and technical support for testing and monitoring the migration of bisphenols in food contact materials.

## 2. Experimental and Methods

### *2.1. Materials and reagents*

4,4'-dihydroxydiphenyl methane (BPF), 2,2-bis (4-hydroxyphenyl) hexafluoropropane (BPAF), 2,2-bis (2,6-dibromo-4-hydroxyphenyl) propane (BADGE), 2,2-bis (2,6-dibromo-4-hydroxyphenyl) propane (TBBPA), they were all supplied from Dr. Ehrenstorfer (Augsburg, Germany). HPLC grade methanol, isopropanol, acetonitrile, methylene chloride, ethyl acetate and n-hexane were purchased from Merck (Darmstadt, Germany).

### *2.2. Instruments and equipment*

ACQUITY UPC$^2$ ultra performance convergence chromatography (Waters, USA); N-EVAP-112 nitrogen blower (Organomation, USA); BILON-2000CT ultrasonic cleaner (Bilang, Shanghai); BSXT-06 soxhlet extractor (Lanyi, Shanghai).

### *2.3. Methods*

#### *2.3.1. Chromatographic conditions*

Column: ACQUITY UPC$^2$ Torus DIOL column (150 mm×30 mm, 1.7 μm); Column temperature: 35°C; Mobile phase: supercritical $CO_2$- methanol; Detection wavelength: 220 nm; Back pressure: 1800 psi; Injection volume: 3 μL. Elution gradient: 0~1 min, 5%~9% B; 1~2 min, 9%~10% B; 2~2.5 min, 10%~20% B; 2.5-5 min, 20% B; 5~5.5 min, 20%~5% B. The flow rate was maintained at 1.3 mL/min.

#### *2.3.2. Migration experiment*

*(1) Soaking treatment*

The plastic food contact materials samples were sealed and soaked at in food simulants (water, 3% acetic acid, 10% ethanol and isooctane) at temperature of 4, 20, 40, 60 or 80°C, and the simulated fluid were taken at the time points of 2, 4, 8, 12, 24, 48, 120 and 240 h, respectively. After the temperature was restored to room temperature, the pretreatment was carried out according to the method as follows.

*(2) Pretreatment of simulants*

Aqueous simulant: 10 mL aqueous simulant and 10 mL organic solvent (toluene, n-hexane, dichloromethane or ethyl acetate) were taken into the separating funnel, oscillated for extraction and repeated twice. After that, combined the organic phase and dried with nitrogen flow at 45°C, then dissolved the residue with methanol and diluted to volume of 1.0 mL for UPC$^2$ analysis.

Isooctane simulant: 10 mL isooctane simulant was directly dried with nitrogen flow at 45°C, the residue was dissolved with methanol and diluted to volume of 1.0 mL for UPC$^2$ analysis.

## 3. Results and Discussion

### *3.1. Linearity, LODs, and LOQs*

Standard mixture solutions with concentrations of 0.2, 1.0, 5.0, 20.0, 50.0 and 100.0 mg/L were prepared for constructed the calibration curves. External standard method was used for quantitative analysis. The chromatographic peak area Y of each substance was used for linear regression of mass concentration X. The results showed that the correlation coefficients of the four bisphenols were all greater than 0.999. In the blank food simulants without bisphenols, standard mixture solutions of different concentrations were successively added, and the detection limits (LODs) were calculated according to three times signal-to-noise ratio (S/N=3). The LODs of the four bisphenols in the aqueous simulants (water, 3% acetic acid and 10% ethanol) was 0.10 mg/L, and the LODs in the isooctane simulants was 0.05 mg/L.

### *3.2. Migration experiment*

#### *3.2.1. The Selection of extraction solvent for food simulants*

In the experiment, the extraction solvent for the four bisphenols in aqueous food simulants (water, 3% acetic acid and 10% ethanol) were firstly investigated. The blank aqueous simulants were added with high, medium and low concentrations of standard mixture, respectively. Toluene, n-hexane, dichloromethane and ethyl acetate were used as extraction solvents. The extraction efficiency in acidic and alcoholic food simulants were similar to those in pure water substrate. As shown in Table 1, the recoveries of bisphenols extracted by n-hexane were the lower than 65.0%, while the recoveries extracted by toluene were the highest (>85.0%). Under the condition of toluene extraction, the recoveries of the four bisphenols in the pure water simulant were ranged from 80.8% to 94.5%, and the relative standard deviations (RSDs) were in the range of 2.69%~6.17%, indicating that the toluene extraction method for food simulants in this experiment can meet the actual test requirements. Therefore, toluene was selected as the optimal solvent for extraction of bisphenols in aqueous simulants.

Table 1. Recoveries and relative standard deviations of the four bisphenols in pure water simulant.

| Extracting solvent | Spiked level/（mg/L） | R/% | | | | RSD/% | | | |
|---|---|---|---|---|---|---|---|---|---|
| | | BPF | BPAF | BADGE | TBBPA | BPF | BPAF | BADGE | TBBPA |
| toluene | 1 | 84.3 | 87.6 | 80.8 | 81.1 | 6.07 | 5.19 | 6.17 | 4.68 |
| | 50 | 89.4 | 90.8 | 86.4 | 84.0 | 4.29 | 5.03 | 4.81 | 3.15 |
| | 100 | 94.5 | 91.6 | 89.7 | 90.2 | 2.94 | 3.58 | 3.10 | 2.69 |
| n-hexane | 1 | 41.3 | 46.4 | 51.8 | 49.6 | 8.16 | 7.49 | 6.40 | 6.84 |
| | 50 | 50.5 | 56.7 | 59.4 | 60.1 | 6.08 | 4.28 | 5.64 | 4.36 |
| | 100 | 61.9 | 59.2 | 64.5 | 63.4 | 4.15 | 3.16 | 3.49 | 3.30 |
| dichloro-methane | 1 | 66.8 | 71.0 | 73.6 | 69.4 | 7.11 | 6.18 | 5.66 | 5.94 |
| | 50 | 75.9 | 78.4 | 80.2 | 73.1 | 5.02 | 5.34 | 4.09 | 3.27 |
| | 100 | 84.9 | 82.2 | 85.1 | 80.6 | 3.55 | 2.99 | 4.18 | 3.05 |
| ethyl acetate | 1 | 70.8 | 75.1 | 77.4 | 69.2 | 6.22 | 6.67 | 5.61 | 5.06 |
| | 50 | 73.5 | 76.0 | 80.7 | 78.8 | 3.58 | 5.44 | 5.03 | 4.39 |
| | 100 | 81.9 | 83.3 | 85.1 | 82.6 | 2.87 | 4.21 | 3.47 | 3.77 |

*3.2.2. The effect of simulants on the migration amount of bisphenols*

In this experiment, plastic food contact maerials samples were immersed in food simulants at 40°C for 24 h, respectively, the migration amount of bisphenols in different food simulants were shown in Figure 1. The results showed that only BPAF migrated in the pure water simulant, while BPF, BADGE and TBBPA were not dissolved. In the ethanol and isooctane food simulants, all the four bisphenols were dissolved in different degrees, and the maximum migration was obtained in isooctane. Therefore, it is not recommended to add bisphenols to plastic food contact materials with ethanol and fatty food.

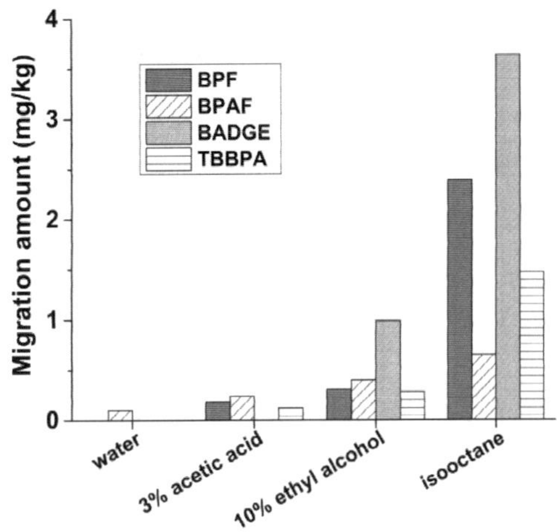


Figure 1. Migration of bisphenols in different simulants (40°C, 24 h).

*3.2.3. The effect of time on migration amount of bisphenols*

Because the migration behavior of bisphenols in isooctane was the most obvious, isooctane was chosen as the simulant in this experiment. Under the contact temperature of 40°C, the migration amount of four bisphenols in the contact materials of plastic food was investigated from 1 to 48 h. As shown in Figure 2, with the extension of contact time, the migration amount of bisphenols also increased. BPAF reached the migration balance at 36 h, while BPF, BADGE and TBBPA all reached the migration balance at 24 h.

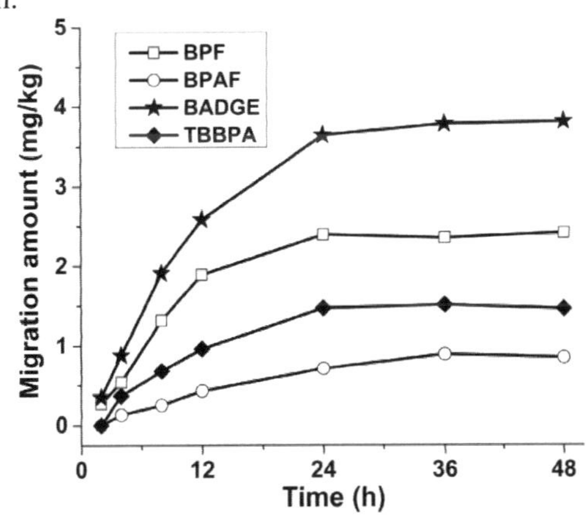


Figure 2. The effect of time on migration amount of bisphenols

## 4. Conclusions

UPC$^2$ was used to establish the detection methods of four bisphenols (BPF, BPAF, BADGE and TBBPA) in plastic food contact materials. The influence of extraction solvent on recoveries of bisphenols in aqueous simulants was investigated. It was found that the extraction efficiency of toluene was the most satisfactory. The effects of simulants, migration time and migration temperature on the migration amount of bisphenols were investigated. Experimental results showed that the migration behaviors of the four bisphenols were the most obvious in isooctane. At the same temperature, the migration of bisphenols increased with the extension of time. The migration speed and migration amount of bisphenols increased significantly with the increase of migration temperature.

**Acknowledgments**

This work was funded by Chongqing scientific research institute performance incentive and guidance special project (cstc2018jxj100007).

**References**

[1] Biedermann-Brem S., Grob K., Fjeldal P. (2008) Release of bisphenol A from polycarbonate baby bottles: mechanisms of formation and investigation of worst case scenarios. Eur. Food Res. Technol., 227: 1053-1060.
[2] Shao B., Han H., Li D., et al (2007) Analysis of alkylphenol and bisphenol A in meat by accelerated solvent extraction and liquid chromatography with tandem mass spectrometry. Food Chem., 105: 1236-1241.
[3] Sun X., Peng J., Wang M., et al (2018) Determination of nine bisphenols in sewage and sludge using dummy molecularly imprinted solid-phase extraction coupled with liquid chromatography tandem mass spectrometry. J. Chromatogr. A, 1552: 10-16.
[4] Cacho J. I., Campillo N., Viñas P., et al (2012) Stir bar sorptive extraction coupled to gas chromatography–mass spectrometry for the determination of bisphenols in canned beverages and filling liquids of canned vegetables. J. Chromatogr. A, 1247: 146-153.
[5] Vandenberg L. N., Colborn T., Hayes T. B., et al (2012) Hormones and endocrine-disrupting chemicals: low-dose effects and nonmonotonic dose responses. Endocr. Rev., 33: 378-455.
[6] Huang Y. Q., Wong C. K., Zheng J. S., et al (2012) Bisphenol A (BPA) in China: a review of sources, environmental levels, and potential human health impacts. Environ. Int., 42: 91-99.
[7] Chen D., Kannan K., Tan H., et al (2016) Bisphenol analogues other than BPA: environmental occurrence, human exposure, and toxicity-a review. Environ. Sci. Technol., 50: 5438-5453.
[8] Clayton E. M., Todd M., Dowd J. B., et al (2011) The impact of bisphenol A and triclosan on immune parameters in the U.S. population, NHANES 2003-2006. Environ. Health Perspect., 119: 390-396.
[9] Zeng J., Kuang H., Hu C., et al (2013) Effect of bisphenol A on rat metabolic profiling studied by using capillary electrophoresis time-of-flight mass spectrometry. Environ. Sci. Technol., 47: 7457-7465.
[10] Matsuda S., Matsuzawa D., Ishii D., et al (2013) Perinatal exposure to bisphenol A enhances contextual fear memory and affects the serotoninergic system in juvenile female mice. Horm. Behav., 63: 709-716.
[11] Chen M., Ike M., Fujita M. (2002) Acute toxicity, mutagenicity, and estrogenicity of bisphenol-A and other bisphenols. Environ. Toxicol., 17: 80-86.
[12] Li Y., Burns K. A., Arao Y., et al (2012) Differential estrogenic actions of endocrine-disrupting chemicals bisphenol A, bisphenol AF, and zearalenone through estrogen receptor alpha and beta in vitro. Environ. Health Perspect., 120: 1029-1035.
[13] Cunha S. C., Fernandes J. O. (2010) Quantification of free and total bisphenol A and bisphenol B in human urine by dispersive liquidliquid microextraction (DLLME) and heart-cutting multidimensional gas chromatography-mass spectrometry (MD-GC/MS). Talanta, 83: 117-125.
[14] Pozdnyakov I. P., Guo L., Glebove M., et al (2010) Aqueous photochemistry of bisphenol E in the presence of beta-cyclodextrin. High Energy Chem., 45: 214-221.

[15] Helies-Toussaint C., Peyre L., Costanzo C., et al (2014) Is bisphenol S a safe substitute for bisphenol A in terms of metabolic function? An in vitro study. Toxicol. Appl. Pharmacol., 280: 224-235.

[16] Kinch C. D., Ibhazehiebo K., Jeong J. H., et al (2015) Low-dose exposure to bisphenol A and replacement bisphenol S induces precocious hypothalamic neurogenesis in embryonic zebrafish. Proc. Natl. Acad. Sci., 112: 1475-1480.

[17] Eladak S., Grisin T., Moison D., et al (2015) A new chapter in the bisphenol A story: bisphenol S and bisphenol F are not safe alternatives to this compound. Fertil. Steril., 103: 11-21.

[18] Munguia-Lopez E. M., Soto-Valdez H. (2001) Effect of heat processing and storage time on migration of bisphenol A (BPA) and bisphenol A-diglycidyl ether (BADGE) to aqueous food simulant from Mexican can coatings. J. Agri. Food Chem., 49: 3666-3671.

[19] Li Y., Burns K. A., Arao Y., et al (2012) Differential estrogenic actions of endocrine-disrupting chemicals bisphenol A, bisphenol AF, and zearalenone through estrogen receptor alpha and beta in vitro. Environ. Health Perspect., 120: 1029-1035.

[20] Wang J., Xiao X., Chen T., et al (2016) High-performance liquid chromatography-ultraviolet method for the determination of total specific migration of nine ultraviolet absorbers in food simulants based on 1,1,3,3-tetramethylguanidine and organic phase anion exchange solid phase extraction to remove glyceride. J. Chromatogr. A, 1451: 58-66.

# Thermodynamics of CuO/AC Adsorbent Deactivating on $PH_3$-containing off gas

**Ming Jiang[1], Qi Jiang[1], Tianguo Li[1], Bo Li[1], Fangdong Zhan[1], Yongmei He[1]***

[1]College of Resources and Environment, Yunnan Agricultural University, Kunming, Yunnan, 650201, China

*Corresponding author: heyongmei06@126.com

**Abstract.** The thermodynamics of activated carbon modified with CuO (CuO/AC) adsorbent deactivation by $PH_3$ was simulated using an HSC thermodynamics database. Chemical reactions were analyzed at adsorption temperatures ranging from 0 to 100°C, and the Gibbs free energy values of all reactions were compared. The results showed that impurities ($O_2$, $H_2$, CO) in off-gas can deactivate CuO in the adsorbent. $O_2$ was primarily responsible for CuO deactivation, and the thermodynamic competitiveness of the reaction involving $O_2$ was the strongest. The most easily generated deactivation products were $Cu(OH)_2$ and $P_4O_{10}$. The adsorption temperature did not obviously affect the deactivation of CuO/AC.

## 1. Introduction

Phosphine ($PH_3$) is a highly toxic gas which commonly originates from the fumigation of grain warehouses [1], the manufacture of semiconductor materials [2], and phosphorus-containing chemical products [3]. $PH_3$ is a gaseous impurity which seriously restricts the comprehensive utilization of industrial off-gas, and it is currently removed using a variety of methods, including absorption [4], adsorption [5], catalytic oxidation [6], and combustion [7]. Among them, adsorption is a low-cost and high-efficiency purification method that is suitable for the thorough purification of off-gas. Adsorbents used to remove $PH_3$ mainly include activated carbon, activated carbon fibers, and molecular sieves.

A preliminary study showed that a modified activated carbon adsorbent demonstrated good adsorption performance for $PH_3$ and that CuO was the main active component [8-11]. However, the inactivation mechanism of $PH_3$ on CuO has not been reported. In this paper, HSC 6.0 thermochemical software was used to simulate possible chemical reactions between $PH_3$ and CuO when $PH_3$ coexists with $O_2$, $N_2$, $H_2$, and CO in the off-gas. The deactivation products are also discussed.

## 2. Thermodynamic calculation method

According to the physicochemical properties of pure substances, the likelihood of a spontaneous reaction when a chemical reaction reaches thermodynamic equilibrium was calculated using the minimum Gibbs free energy, and the deactivation products were analyzed. Before the simulation, the assumed conditions and parameters were as follows: 1) the reaction was assumed to be in thermodynamic equilibrium, 2) the influence of chemical reaction kinetics was not considered; and 3) the adsorption temperature range was maintained between 10-100 °C, with a heating rate of 10 °C.

## 3. Results and discussion

### 3.1. $CuO-PH_3$ system

The possible reactions between CuO in the adsorbent and $PH_3$ in the gas stream are shown in equations (1)-(15). Figure 1 shows the relationship between Gibbs free energy ($\Delta G$) and temperature ($T$) calculated by HSC software. Figure 1 shows that the Gibbs free energy of reactions (1)-(3) and (9) -(15) was less than zero ($\Delta G<0$) from 0-100 °C, indicating that these reactions were spontaneous. The solid products included $CuP_2$, $Cu_2O$, P, $Cu_3P$, Cu, $Cu_2P_2O_7$, $Cu_3(PO_4)_2$, $CuHPO_4$, $Cu(H_2PO_4)_2$, $H_3PO_4$, $Cu(OH)_2$, and $Cu_2O$. The absolute value of Gibbs free energy of reaction (13) was the largest, which indicates that this reaction was the most spontaneous. In addition, the adsorption temperature had no apparent effect on the spontaneity of these reactions.

$4PH_3(g)+2CuO=2CuP_2+6H_2(g)+O_2(g)$ (1)

$PH_3(g)+3CuO=1.5Cu_2O+P+1.5H_2O$ (2)

$1.333PH_3(g)+4CuO=1.333Cu_3P+2H_2O+O_2(g)$ (3)

$2PH_3(g)+6CuO=6CuH(g)+2PO_2+O_2(g)$ (4)

$4PH_3(g)+12CuO=12CuH(g)+2P_2O_5+O_2(g)$ (5)

$4PH_3(g)+12CuO=12CuH(g)+P_4O_{10}+O_2(g)$ (6)

$4PH_3(g)+12CuO=12CuH(g)+P_4O_8+2O_2(g)$ (7)

$4PH_3(g)+10CuO=10CuH(g)+P_4H_2+5O_2(g)$ (8)

$PH_3(g)+1.5CuO=1.5Cu+P+1.5H_2O(g)$ (9)

$4PH_3(g)+16CuO=2Cu_2P_2O_7+12CuH(g)+O_2(g)$ (10)

$2PH_3(g)+8CuO=Cu_3(PO_4)_2+3H_2(g)+5Cu$ (11)

$PH_3(g)+4CuO=CuHPO_4+H_2(g)+3Cu$ (12)

$2PH_3(g)+8CuO=Cu(H_2PO_4)_2+H_2(g)+7Cu$ (13)

$PH_3(g)+4CuO=H_3PO_4+4Cu$ (14)

$PH_3(g)+4.5CuO=1.5Cu(OH)_2+P+1.5Cu_2O$ (15)

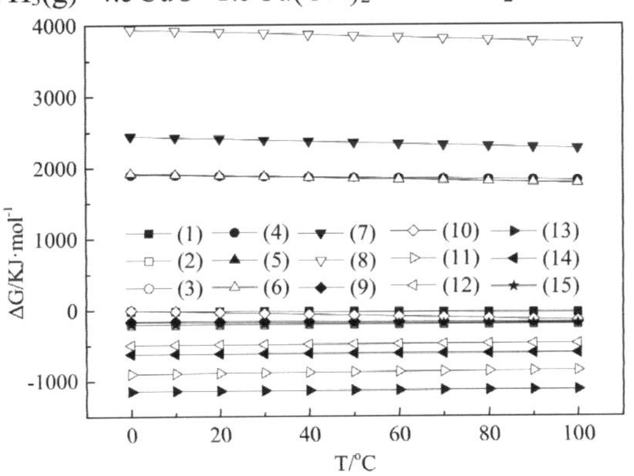


Figure 1. Relationship between Gibbs energy and temperature of equations (1)-(15).

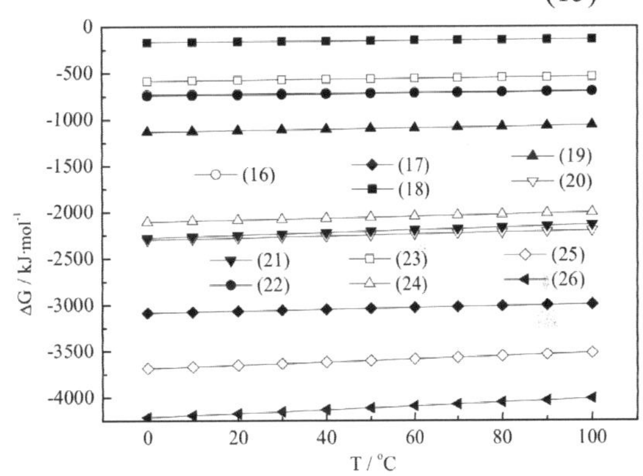


Figure 2. Relationship between Gibbs energy and temperature of equations (16)-(26).

### 3.2. $CuO$-$PH_3$-$O_2$ system

The possible reactions between CuO and $PH_3$ when $O_2$ is present in the gas stream are shown in equations (16)-(26). Figure 2 shows that the Gibbs free energy of all reactions is less than zero, indicating that these reactions are spontaneous at adsorption temperatures from 10-100 °C. This shows that the presence of $O_2$ significantly affects the deactivation of CuO. By comparing the absolute Gibbs free energy values of these chemical reactions, it can be concluded that the absolute Gibbs free energy of reaction (26) is the largest, which indicates that this reaction is the easiest to carry out. This is also the main adsorbent deactivation reaction occurring in this system, which produces $Cu(OH)_2$ and $P_4O_{10}$, in agreement with previous reports[2].

$2PH_3(g)+CuO+O_2(g)=CuP_2+3H_2O$ (16)

$2PH_3(g)+2CuO+4O_2(g)=Cu_2P_2O_7+3H_2O$ (17)

$4PH_3(g)+2CuO+O_2(g)=2CuH(g)+P_4H_2+4H_2O$ (18)

$$PH_3(g)+CuO+2O_2(g)=CuHPO_4+H_2O \tag{19}$$
$$2PH_3(g)+CuO+4O_2(g)=Cu(H_2PO_4)_2+H_2O \tag{20}$$
$$2PH_3(g)+3CuO+4O_2(g)=Cu_3(PO_4)_2+3H_2O \tag{21}$$
$$2PH_3(g)+CuO+1.5O_2(g)=Cu(OH)_2+2P+2H_2O \tag{22}$$
$$PH_3(g)+1.5CuO+1.75O_2(g)=1.5Cu(OH)_2+PO_2 \tag{23}$$
$$2PH_3(g)+3CuO+4O_2(g)=3Cu(OH)_2+P_2O_5 \tag{24}$$
$$4PH_3(g)+6CuO+7O_2(g)=6Cu(OH)_2+P_4O_8 \tag{25}$$
$$4PH_3(g)+6CuO+8O_2(g)=6Cu(OH)_2+P_4O_{10} \tag{26}$$

### *3.3. CuO-$PH_3$-$N_2$ system*
The possible reactions between CuO and $PH_3$ when $N_2$ is present in the gas stream are shown in equations (27)-(34). Figure 3 shows that the Gibbs free energy of all reactions was greater than zero, indicating that these reactions were nonspontaneous. These results show that the addition of $N_2$ inhibited the CuO deactivation in the adsorbent.

$$PH_3(g)+2CuO+3N_2(g)=2CuN_3+1.5H_2(g)+PO_2 \tag{27}$$
$$2PH_3(g)+5CuO+7.5N_2(g)=5CuN_3+3H_2(g)+P_2O_5 \tag{28}$$
$$4PH_3(g)+8CuO+12N_2(g)=8CuN_3+6H_2(g)+P_4O_8 \tag{29}$$
$$4PH_3(g)+10CuO+15N_2(g)=10CuN_3+6H_2(g)+P_4O_{10} \tag{30}$$
$$PH_3(g)+1.5CuO+2.25N_2(g)=1.5CuN_3+P+1.5H_2O \tag{31}$$
$$PH_3(g)+2CuO+3N_2(g)=2CuN_3+P+1.5H_2(g)+O_2(g) \tag{32}$$
$$PH_3(g)+2CuO+3.5N_2(g)=2CuN_3+PN(g)+1.5H_2(g)+O_2(g) \tag{33}$$
$$PH_3(g)+1.5CuO+2.75N_2(g)=1.5CuN_3+PN(g)+1.5H_2O \tag{34}$$

### *3.4. CuO-$PH_3$-$H_2$ system*
The possible reactions between CuO and $PH_3$ when $H_2$ is present in the system are shown in equations (35)-(40). Figure 3 depicts only the Gibbs free energy of reaction (35), which is less than zero at adsorption temperatures between 10-100°C, indicating that the reaction can proceed spontaneously to produce the deactivation product $Cu_3P$. In addition, the $\Delta G$ value of formula equation (35) did not significantly change as the adsorption temperature increased, indicating that the adsorption temperature did not significantly affect the spontaneous nature of the reaction.

$$PH_3(g)+3CuO+1.5H_2(g)=Cu_3P+3H_2O \tag{35}$$
$$PH_3(g)+3CuO+H_2(g)=3CuH(g)+PO_2+H_2O \tag{36}$$
$$2PH_3(g)+7CuO+2.5H_2(g)=7CuH(g)+P_2O_5+2H_2O \tag{37}$$
$$4PH_3(g)+12CuO+4H_2(g)=12CuH(g)+P_4O_8+4H_2O \tag{38}$$
$$4PH_3(g)+14CuO+5H_2(g)=14CuH(g)+P_4O_{10}+4H_2O \tag{39}$$
$$4PH_3(g)+4CuO+H_2(g)=4CuH(g)+P_4H_2+4H_2O \tag{40}$$

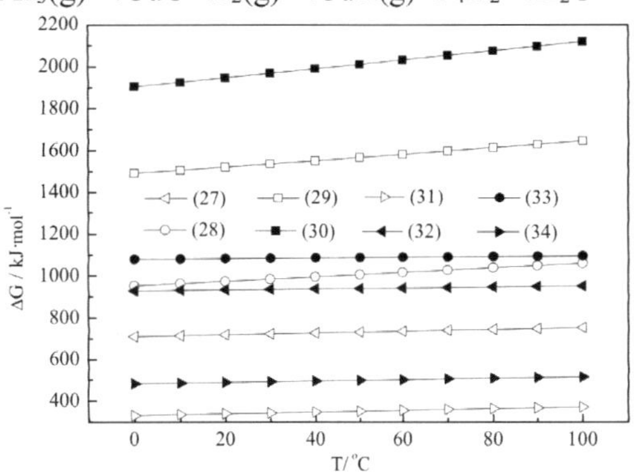


Figure 3. Relationship between Gibbs energy and temperature of equations (35)-(40).

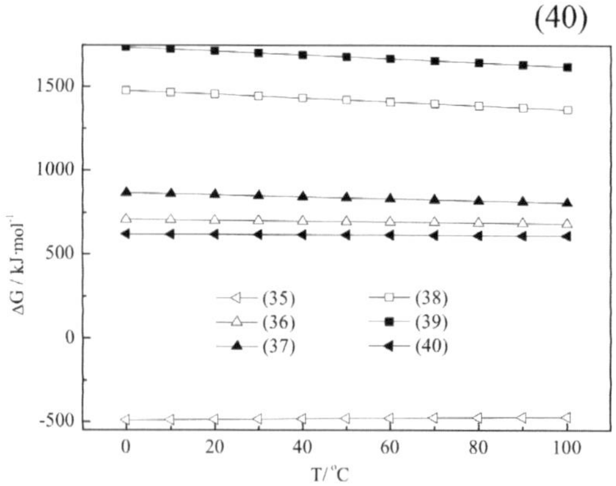


Figure 4. Relationship between Gibbs energy and temperature of equations (27)-(34).

### *3.5. CuO-PH3-CO system*

The possible reactions between CuO and $PH_3$ when CO is present in the gas stream are shown in equations (41)-(74). Figure 5 and Figure 6 show that the Gibbs free energy of the reactions in equations (48), (55), and (69) were positive, while others may occur spontaneously at adsorption temperatures from 10-100 °C. The absolute value of Gibbs free energy of reaction (59) was the largest, indicating that this reaction is the most spontaneous, and the deactivation products are $Cu_2P_2O_7$ and $P_4H_2$.

$4.8PH_3(g)+1.4CuO+3.6CO(g)=1.4CuP_2+3.6CH_4(g)+P_2O_5$ (41)
$4PH_3(g)+CuO+3CO(g)=CuP_2+3CH_4(g)+2PO_2$ (42)
$8PH_3(g)+2CuO+6CO(g)=2CuP_2+6CH_4(g)+P_4O_8$ (43)
$9.6PH_3(g)+2.8CuO+7.2CO(g)=2.8CuP_2+7.2CH_4(g)+P_4O_{10}$ (44)
$5.867PH_3(g)+5.6CuO+4.4CO(g)=1.867Cu_3P+4.4CH_4(g)+P_4O_{10}$ (45)
$5.333PH_3(g)+4CuO+4CO(g)=1.333Cu_3P+4CH_4(g)+P_4O_8$ (46)
$4PH_3(g)+3CuO+3CO(g)=Cu_3P+3CH_4(g)+3PO_2$ (47)
$10PH_3(g)+3CuO+15CO(g)=Cu_3P+7.5C_2H_4(g)+9PO_2$ (48)
$5.5PH_3(g)+3CuO+8.25CO(g)=Cu_3P+4.125C_2H_4(g)+2.25P_2O_5$ (49)
$10PH_3(g)+3CuO+15CO(g)=Cu_3P+7.5C_2H_4(g)+2.25P_4O_8$ (50)
$5.5PH_3(g)+3CuO+8.25CO(g)=Cu_3P+4.125C_2H_4(g)+1.125P_4O_{10}$ (51)
$6PH_3(g)+CuO+9CO(g)=CuP_2+4.5C_2H_4(g)+P_4O_{10}$ (52)
$10PH_3(g)+CuO+15CO(g)=CuP_2+7.5C_2H_4(g)+2P_4O_8$ (53)
$6PH_3(g)+CuO+9CO(g)=CuP_2+4.5C_2H_4(g)+2P_2O_5$ (54)
$10PH_3(g)+CuO+15CO(g)=CuP_2+7.5C_2H_4(g)+8PO_2$ (55)
$3.333PH_3(g)+2CuO+5CO(g)=Cu_2P_2O_7+2.5C_2H_4(g)+1.333P$ (56)
$6.667PH_3(g)+2CuO+5CO(g)=Cu_2P_2O_7+5CH_4(g)+4.667P$ (57)
$6.667PH_3(g)+3CuO+5CO(g)=Cu_3(PO_4)_2+5CH_4(g)+4.667P$ (58)
$9PH_3(g)+5CuO+12.5CO(g)=2.5Cu_2P_2O_7+6.25C_2H_4(g)+P_4H_2$ (59)
$7.6PH_3(g)+2CuO+5CO(g)=Cu_2P_2O_7+5CH_4(g)+1.4P_4H_2$ (60)
$3.333PH_3(g)+3CuO+5CO(g)=Cu_3(PO_4)_2+2.5C_2H_4(g)+1.333P$ (61)
$9PH_3(g)+7.5CuO+12.5CO(g)=2.5Cu_3(PO_4)_2+6.25C_2H_4(g)+P_4H_2$ (62)
$7.6PH_3(g)+3CuO+5CO(g)=Cu_3(PO_4)_2+5CH_4(g)+1.4P_4H_2$ (63)
$2PH_3(g)+2CuO+8CO(g)=Cu_2P_2O_7+8C+3H_2O$ (64)
$2PH_3(g)+2CuO+5CO(g)=Cu_2P_2O_7+5C+3H_2(g)$ (65)
$PH_3(g)+CuO+3CO(g)=CuHPO_4+3C+H_2(g)$ (66)
$PH_3(g)+CuO+4CO(g)=CuHPO_4+4C+H_2O$ (67)
$PH_3(g)+10CuO+CO(g)=5Cu_2O+H_3PO_4+CO_2(g)$ (68)
$PH_3(g)+3CuO+ 3CO(g)=3CuH(g)+P+3CO_2(g)$ (69)
$2PH_3(g)+8CuO+CO(g)=6CuH(g)+Cu_2P_2O_7+CO_2(g)$ (70)
$PH_3(g)\ 6.5CuO+CO(g)=1.5Cu(OH)_2+P+2.5Cu_2O+CO_2(g)$ (71)
$PH_3(g)+5CuO+CO(g)=H_3PO_4+5Cu+CO_2(g)$ (72)
$2PH_3(g)+10CuO+2CO(g)= Cu(H_2PO_4)_2+9Cu+H_2(g)+2CO_2(g)$ (73)
$PH_3(g)+2.5CuO+CO(g)=P+2.5Cu+1.5H_2O+CO_2(g)$ (74)

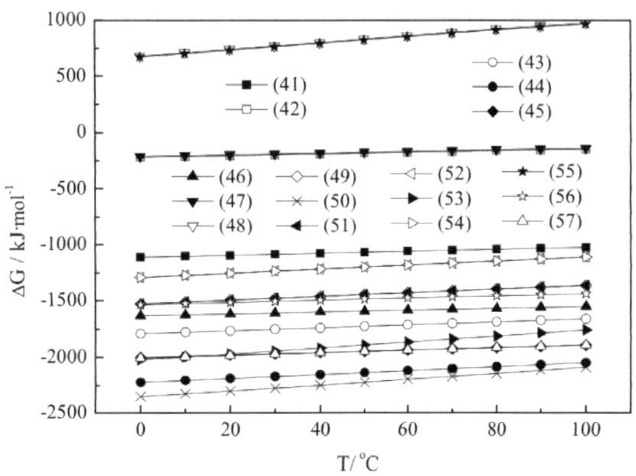


Figure 5. Relationship between Gibbs energy and temperature of equations (41)-(57).

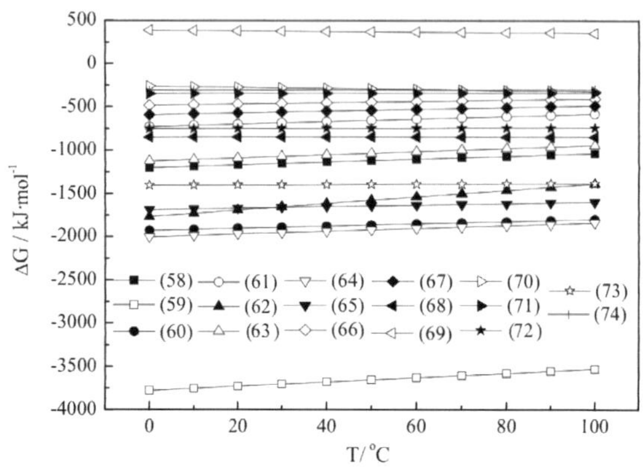


Figure 6. Relationship between Gibbs energy and temperature of equations (58)-(74).

## 4. Conclusion

(1) $PH_3$ can lead to the deactivation of CuO in activated carbon under certain reaction conditions.

(2) $N_2$ in the gas stream did not participate in chemical reactions between $PH_3$ and CuO, but $O_2$, $H_2$, and CO can accelerate the deactivation of CuO. $O_2$ was the predominant agent CuO deactivation in the adsorbent and produced the deactivation products $Cu(OH)_2$ and $P_4O_{10}$.

**Acknowledgments**

The authors gratefully acknowledge the financial support from the National Natural Science Foundation of China (No. 51768074, 41701362), the Natural Science Foundation of Yunnan Province (No. 2017FG001-014, 2018FD020), and the Scientific Research Foundation of Yunnan Agricultural University (No. A2002350).

**References**

[1] Woodman J D, Haritos V S and Copper P D 2008 *Comparative Biochemistry and Physiology Part C: Toxicology & Pharmacology* **147** 271
[2] Li W C, Bai H L, Hsu J N, Li S N and Chen C 2008 *Industrial & Engineering Chemistry Research* **47** 1501
[3] Ning P and Wang X Y 2015 *Frontiers of Environmental Science & Engineering* **9** 181
[4] Wei Y F, Huang X F, Li W Q, Gao K Y, Qin Y S and Zhou T 2011 *Fresenius Environmental Bulletin* **20** 3120
[5] Wang X Q, Ning P, Shi Y and Jiang M 2009 *Journal of Hazardous Materials* **171** 588
[6] Ma L P, Ning P, Zhang Y Y and Wang X Q 2008 *Chemical Engineering Journal* **137** 471
[7] Elliot B, Balma F and Johnson F 1990 *Solid State Technology* **33** 89
[8] Yu Q F, Li M, Ning P, Yi H H and Tang X L 2019 *Journal of Wuhan University of Technology-Mater. Sci. Ed.* **34** 487
[9] Yu Q F, Ning P, Yi H H, Tang X L, Li M and Yang L P 2012 *Separation Science and Technology* **47** 527
[10] Ren Z D, Quan S S, Zhu Y C, Chen L, Deng W X and Zhang B 2015 *RSC Advances* **5** 29734
[11] Yi H H, Yu Q F, Tang X L, Ning P, Yang L P, Ye Z Q and Song J H 2011 *Industrial & Engineering Chemistry Research* **50** 3960

IOP Conf. Series: Materials Science and Engineering **738** (2020) 012026 doi:10.1088/1757-899X/738/1/012026

# Study on effects of hot isostatic pressing process on microstructure and properties of CNTs/Al2009 composites

**X L He[1,2*], T B He[1,2], P J Tang[1,2], X Y Wang[1,2] and P Y Li[1,2]**

[1]AECC Beijing Institute of Aeronautical Materials, Beijing 100095, China

[2]Beijing Engineering Research Center of Advanced Aluminum Alloys and Application, Beijing 100095, China

*Corresponding author's e-mail: okhere007@126.com

**Abstract.** The CNTs/Al2009 composite powders were prepared by cryogenic milling method and then the CNTs/ Al2009 matrix composites were fabricated by hot extrusion process or hot isostatic pressing (HIP)/hot extrusion process. The effects of HIP process on microstructure and interface structure of CNTs/Al2009 composites were studied by X-ray diffraction, raman spectroscopy, scanning electron microscopy and transmission electron microscope. Finally, the effects of HIP process on the CNTs/2009Al composites were characterized by tensile strength at room temperature. The results showed that the cryogenic milling process could disperse 1.0 wt. % of CNTs into the aluminum matrix, and the composites prepared by HIP and hot extrusion process had uniform microstructure and excellent tensile properties at room temperature. The tensile strength, yield strength and elongation of the 1.0 wt.% CNTs/Al2009 composites prepared by HIP and hot extrusion process were 560 MPa, 443.3 MPa and 10.2%, respectively. Hot isostatic pressing process had an important influence on the interfacial bonding strength of CNTs/Al2009 composites. The strength of the composite prepared by HIP and hot extrusion process was 20 MPa higher than that of direct hot extrusion process, and the elongation was basically the same. The main reason was that the high temperature and high pressure conditions of hot isostatic pressing process could promote the reaction of carbon and aluminum at the interface between CNTs and aluminum matrix, so that a few CNTs with severe damage and defects in some CNTs formed $Al_4C_3$ phase with aluminum, thereby improving two compatibility, and enhanced interface bonding.

## 1. Introduction

With high specific modulus and strength, aluminum matrix composites are effective to improve fuel efficiency by weight reduction in aircraft and transportation industries [1–3]. Carbon nanotubes (CNTs) have recently emerged as materials with exceptional properties exceeding those of any conventional material (e.g. Young's modulus up to 1-1.8 TPa, tensile strength up to 150 GPa, low density of 1.2~1.8 g/cm$^3$, and neglected coefficient of thermal expansion, as well as good ductility and plastic deformation capacity [4-6]). Key requirements to achieve both properties are uniform dispersion of the CNTs in the matrix, and good interfacial bonding between the reinforcement and the matrix. Powder metallurgy techniques have been used extensively due to the ease of incorporating the CNTs in the matrix powder [7-14]. The mechanical dispersion processes (like ball milling, friction stir processing) encounter the problem of CNTs structural damage. The low temperature requirements help minimize possible thermal damage to the CNTs. Some researchers reported manufacturing CNTs reinforced aluminum matrix composites through cryogenic milling [15-17]. The interface acts as a link between the reinforcement and the matrix, and has a critical impact on the overall performance of the composites. The interface of CNTs/Al matrix composites is affected by many factors, including the composites preparation process and parameters, the wettability of CNTs and aluminum matrix and the chemical reaction at the interface. These factors together determine the interface state of the interface.

In this paper, the CNTs/2009Al composite powders were prepared by cryogenic milling method and then the CNTs/2009Al matrix composites were fabricated by hot extrusion process. In order to improve the interface bonding between CNTs and aluminum matrix, the preparation process of CNTs/2009Al matrix composites was improved and hot isostatic pressing (HIP) proce ss was added before hot extrusion process. The effects of HIP process on microstructure and interface structure of CNTs/Al2009 composites were studied by X-ray diffraction, Raman spectroscopy, Scanning electron microscopy and Transmission electron microscope. Finally, the effects of HIP process on the CNTs/2009Al composites were characterized by tensile strength at room temperature.

## 2. Experimental details

### *2.1. Materials*

Multiwalled carbon nanotubes with nominal diameter of 40-60 nm, length of 5-15 μm, and purity more than 97 wt.% were purchased from Shenzhen Nanotech Port Co. The Al2009 alloy powders were produced by Beijing Institute of Aeronautical Materials. CNTs/2009Al composite powders were prepared by cryogenic milling process for 2h milling time and the CNTs content was 1.0 wt.%. And then the CNTs/2009Al composites were fabricated by hot extrusion processor by HIP/hot extrusion process.

### *2.2. Analysis and characterization*

The X-ray diffraction was measured using a Bruker D8 Advance diffractometer with a Cu Kα radiation source (50 kV and 30 mA) in the 2θ range from 20° to 90°, the step size and scan rate were 0.02 and 4°/min. Raman spectroscopy (Jobin Yvon HR800) with a laser wavelength of 532 nm was used to evaluate the disorder and structure evolution in CNTs. The microstructure and interface structure of composites were characterized by scanning electron microscope (SEM, Nova Nano SEM 450) and transmission electron microscope (TEM, JEOL 2100). Tensile test of dog-bone cylindrical sample with a diameter of 5 mm and gauge length of 25 mm was carried out using an Instron testing machine at room temperature.

## 3. Results and discussion

### *3.1. Effects of hot isostatic pressing process on microstructure of composites*

Figure 1 showed XRD pattern of CNTs/Al2009 composites prepared by hot extrusion process and by HIP and hot extrusion process. As shown in Figure 1, no other diffraction peaks were found in the composite except for the α(Al) and S($Al_2CuMg$) phases, regardless of whether or not HIP process was used. After quenching, the aluminum alloy formed a supersaturated solid solution and was in an unstable state. It spontaneously decomposed to produce a second phase at room temperature. The main strengthening phase of Al-Cu-Mg alloy was θ($Al_2Cu$) phase and S($Al_2CuMg$) phase. With the difference of Cu and Mg ratio, the phase composition was also different. The higher the Cu content, the smaller the S phase and the more θ phase. On the contrary, the Mg content increased, and the smaller the θ phase, the more S phase. In general, when Cu/Mg $\leq$ 2.6, the precipitated phase was almost all S phase, and when Cu/Mg > 2.6, S + θ phase or θ phase was formed. In this paper, the content ratio of Cu and Mg was 2.6, which was at the critical value, and the main strengthening phase was the S phase. Due to the low content of CNTs (1 wt.%) and the carbon atomic number of 6, which was a light element, this may be the reason why CNTs were not detected.

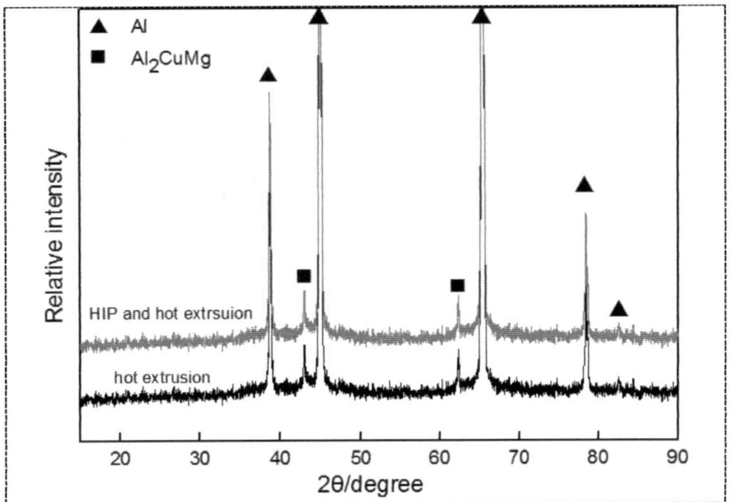


Figure 1. XRD pattern of CNTs/Al2009 composites prepared by hot extrusion and by HIP and hot extrusion process

As seen from Figure 2, there were two distinct characteristic peaks on the Raman spectrum curve, the G peak at around 1580 $cm^{-1}$ and the D peak near 1350 $cm^{-1}$. Composite powders, composites prepared by hot extrusion process and prepared by HIP/hot extrusion process had ID/IG values of 1.25, 1.33 and 1.17, respectively. And CNTs had the highest defects in composites obtained without HIP process. In addition, the G peak in the composite material moved toward the high frequency direction, indicating that a certain internal stress was generated in the CNTs due to the hot extrusion process. The ID/IG value of the material obtained by HIP process was slightly lower than that of the ball milled powder. The defects of CNTs in the composite would be consistent with the CNT state in the powder, or slightly deteriorated. In the composites by HIP and hot extrusion process, there was a small peak near 850 $cm^{-1}$, which was $Al_4C_3$.

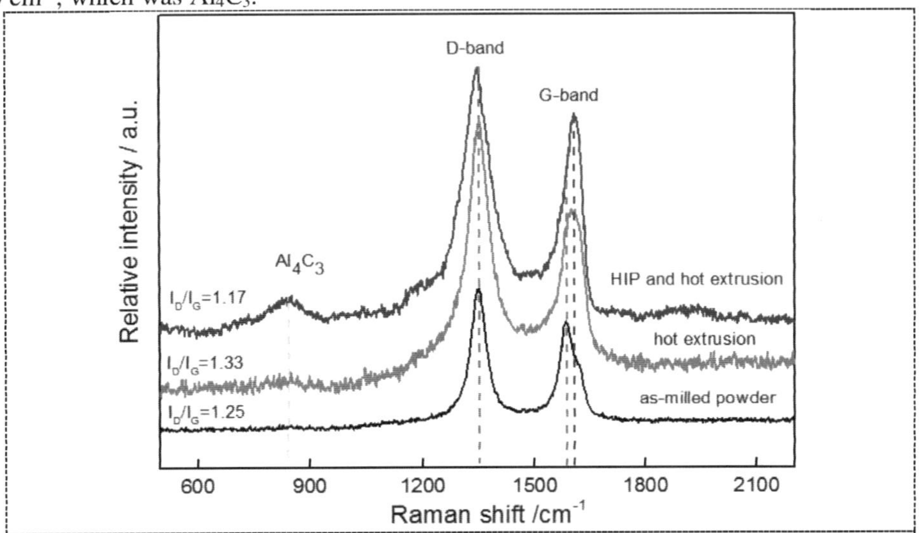


Figure 2. Raman spectra of CNTs/Al2009 composite powders and composites

Figure 3 showed grain structure of CNTs/Al2009 composites prepared by hot extrusion process and by HIP and hot extrusion process. It could be found from the figure that the composites maintained a fine grain size regardless of whether it was subjected to HIP process and the grain length was about 1 to 2 μm and the width was about several hundred nanometers. A large number of second phase particles were precipitated in both the crystal and the grain boundaries of the composite. In addition,

IOP Conf. Series: Materials Science and Engineering **738** (2020) 012026 doi:10.1088/1757-899X/738/1/012026

regions in which dislocations strongly interacted with the second phase particles were also observed from within the grains of the composites.

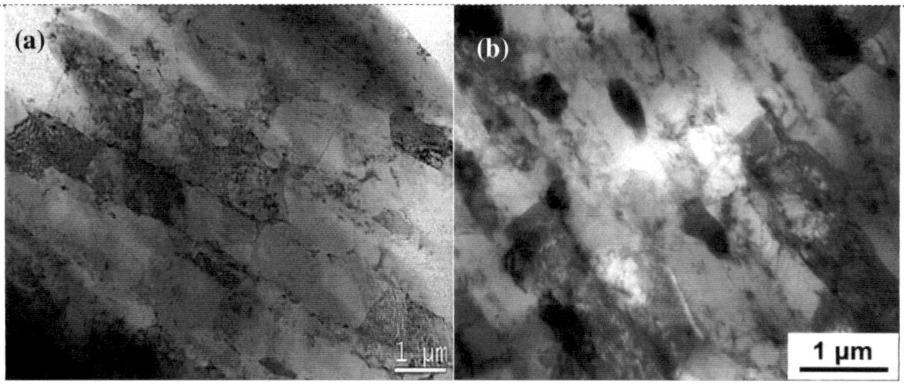


Figure 3. Grain structure of CNTs/Al2009 composites prepared by hot extrusion (a) and by HIP and hot extrusion process(b)

*3.2. Effects of hot isostatic pressing on interface structure of composites*

Figure 4 showed interface structure of CNTs/Al2009 composites prepared by hot extrusion directly. As shown in Figure 4, CNTs basically maintained a layered structure, but some atomic layers were disordered, indicating an increase in defects, because strong mechanical collisions in the high-energy ball milling process caused certain damage to the CNTs structure. The interface between the CNTs and the aluminum matrix was smooth and had no interfacial reactants. Figure 4 (c) showed the interface between the CNTs tip and the aluminum matrix, and aluminum atoms enter between the tube walls of the CNTs. Graphite and aluminum were not wetted. At high temperature, the atom became active, and the diffusion rate of Al was accelerated. During the hot extrusion process, the high temperature and high pressure closely combined the CNTs with aluminum through the interatomic force. As shown in Figure 4 (d), there was also a crystalline transition layer between the CNTs and the aluminum, which had a thickness of about 2 nm and was completely coherent with the aluminum matrix. Although the material type of the transition layer was not clear, it was obvious that its existence would be able to improve interface wettability and improve interface bonding strength.

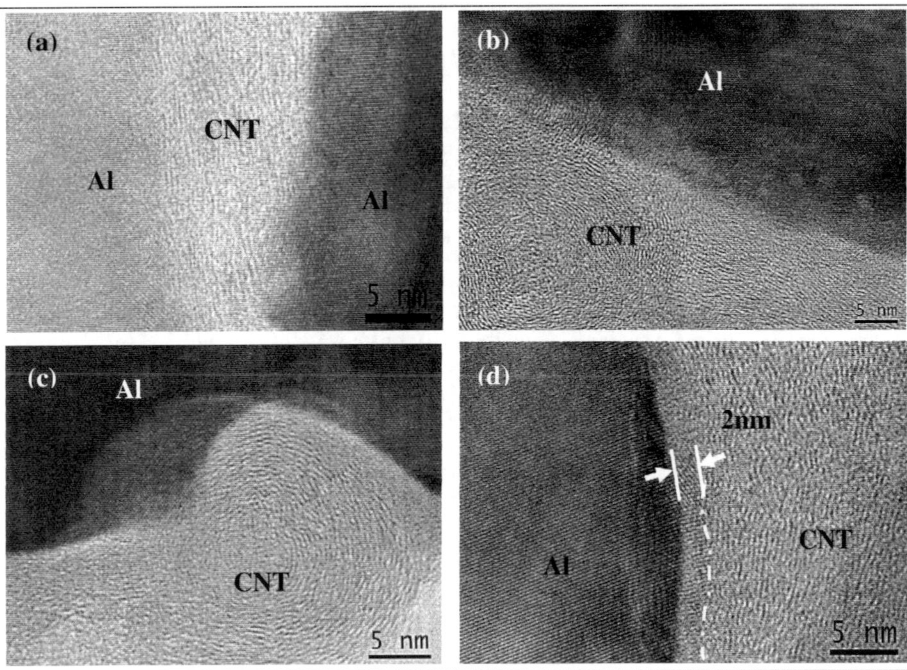


Figure 4. Interface structure e of CNTs/Al2009 composites prepared by hot extrusion process

Figure 5 showed interface structure of CNTs/Al2009 composites prepared by HIP and hot extrusion process. The tubular CNTs could be clearly seen in Figure 5 and it was indicated that the CNTs structure was intact and the tube end was closed. As shown in previous research work [15], the presence of CNTs (on the side of the rod phase), but they were gathered together. Part of the rod-like phase was amorphous, which could be presumed to be amorphous carbon, and the crystalline region was $Al_4C_3$ phase formed by a partial reaction of CNTs with aluminium, which was in the presence of $Al_4C_3$ phase as described in Figure 2.

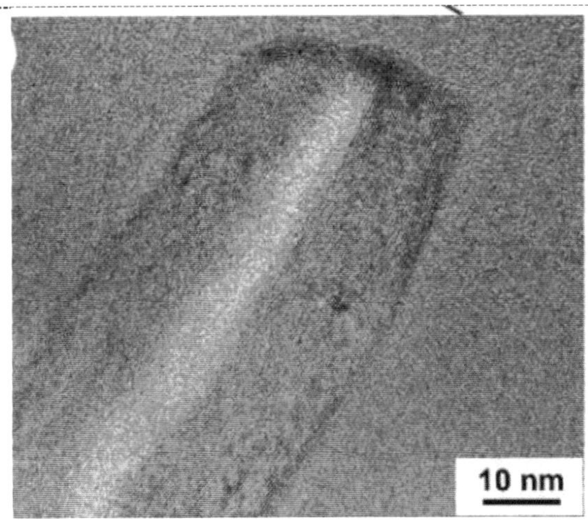


Figure 5. Interface structure e of CNTs/Al2009 composites prepared by HIP and hot extrusion process

IOP Conf. Series: Materials Science and Engineering **738** (2020) 012026 doi:10.1088/1757-899X/738/1/012026

### *3.3. Effects of hot isostatic pressing process on properties of composites*

Table 1 showed the tensile properties of CNTs/Al2009 composites at room temperature. As seen from Table 1, the tensile strength and yield strength of the composites prepared by HIP and hot extrusion process were higher than those prepared by hot extrusion process, which were increased by about 20 MPa. The elongation of the two materials was basically the same. The main reason of the increase in tensile strength was also the increase in interfacial bonding strength. Combined with XRD, TEM and Raman spectroscopy, it was indicated that a small amount of $Al_4C_3$ appeared in the CNTs/Al2009 composites obtained by HIP and hot extrusion process. During the HIP process, the composite powders were in a high temperature and high pressure state for a long time, which provided conditions for the reaction of carbon and aluminum at the interface between the CNTs and the aluminum matrix. The stability of the CNTs was very high, but many openings were formed by the chopping of the CNTs in the ball milling, or some defects occured in the tube walls, and the reactivity in these places was increased, thereby forming an intermetallic compound with aluminum matrix. The severely damaged CNTs reacted with aluminum to form carbides and the number of high-defective CNTs decreased, so the relative amount of low-defective CNTs increased, and the Raman scattering intensity also increased. Therefore, the ID/IG value of the material obtained by HIP process was decreased. Among composites without HIP process, some of the CNTs with higher defects were still retained, and the influence of plastic deformation during the extrusion process was affected, so the ID/IG value of the composite showed a slight increase.

CNTs and aluminum matrix were generally non-wetting (wetting angle was about 127°), while $Al_4C_3$ formation could significantly reduce the wetting angle, thereby increasing the wettability between the two. A very small amount of nano-scale reaction product generated by the interface between CNTs and aluminum matrix could improve the interfacial bonding force between the two, so that the load acting on the matrix could be effectively transmitted to the reinforcement and improved the performance of the composites. Obviously, the positive effect of the HIP process was twofold, which converts the severely damaged CNTs and aluminum matrix reaction into $Al_4C_3$, which not only improved the interfacial bonding strength, but also reduced the defects in the CNTs.

Table 1. Tensile properties of CNTs/Al2009 composites at room temperature.

| Process | Tensile strength/MPa | Yield Strength/MPa | Elongation/% |
|---|---|---|---|
| Hot extrusion | 540 | 423.3 | 10.0 |
| HIP and hot extrusion | 560 | 443.3 | 10.2 |

## 4. Conclusions

Following conclusions were drawn from this work:

(1) The cryogenic milling process could disperse 1.0 wt. % of CNTs into the aluminum matrix, and the composites prepared by HIP and hot extrusion process had uniform microstructure and excellent tensile properties at room temperature. The tensile strength, yield strength and elongation of the 1.0 wt.% CNT/Al2009 composites prepared by HIP and hot extrusion process were 560 MPa, 443.3 MPa and 10.2%, respectively.

(2) Hot isostatic pressing process had an important influence on the interfacial bonding strength of CNTs/Al2009 composites. The strength of the composite prepared by HIP and hot extrusion process was 20 MPa higher than that of direct hot extrusion process, and the elongation was basically the same. The main reason was that the high temperature and high pressure conditions of HIP process could promote the reaction of carbon and aluminum at the interface between CNTs and aluminum matrix, so that a few CNTs with severe damage and defects in some CNTs formed $Al_4C_3$ phase with aluminum, thereby improving two compatibility, and enhanced interface bonding.

## References

[1] Miracle D B. (2005) Metal matrix composites - from science to technological significance, *Compos. Sci. Technol.* 65: 2526-40.

[2] Liu Z Y, Wang Q Z, Xiao B L, Ma Z Y and Liu Y. (2010) Experimental and modelling investigation on SiCp distribution in powder metallurgy processed SiCp/2024 Al composites, *Mater. Sci. Eng.* A 527: 5582-91.

[3] Torralba J M, Costa C E and Velasco F. (2003) P/M aluminum matrix composites: an overview, *J. Mater Proce Technol.* 133: 203.

[4] George R, Kashyap K T and Rahul R. (2005) Strengthening in carbon nanotube/aluminium (CNT/Al) composites. *Scripta. Mate.* 53(10): 1159-63.

[5] Thostenson E T, Ren Z F and Chou T W. (2001) Advances in the science and technology of carbon nanotubes and their composites: a review. *Compos. Sci. Technol.* 61: 1899-1912.

[6] Xie X L, Mai Y W and Zhou X P. (2005) Dispersion and alignment of carbon nanotubes in polymer matrix: A review. *Mater. Sci. Eng.* R: Reports, 49(4): 89-112.

[7] Kuzumaki T, Miyazawa K and Ichinose H. (1998) Processing of carbon nanotubes aluminium composite. *Mater. Res.* 13: 2445-9.

[8] Morsi K, Esawi A M K, Lanka S, Sayed A and Taher M. Spark plasma extrusion (SPE) of ball-milled aluminium and carbon nanotube reinforced aluminium composite powders. *Compos A: Appl Sci Manuf.* 41: 322-6.

[9] George R, Kashyap K T, Rahul R and Yamdagni S. (2005) Strengthening in carbon nanotube/aluminium (CNT/Al) composites. *Scripta. Mater.* 53: 1159-63.

[10] Choi H J, Kwon G B, Lee G Y and Bae D H. (2008) Reinforcement with carbon nanotubes in aluminium matrix composites. *Scripta. Mater.* 59: 360-3.

[11] Tokunaga T, Kaneko K and Horita Z. (2008) Production of aluminium-matrix carbon nanotube composite using high pressure torsion. *Mater. Sci. Eng.* A 490: 300-4.

[12] Kwon H, Estili M, Takagi K, Miyazaki T and Kawasaki A. (2009) Combination of hot extrusion and spark plasma sintering for producing carbon nanotube aluminium matrix composites. *Carbon* 47: 570–7.

[13] Morsi K, Esawi AMK, Borah P, Lanka S, El-Sayed A. (2010) Characterization and spark plasma sintering of mechanically milled aluminium-carbon nanotube (CNT) composite powders. *J. Compos. Mater.* 44: 1991-2003.

[14] Perez B R, Estrada- G I and Ferreira P J. (2008) Novel Al-matrix nanocomposites reinforced with multi-walled carbon nanotubes. *J. Alloys. Compd.* 450: 323–6.

[15] He T B, He X L and Tang B J. (2017) The use of cryogenic milling to prepare high performance Al2009 matrix composites with dispersive carbon nanotubes. *Mater. Des.*114(15): 373-82.

[16] Azeez A A, Rhee K Y and Park S J. (2013) Application of cryomilling to enhance material properties of carbon nanotube reinforced chitosan nanocomposites. *Compos. Part B: Eng.* 50(0): 127-34.

[17] Woo D J, Hooper J P and Osswald S. (2014) Low temperature synthesis of carbon nanotubereinforced aluminum metal composite powders using cryogenic milling. *J. Mater. Res.* 29(22): 2644-56.

IOP Conf. Series: Materials Science and Engineering **738** (2020) 012027 doi:10.1088/1757-899X/738/1/012027

# Effect of Trees Knot Defects on Wood Quality: A Review

**Hongyue Qu, Ming Chen, Yuan Hu and Jianhua Lyu***

Forestry College, Sichuan Agricultural University, Chengdu, Sichuan, 611130, China

*Corresponding author's e-mail: ljh@sicau.edu.cn

**Abstract.** This paper reviews the research progress of effect of trees knot defects on wood quality, including the research strategy, technology, model, and research directions. Firstly, the effect of the knot on the appearance, mechanical properties, mechanical processing properties, chemical composition and bonding properties of wood were summarized. Secondly, the relationship between the influence of the knot on the quality of the wood and the size, position, color, type, shape and distribution of the knot were analyzed. Then, the advantages and disadvantages of different non-destructive testing methods and different prediction models were analyzed and compared. The following conclusions on future research trends can be drawn from the literature review and analysis: 1) Improve the fitting degree of the prediction model of the influence of the knot on the wood quality; 2) Optimize the single detection method and multiple detection methods to combine organically, and the test results are combined with the artificial cultivation of trees to fundamentally reduce the impact of knots on wood quality; 3) Develop on-line technology to measure the effect of knots on wood quality, and make different wood products based on the test results to maximize the utilization of wood.

## 1. Introduction

Wood is a renewable natural green material. It has the characteristics of light weight, high strength, good elasticity, heat insulation and etc. It is a high-quality raw material for making furniture. But woods have different grades, and 90% of wood grades depend on the knot. In wood processing, knot is considered as a wood defect, will make wood quality, use value, strength, durability reduction. In order to meet people's demand for wood products, many scholars at home and abroad began to carry out relevant studies on improving wood quality. The improvement of wood quality is closely related to the state, size, position and tilt angle of knot, and optimizing knot is equivalent to improving the quality of wood [1].

## 2. Research status of knot features

Longuetaud F et al. [2] showed that, the knot refers to the branches buried inside the trunk and on the surface. In plant physiology, knots play an important role, and external branches are the main source of nutrients needed for tree growth. The leaves on the live branches transform their external $CO_2$ into organic matter and $O_2$ through their own photosynthesis. The organic matter is transported to the roots through the screen of the phloem of the trees and absorbed to promote the growth of the trees. From a physiological point of view, knots are necessary for tree growth, while in wood science, knots reduce the quality of wood. In order to improve the quality of wood, the law of the existence of the festival has become a basic research.

The basic features of the knot include its type, shape, distribution and color. Kong F C [3] and so on, from the texture of knot and the conjunctive morphology of surrounding tissues, which can be classified into living knot and dead knot. From the perspective of the material of the knot, they can be divided into leakage knot, decay knot and sound knot. According to the section of the knots, they are divided into circular knot, strip knot and palmate knot. Common knot types and features are shown in figure 1. The knot prediction model established by Trincado G [4] shows that most of the knots are conical, and only a small number of knots appear cylindrical. Huang S Y [5] and others found that the distribution and formation of knots are not only related to environmental factors and site conditions, but also related to genes. For example, there is a large difference in the number of knots in different tree species. Larch, Camphor pine and other knots have a large number, while Poplar, Eucalyptus and other knots have a small number. Li J [6] research shows that the color of knot is darker than that of normal wood. Because the dead knot has been separated from the surrounding tissue, the color of the edge changes greatly, presenting a hard black scar. There is a clear boundary between the color and the surrounding, and the color of the living knot gradually changes and deepens.

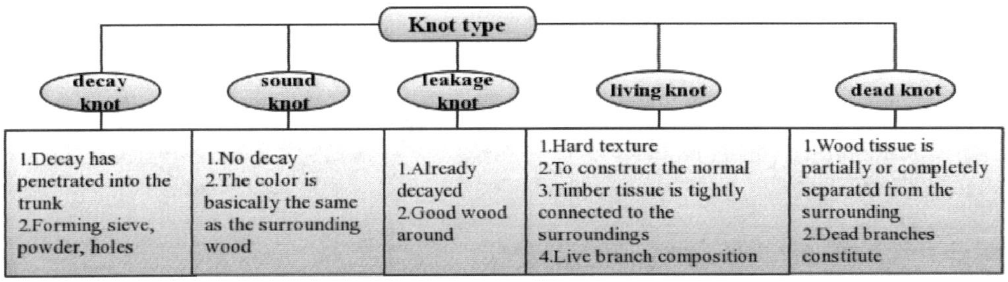


Figure 1. Common Section Types and Characteristics

## 3. Research status of the influence of knot on wood quality

### *3.1. Effect of knot on the appearance of wood*

In the research of Li J [6], the wood color of the knot is deeper than the normal material nearby. The unique pattern and deep color give people a beautiful stimulation, a dynamic floating feeling, giving the environment a sense of luxury. Crafts can be processed and utilized. However, the knot has destroyed the appearance consistency of the whole piece of plate. The research by Zhao P J [7] and others shows that the texture of the knot is obviously different from the surrounding, and it is distorted to varying degrees, showing a unique texture and changing physical and mechanical properties of wood.

### *3.2. Effect of knot on mechanical properties of wood*

Jin R L [8] found that the density and water content of the knots are significantly different from the surrounding wood tissue. During the drying process, the knots may fall off and cause holes. In the mechanical properties test, Cao Y [9] and other studies have shown that the knots reduce the tensile and compressive strength of the wood, but improve the horizontal compressive strength and the shear strength. Zhang X Y [10] found that the knots of larch increased their flexural strength. In 2018, Rocha [11] et al. found in their study on the influence of eucalyptus' knots on its modulus of elastic (MOE) and the compressive strength (Fc) that the greater the number of knots, the greater the difference between MOE and Fc, while the effect of small knots (8.31mm$^2$) and normal wood was almost equal. Eucalyptus knots are more sensitive to hardness than strength.

### *3.3. Effect of knot on wood machining performance*

Jiang H [12] and others believe that the hardness of knots is greater than the hardness of wood itself, and the knots may be carbonized due to high temperature during sawing. At the same time, the angle of the knot is large, which greatly shortens the service life of the wood tools when processing wood. In the processing of veneers, the boiling and softening of the knots before the wood is made is particularly important.

### *3.4. Effect of knot on chemical composition of wood*
The study by Lukacevic M [13] microscopically shows that the internal fibers of the wood deviate from the knot, resulting in imperfections in the fibers inside the wood. Studies by Vek V [14] have shown that extracts from the knot are more abundant than other parts, and the essential oil content is significantly higher than other parts.

### 3.5. *Effect of knot on wood glue performance*
Lin L Y [15] and other studies have shown that the knot will reduce the gluing properties of the wood, the size of the knot has a significant impact on it, and the type of the knot is not significant. Xie L S [16] and others found that the knot has a significant effect on the normal shear strength, wood breakage rate, immersion peeling rate and boiling stripping rate of the seam.As the diameter of the knot increases, the glue seam shear strength and wood breakage rate can be reduced by 40% and 67%.

Generally, dead knot has a greater effect on wood quality than live knot. However, in the study of Ke B F [17], it was found that the effect of the dead knot on the bending strength of the live knot is small, (1) the texture slope of living knot is greater than that of dead knot; (2) cracking due to drying of the living knot and surrounding wood. Considering again that the live knot can resist several kinds of stress, while the dead knot cannot, it can be concluded that the influence of the dead knot and the live knot on the stress is basically equal.

## 4. Research status of knot detection
External knot measurements are relatively easy, while internal knot is not. Traditional knot detection destroyed the logs, reducing the economic value of the wood. Now, most of the use of non-destructive testing. Rummukainen H [18] et al. and Zhang H J [19] et al. used X-ray CT scanning technology to detect the internal conditions of trees. Burcham D C [20] et al., Zhang T [21] et al., Qin J C [22] et al.using ultrasonic detection technology to obtain internal knot parameters. Liu S T [23] et al. and Wang Z C [24] et al. used stress wave technology to detect internal knots, Wu F M [25] et al. and Bogosanovic M [26] et al. used microwave technology to obtain internal knots. Yu H Z [27] et al. and Ding W J [28] determined the position and size of the knot by near-infrared spectroscopy. But these methods have their own advantages and disadvantages, as shown in Table 1.

Table 1. Nondestructive testing methods, principles, advantages and disadvantages

| Detection method | Principle | Ddvantage | Disadvantage |
|---|---|---|---|
| X - ray CT scan[18,19] | The xylem inside the wood and the part of the knot have different absorption capacity to the X-ray, which presents different images on the negative film, and determines the shape and size of the knot | (a)Easy to operate<br>(b)Imaging resolution is high, the image is clear and intuitive | (a)Insensitive to cracks parallel to the surface<br>(b)Fluorescent materials required for imaging, with radioactivity<br>(c)Cost too high |
| Ultrasonic testing[20,21,22] | The type, size and position of the knot will be determined by receiving the reflected ultrasonic wave and measuring the corresponding wave velocity and waveform | Good controllability and repeatability | (a)easy to be interfered during detection<br>(b)The qualitative and quantitative determination of defects is not very accurate<br>(c)The coupling agent required for testing may cause contamination problems |
| Stress wave detection[23,24] | The tested wood is struck by pulse to produce stress wave, and whether the tested wood has defects is judged by the change rule of time or speed of stress wave propagation | (1)More suitable for complex and severe outdoor environment, with strong practical application<br>(2)The testing | (1)The instrument is heavy and difficult to carry and transport<br>(2)The physical properties of different parts of the tree and directions lead to different |

| | | | |
|---|---|---|---|
| | | machine is cheap and easy to carry<br>(3)No uncontrollable factors and unknown pollution | propagation mechanisms of stress waves, affecting the results of stress wave detection on the knot<br>(3)Unable to detect knot types |
| Microwave detection[25,26] | When the wood is scanned, the fibers in the knots are different from the surrounding wood, which causes the variation of the microwave to be very prominent and obvious, thereby measuring the position of the knot. | (1)High sensitivity of detection equipment<br>(2)High operational safety | The moisture content of wood has a serious impact on the detection of knots |
| Near infrared spectroscopy[27,28] | When the infrared light passes through the knot, it will cause a difference in heat transfer, thus determining the position and size of the knot | The speed of propagation is very fast<br>(3)Ensure that the tested wood structure is complete and abundant | (1)There is a lot of redundant and non-linear information in the detection, which leads to the inaccuracy of the later knot reduction model<br>(2)In the process of identification, it takes too long and is not practical |

To improve the effect of knot detection, we can better study the effect of knots on wood quality. In addition to the above methods, non-destructive testing methods for wood knots include neural network technology [29], geological radar (GPR) technology [30], and resistance technology [31], but these methods also have limitations and need Further improvement.

## 5. Application of the models in the study of knots

The current research involves a variety of prediction models for wood knots. In 2014, Gerhard [32] et al. used the morphological characteristics of local materials to analyze the properties of local materials. For example, a non-destructive tensile test is performed on wood containing knots, and the stress of the corresponding knot and the tensile stiffness of the wood are measured by an optical imaging device, thereby establishing a regression equation and analyzing the tensile stiffness of the other knots; The knot was subjected to a destructive tensile test to measure the tensile strength of the wood. The regression analysis was also used to determine the tensile strength of other knot woods.

In 2017, Hao [33] J et al. established a multivariate regression model to predict the length of the knot by three factors: the diameter of the branches, the angle of the branches and the age of the branches. The purpose of this is to predict the influence of the formation of the wood on the wood material by the multivariate regression equation.

In the artificial red pine knot attribute based on the mixed effect model studied by Jia W W [34] et al. in 2018, different mixed effect models were established by using various parameters of the red pine knot, and the fitting accuracy was higher than that of the traditional regression model. For example, the basic model with the optimal age of death is introduced into the random parameters of the pitch base diameter, and the generalized linear mixed model has the highest fitting degree. The model is based on the common model and introduces random parameters of the tree effect level. Construct a single-level generalized linear mixed model.

Compare the prediction model, found that multivariate regression equation than a singal regression equation add more independent random parameters, makes the predicted parameters data closer to the actual data, and generalized linear mixed model compared with the multiple regression equation can reflect the stand growth in average change trend also can reflect individual differences.

## 6. Discussion on future research direction

It can be seen from the above that the influence of the knot on the quality of the wood is closely related to the size, position, color, type, shape and distribution of the knot. 1) The effect on the appearance is mainly that the color and texture of the knot destroy the consistency of the appearance of the sheet. 2) The influence on the mechanical properties of wood is mainly due to the difference in density and water content of the knots. Reduce the tensile strength, compressive strength and flexural strength of parallel to the grain, increase the compression strength of perpendicular to the grain and the shear strength of parallel to the grain. 3) The influence on the mechanical properties of wood is mainly due to the increase of wood hardness and greatly shorten the service life of wood tools. 4) The influence on the chemical composition of wood is mainly due to the obvious increase of the extract of the knot and the fiber breakage. 5) The effect on the gluing properties of wood is mainly due to the positive correlation between the size of the knot and the gluing performance.

Because the current destructive test method is used in the study of wood quality in the section, the test wood can not be used for processing in the later stage, which reduces the use value and economic value of the wood. At the same time, the prediction model of the influence of knots on wood quality is not high in fitting; the non-destructive testing method of the knot, the detection result is not accurate enough, and the detection process is long.

In the future, more scientific and technological means will be used to optimize the research methods of the influence of knots on wood quality. For example, without destroying the logs, the prediction model of the knots is used to know the influence of the knots on the quality of the wood. However, the current prediction model of the knots is not high enough. In the future, it is possible to improve the fitness by reducing errors and finding optimal prediction models for different parameters.

1) Optimize the method of detecting the knot, for example, optimize the single detection method and improve the detection result. organically combine various detection methods, exploit the advantages of each detection method, and make up for the deficiency of the single detection method. Combining the test results with the artificial cultivation of trees fundamentally reduces the impact of knots on wood quality.

2) At the current level of science and technology, the development of online detection technology for the impact of knots on wood quality provides an important basis for post-production wood products. At the same time, select different wood according to different purposes to maximize the utilization rate of wood. For example, when sawing plywood try to avoid knots. The wood with more knots can be used for making paper etc.

**Acknowledgments**

The authors are grateful for the support of Opening Foundation for Industrial Design Industry Research Center, Key Research Base of Humanities and Social Sciences, Sichuan Education Department (GYSJ18-037), the Project of Modern Design and Culture Research Center, Sichuan Key Research Base of Philosophy and Social Sciences (Grant No. MD18Z002), and Research Project of Double Support Program for Art Special Project in Sichuan Agricultural University of 2019.

**References:**

[1] Niu S G and Chen L 2014 Analysis of the influence of wood defects on material quality *Heilongjiang Science and Technology Information.* 04 252

[2] Longuetaud F, Mothe F Kerautret B and Leban J M 2012 Automatic knot detection and measurements from X-ray CT images of wood: A review and validation of an improved algorithm on softwood samples *Computers and Electronics in Agriculture.* 85

[3] Kong F C 2015 Application of wood defects and wood inspection technology *Private Technology.* 06 229

[4] Trincado G and Burkhart H E 2008 A model of knot shape and volume in Loblolly pine trees *Wood Science and Technology.*40(4) 634-46

[5] Huang S Y, Wang J H, Lyu J X and Lei Y C 2011 Progress in World knot Research *China Forest Products Industry.* 38 3-7

[6] Li J 2010 Ecological properties of wood-wood is a contributor to human health in green environment *Journal of Northeast Forestry University.* 38 (05) 1-8

[7] Zhao P J, Zhou S R and Ye L X 2016 Discussion on the influence of knot on material quality in wood inspection *Friends of Farmers' Wealth.* 04 102
[8] Jin R L 2016 Discussion on the main points of wood defect and wood inspection technology *Friends of Farmers' Wealth.* (02) 140
[9] Cao Y, Street J and Li M 2019 Evaluation of the effect of knots on rolling shear strength of cross laminated timber (CLT) *Construction and Building Materials.* 222 579-87
[10] Zhang X Y, Jiang X M and Yin Y F 2018 Effect of node on acoustic ultrasonic parameters and bending resistance of larch wood *Wood Processing Machinery.* 29(01) 13-17+33
[11] Rocha M F V, Costa, L R and Costa L J 2018 Wood knots influence the modulus of elasticity and resistance to compression *Floresta e Ambiente.* 25(4)
[12] Jiang H, Lin L J, Zheng H P, Zhou D Q and Fang Y M 2019 Detection of knot defects in sawn timber based on air-coupled ultrasonic wave *China Forest Products Industry.* 33(01) 46-49
[13] Lukacevic M and Kandler G 2019 A 3D model for knots and related fiber deviations in sawn timber for prediction of mechanical properties of boards *Materials & Design.* 166: 107617
[14] Vek V, Oven P and Ters T 2013 Extractives of mechanically wounded wood and knots in beech *Holzforschung.* 68(5) pp. 529-39
[15] Lin L Y, He S and Fu F 2012 Effects of knot on wood Gluing properties *China Forest Products Industry.*26(04) 12-15
[16] Xie L S, Liu J Q and Qiao L T 2016 Effect of knot on gluing property of larch gluing wood *Journal of Forestry Engineering.* 1(01) 21-24
[17] Ke B F 1982 Effect of Knot on Flexural Strength and Flexural Elastic Modulus of Masson Pine Components *Journal of Anhui Agricultural College.* (02) 11-21+120
[18] Rummukainen H, Makkonen M and Uusitalo J 2019 Economic value of optical and X-ray CT scanning in bucking of Scots pine *Wood Material Science & Engineering.* 1-10
[19] Zhang H J, Guan C and Wen J 2016 Application and research progress of non-destructive testing of wood materials *Journal of Forestry Engineering.* 1(06) 1-9
[20] Burcham D C, Brazee N J and Marra R E 2019 *Trees.* 33 681
[21] Zhang T, Cheng X W and Lu W D 2016 Study on the detection of internal cavity defects in wood by ultrasonic method *Journal of Southwest Forestry University.*36(01) 121-125+130
[22] Qin J C, Liu X Z, Koen Van Den A and Cui G Y 2018 The study of wood knots using acoustic nondestructive testing methods *Ultrasonics.* 88
[23] Liu S T, Cheng, S G, Chen X, B R Y, Wang J L and Ge X H 2019 Detection of Complex Defects in Trees *Electronic Test.* (02) 57-59
[24] Wang Z C Li G H, Feng H L, Fang Y M and Fei H 2015 Identification and classification of wood defects based on stress wave and support vector machine *Journal of Nanjing forestry University* (Natural Science Edition). 39(03) 130-6
[25] Wu F M 2018 Research progress of microwave nondestructive testing technology for living wood *Forestry Machinery & Woodworking Equipment.* 46(12) 9-14
[26] Bogosanovic M, Al Anbuky A and Emms G W J 2010 *Wood Sci.* 56 357
[27] Yu H L, Zhang M and Hou H Y 2019 Near infrared spectroscopy analysis of solid wood plate knot shape inversion *Spectroscopy and Spectral Analysis.* 39(08) 2618-23
[28] Ding W B and Wang B J 2019 Study and Development Trend of Wood Defect Detection Technology *Forestry Machinery & Woodworking Equipment.* 47(01) 4-8
[29] Liu J M, Xu k H and Wang L H 2018 Research on wood internal defect diagnosis based on mea-bp neural network *Forest industry.* 45(02) 19-24
[30] Di X H and Wang L H 2013 Discussion on the feasibility of GPR application in wood nondestructive testing *Nondestructive Testing.* 35(11) 51-54
[31] Yue X Q 2017 Quantitative detection of wood internal decay based on electrical resistance method *Northeast Forestry University.*
[32] Fink G, Kohler J and Eur J 2014 *Wood Prod.* 72 331

IOP Conf. Series: Materials Science and Engineering **738** (2020) 012027 doi:10.1088/1757-899X/738/1/012027

[33] Hao J, Meng M J, Huang D W and Wei J L 2017 Distribution characteristics and prediction model of knots in plantation of lattice wood J*ournal of Nanjing Forestry University* (Natural Science Edition). 41(03) 100-4
[34] Jia W W, Cui C and Li F R 2018 The attribute of artificial red pine knots based on mixed effect model *Chinese Journal of Applied Ecology*. 29(01) 33-43

IOP Conf. Series: Materials Science and Engineering **738** (2020) 012028 doi:10.1088/1757-899X/738/1/012028

# A research on the relationship between the structure of nuclear energy and string theory, energy, wormhole

**Zhang Tiancheng**

Faculty of science, University Malaya, Kuala Lumpur, Malaysia

**Abstract:** In the 1870s, Europe entered the era of electric revolution. Not only large enterprises, but also small enterprises have adopted new power - electric energy. At first, an engine equipment only supplies lighting power for a house or a street. People call this kind of power station a "household type" power station, which produces very little power.
The power plant was originally direct current. Edison, a famous American inventor, started to build the central power plant in 1881. In 1882, two power plants with initial scale were put into operation. In January 1882, the Edison Company of Holland bridge in London began to generate electricity to supply the city cathedral and bridgehead hotel in the west of the post office bridge of the San Marco factory. Later, a water power station appeared on the fintan River in Petersburg, Russia. The power station was built on a barge to provide power for the lighting of Nevsky street. In the early stage of power production and transmission, there was a long and fierce debate about whether it was DC or AC. Edison advocated the use of DC, and people have thought about various ways to expand the scope of DC power supply, so that the power supply situation in small and medium-sized cities has been significantly improved. The use of nuclear energy to produce power plants, also known as nuclear power plants (NPPs). There is a strong binding force between the nuclei (neutrons and protons). When heavy nucleus splits and light nucleus polymerizes, it will release huge energy, which is called nuclear energy.
The technology has been relatively mature, forming a scale of operation, but the nuclear power plant that produces electric energy from the nuclear energy released by heavy nuclear fission is analyzed from the point of view of energy conversion, which is a conversion process from heavy nuclear fission nuclear energy to heat energy to mechanical energy to electric energy. Now that we have more advanced nuclear power technology, this paper introduces a new power plant concept and its relationship with microphysics

## 1. Introduction

In the past, we paid great attention to the development of traditional energy, such as coal and steam. In recent years, due to the development of technology, the nuclear power industry began to rise gradually But we are still stagnant in the traditional extraction of nuclear materials, such as the use of traditional acid leaching purification. Again, I would like to explain a new idea of nuclear material extraction In addition, we will discuss the nucleon reaction in the microscopic condition, and try to explain the relationship between string theory and wormhole.

*1.1.* In the past, we used to extract nuclear raw materials through mineral materials such as ores. Again, I would like to try to show that we can extract nuclear reaction raw materials with 70% of the earth's coverage material, water Because water is composed of two hydrogen atoms and one water atom, the energy of lithium or deuterium in nuclear reaction is much higher than that of uranium, and its cost is lower. In order to generate the maximum artificial energy with the minimum energy consumption, the

IOP Conf. Series: Materials Science and Engineering **738** (2020) 012028 doi:10.1088/1757-899X/738/1/012028

reactor should be located on the seaside, which can facilitate the extraction of water, and can use sea water for cooling. The energy source of the method of extracting lithium is the abundant sunshine in the coastal area.

*1.2.* The following is the construction concept of coastal nuclear power plant

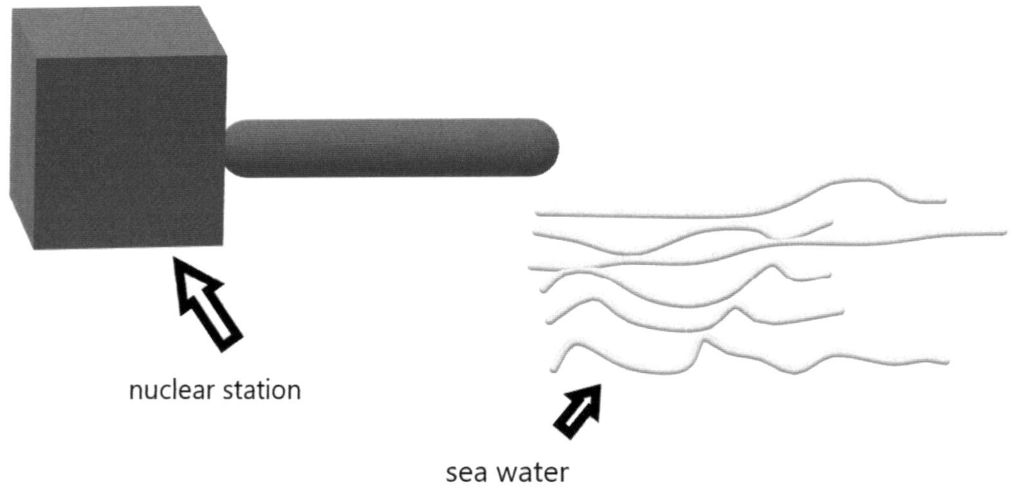


(Illustrate1.1) the location of nuclear power plant

We can use tritium or lithium rich in seawater for controllable nuclear fission or even fusion. The basic power provided by solar energy can be exchanged for unexpected artificial energy

## 2. Electronic positioning at the micro level

Because we can't accurately locate the electron, we can't capture and drain the electron in the conventional way, but I think the electron has the magnetism we know, maybe we can directly drain the electron through similar magnetism to form the current, and the dynamic potential energy of the electron comes from the nuclear gravity to revolve around it. Unfortunately, my existing equipment can't prove this, and I hope to realize this expectation in the future. Nuclear fission and string: when nuclear fission occurs, I think it is not only accompanied by the separation of particles, but also with the expansion of string size, which brings about continuous particle reaction, and if not controlled, the reaction will become more and more intense. It's like the butterfly effect at the micro level. '

*2.1. While considering the nucleon reactions we can observe*

I think there should be more microscopic material activity, string activity. For example, in nuclear fusion, the electrons outside the two particles will separate, so that the two atomic nuclei attract each other to generate huge energy.

I think there is string and membrane activity in the particles When two particles attract, the strings that make them vibrate and produce Wave, the wave contacts each other to produce huge energy and form several new particles.

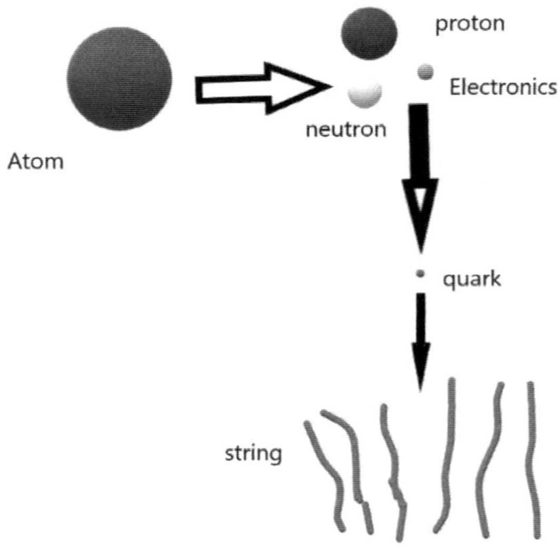


(Illustrate2.1) Nuclear fission and string

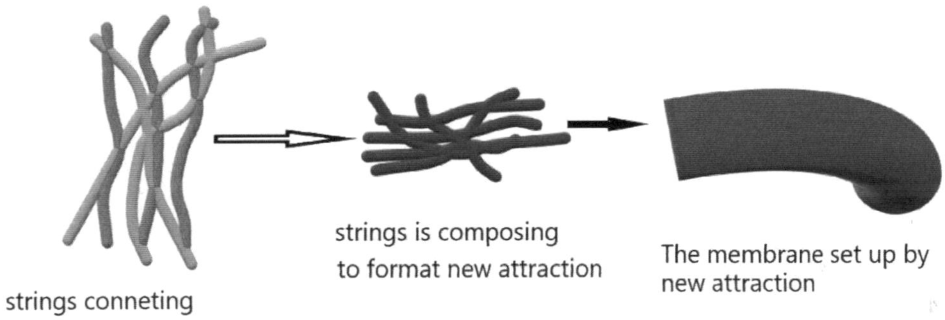


(Illustrate2.2) The changes in chords and membranes

## 3. The deceleration or acceleration of the vibration of a string

I think that when two particles are close, each string will exert a force on each other, which creates a huge amount of energy in whatever form The closer the two particles get to each other, the faster the string vibrates. In fact, I think the acceleration force comes from the membrane in each particle These membranes let the strings shuttle like channels. When several particles get close to each other, both the membranes and the strings will gravitate to each other and accelerate.

| Kin d | u | d | s | c | b | t |
|---|---|---|---|---|---|---|
| (e) | 2/ 3 | -1/3 | - 1/3 | 2/3 | -1/3 | 2/3 |

(Illustrate3) Quark calculation

According to this theory, hadrons are not particles, but quarks. For example, protons are composed of u, u, Du,u,d set up:$p = (uud)$, $n = (udd)$, $\pi^+ = (u\bar{d})$, $\pi^- = (d\bar{u})$, $\bar{d}, \bar{u}$ are Anti quark,

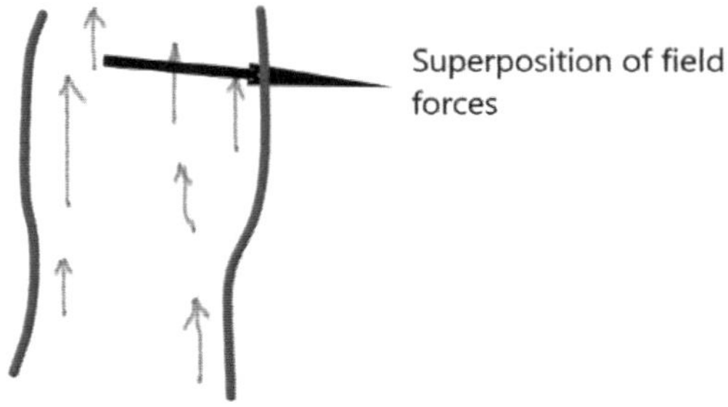


Illustrate 3.1 The forces between the strings after fission

### 3.1. *About the fluidity of string*
As we know, the earth's magnetic field is divided into two poles. But for the micro world, I think the string has the indefinite magnetism, and the magnetism of each string is changing all the time. In some specific cases, the contact keeps the string's properties unchanged for a period of time, and we also get the phenomenon that can be observed

## 4. About the Einstein Ross Bridge
When a particle reacts, especially in a short time, it accumulates a lot of energy. Maybe at some point, we have reached the standard of wormhole opening in the micro world. Like all the strings connected, another channel was built.

## 5. Conclusion
No matter what we know now or what we haven't, energy development is related to the future of mankind. The new type and power station idea described in this paper has its own advantages. This paper also gives some explanations for the particle reactions in the micro world, including the relationship between the reactions and string theory, membrane, wormhole, and string theory changes that we are still in a strange stage. However, we have to make a choice because of economic or other influences. I hope this paper can be a little reference for energy and physics.

## References
[1] Aharony O, Bergman O, Jafferis D L, et al. N=6 superconformal Chern-Simons-matter theories, M2-branes and their gravity duals[J]. Journal of High Energy Physics, 2008.
[2] Dimofte T, Gaiotto D, Gukov S. Gauge Theories Labelled by Three-Manifolds[J]. Communications in Mathematical Physics, 2014, 325(2):367-419.
[3] Woit P. Not Even Wrong: The Failure of String Theory and the Search for Unity in Physical Law[J]. Physics Today, 2006, 59(12):57-58.
[4] Yau S T, Nadis S, Erlich J. The Shape of Inner Space: String Theory and the Geometry of the Universe's Dimensions[J]. Notices of the American Mathematical Society, 2010, 58(8):559-560.
[5] Woit P. Not Even Wrong: The Failure of String Theory and the Search for Unity in Physical Law[J]. Physics Today, 2006, 59(12):57-58.

[6] Yau S T, Nadis S, Erlich J. The Shape of Inner Space: String Theory and the Geometry of the Universe's Dimensions[J]. Notices of the American Mathematical Society, 2010, 58(8):559-560.
[7] Polchinski J. Bouncing Universes in String Theory? [J]. Physica Scripta, 2006, 2005(T117):64.

IOP Conf. Series: Materials Science and Engineering **738** (2020) 012029 doi:10.1088/1757-899X/738/1/012029

# Synthesis of $Fe_3O_4$/PAM Superparamagnetic Nanoparticles by Using Electrostatic Interaction

**Qingyu Huang[1*], Shuzhi Wang[2]**

[1]College of Chemical Engineering, Daqing Normal University, Daqing 163712, China

[2]Organic geochemical laboratory, Daqing Oilfield Company Ltd Exploration and Development Research Institute, Daqing 163712, China

5031416@163.com

**Abstract.** In the study, two kinds of emulsions were prepared, respectively containing nano water droplets (NWDs) with positive and negative charges on the surface. The positive charges nano water droplets, called Fe-NWDs contained $Fe^{2+}$ and $Fe^{3+}$ ions and negative charges nano water droplets, called OH-NWDs, contained $OH^-$ ions. After the above two kinds of emulsions were mixed, $Fe_3O_4$/PAM superparamagnetic nanoparticles (SMNPs) were synthesized in the integrated NWDs by means of electric interaction, co-precipitation and polymerization reaction. The shapes of $Fe_3O_4$/AM SMNPs were spherical or ellipsoid ranged in diameter from 80nm to 360nm tested by TEM and the $Fe_3O_4$/AM SMNPs had superparamagnetic character tested by VSM. The magnetic saturation value of $Fe_3O_4$/PAM SMNPs was 27.38emu/g. The XRD graph indicated $Fe_3O_4$ was contained in the SMNPs. The infrared spectrum proved that $Fe_3O_4$ was coated by PAM exactly.

## 1. Introduction

Recently, functional polymeric nanoparticles have renewed interest due to their potential applications in mechanical, thermal, electrical, magnetic, and smart[1,2]. Magnetic polymeric nanoparticles are considered one of most important functional polymer nanocomposites because of their widespread applications, such as biosensors, bio-separation, site-specific drug delivery, magnetic resonance imaging, and cancer hyperthermia treatments, magnetic fluids, detection and separation in microbiology, drug delivery systems and data storage[3,4].

Magnetic polymeric nanoparticles are often made of the magnetic material coated with polymer base. The magnetic materials include iron oxide, nickel, cobalt, $CoFe_2O_4$, and so on. By far, there are many methods for the preparation of the magnetic polymeric nanoparticles, such as physical mixing, solution polymerization and dispersion polymerization[5,6]. But some products appear magnetic nanoparticles were unevenly dispersed and some approaches were either complicated or costly.

We have known emulsifiers can improve the stability of emulsions, and ionic surfactants can give charge to the surface of the substance they cover. Therefore, in this study, anionic and cationic surfactants were used to make nano water droplets, named NWDs, with positive charges and negative charges on the surface respectively in water-in-oil (W/O) emulsion. The positive charges nano water droplets, named Fe-NWDs contained $Fe^{2+}$ and $Fe^{3+}$ ions and negative charges nano water droplets, named OH-NWDs, contained OH- ions. By means of electrostatic interaction, the Fe-NWDs and OH-NWDs were merged together rapidly and $Fe_3O_4$ nanoparticles were synthesized in the integrated NWDs by means of co-precipitation reaction. After polymerization, $Fe_3O_4$/PAM superparamagnetic nanoparticles (SMNPs) were synthesized in the NWDs. The measurement results of the SMNPs was discussed in the paper.

## 2. Experiment section

Please follow these instructions as carefully as possible so all articles within a conference have the same style to the title page. This paragraph follows a section title so it should not be indented.

### *2.1. Preparation of $Fe_3O_4$/PAM SMNPs*

Experimental drugs included $FeCl_2 \cdot 4H_2O$, Op10, toluene, acrylamide (AM), $NH_4OH$, hexadecyl trimethyl ammonium bromide (CTAB), Span80, sodium dodecyl sulfate (SDS), potassium peroxodisulfate (KPS) and $FeCl_3 \cdot 6H_2O$. All the above chemical raw materials were of analytical grade. At 25°C, the aqueous solution, composed of 1M 2.5ml $FeCl_2$ and 1M 5.0ml $FeCl_3$, 1.85g AM, 0.30g Op10 and 0.0477g ($1.309*10^{-4}$mol) CTAB dissolved in isoamyl alcohol, was slowly added to the oil solution, including 2.0g Span80 and 24ml toluene. After stirring at 2000 rpm for 2 h, the emulsion, named C-E, was formed. Similarly, aqueous solution, composed of 4ml $NH_4OH$, 1.85g AM, 0.20g Op10 and 0.0378g ($1.309*10^{-4}$mol) SDS was slowly added to the oil solution including 2.0g Span80 and 24ml toluene. After stirring at 2000 rpm for 2 h, the emulsion, named S-E, was formed. Then the C-E was slowly added to the S-E when the stirring speed was maintained at 1000 rpm. Based on electrostatic interaction and the coprecipitation, the $Fe_3O_4$ magnetic nanoparticles were prepared in the black emulsion, named CS-E. After 2h, the CS-E was heated to 50°C and injected into with nitrogen for 20 min. The polymerization was initiated by adding 0.04g KPS in the CS-E. After 4 h, PAM was obtained. Now the $Fe_3O_4$/PAM SMNPs were synthesized. The mass ratio of $Fe_3O_4$ to PAM was about 20.78% in $Fe_3O_4$/PAM SMNPs.

The process of synthesizing the $Fe_3O_4$/PAM SMNPs consists of three steps: 1. the Fe-NWDs and OH-NWDs were merged together rapidly and form the integrated NWDs, 2. in the integrated NWDs, $Fe_3O_4$ nanoparticles were obtained through co-precipitation reaction, 3. and then $Fe_3O_4$/PAM SMNPs were obtained using polymerization. In the first step, electrostatic attraction was the core idea for the efficient and rapid synthesis of $Fe_3O_4$ nanoparticles, and this step was also the key step for the synthesis of $Fe_3O_4$/PAM SMNPs. The synthesis mechanism of $Fe_3O_4$/PAM SMNPs was shown in figure 1.

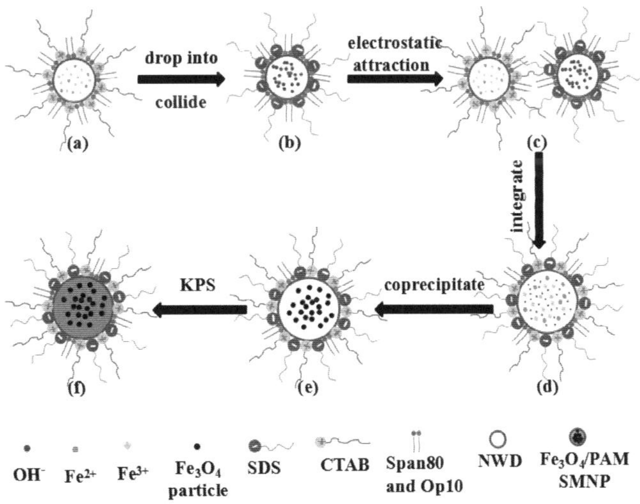


Figure 1. Synthesis mechanism of $Fe_3O_4$/PAM SMNPs

With stirring 2000 rpm, a lot of NWDs as micro-reactors were obtained in the C-E and S-E respectively. In the C-E, many polar groups of CTAB, Span80 and Op10 molecules adhering the surface of Fe-NWD in which $Fe^{2+}$ and $Fe^{3+}$ ions were scattered, shown in figure 1(a). Therefore, the Fe-NWD surface was likely to be positively charged. Similarly, in the S-E, many polar groups of SDS,

Span80 and Op10 molecules adhering the surface of OH-NWD in which $OH^-$ ions were scattered, shown in figure 1(b). The surface of OH-NWD was likely to be negatively charged.

When the C-E was slowly dropped into the S-E showed in figure 1(a) and (b), the Fe-NWD would collides with OH-NWD under the stirring of 1000 rpm. It can be seen from figure 1(c) and (d) that, once OH-NWD and Fe-NWD collided continuously, they would be voluntarily and rapidly merged together under electrostatic attraction. Soon the integrated NWD, in which $OH^-$, $Fe^{2+}$ and $Fe^{3+}$ ions were blended, were formed in figure 1(d). After coprecipitation, the $Fe_3O_4$ nanoparticles were immediately obtained in the integrated NWD shown in figure 1(e).

Figure 1(f) showed the polymerization was initiated by adding KPS and PAM was formed in which $Fe_3O_4$ particles were scattered. Now $Fe_3O_4$/PAM SMNPs were obtained.

In addition, the stabilized emulsion system was very important for synthesis of $Fe_3O_4$/PAM SMNPs. The emulsion system was very stable when the hydrophile lyophile balance, named HLB, was around 5.76 [7]. The HLB values of compound emulsifiers were calculated with the following formula [8, 9]:

$$HLB = \sum (HLB_i \times m_i) / \sum m_i$$

So the HLB of the S-E emulsion including 0.0378g SDS (HLB=40), 2.0g Span80 (HLB=4.3) and 0.20g Op10 (HLB=13.9) was about 5.76. Similarly, the HLB of the C-E emulsion including 0.0477g CTAB (HLB=15.8), 2.0g Span80 and 0.30g Op10 was about 5.76. The S-E or C-E were very stable.

### *2.2. Analytical methods*

The photograph of $Fe_3O_4$/PAM SMNPs was obtained by TEM (Hitachi H-600-II, Japan). The magnetic hysteresis loops of samples were tested by a VSM (Lake Shore 7410, USA). Using KBr wafer, IR was tested by a Spectrum400 FT-IR spectrometer (PerkinElmer, USA). The XRD of $Fe_3O_4$/PAM SMNPs was measured by a X ray diffractometer using Cu Kα radiation (λ=1.54056 Å) (D8 Discover, Bruker, Germany).

## 3. Results and discussions

### *3.1. XRD results:*

The XRD graph of sample was shown in figure 2. The strong peaks in the pattern could be indexed to be the (220), (311), (400), (422), (440), and (511) crystal face in the cubic $Fe_3O_4$ (19-0629). It can be sure that the SMNPs consist of $Fe_3O_4$.

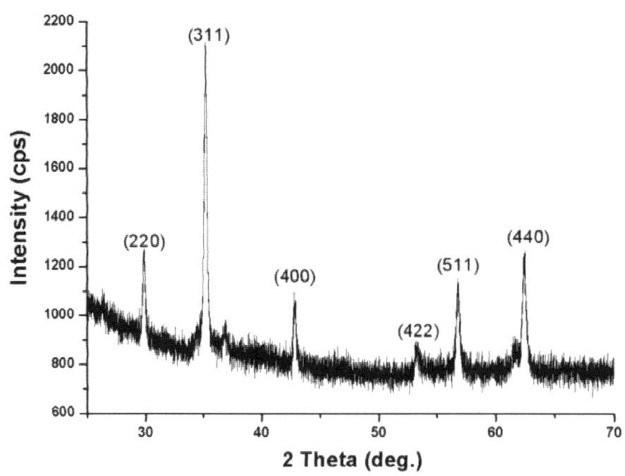


Figure 2. The XRD graph of sample.

### *3.2. FT-IR results:*

Figure 3 showed the FT-IR spectrum of sample. The peak at 3376cm$^{-1}$ and 3193cm$^{-1}$ were due to N-H stretching vibration. The antisymmetric and symmetric vibrations at 2923cm$^{-1}$ and 2853cm$^{-1}$, respectively, were clearly observed due to the aliphatic alkyl chains. The peak at 1664cm$^{-1}$, 1412cm$^{-1}$ and 1112cm$^{-1}$ were due to C=O stretching vibration, C-N stretching vibration and O-C-O stretching vibration respectively. The peak at about 588cm$^{-1}$ was assigned to the Fe-O bonds. The spectrum of sample proved that $Fe_3O_4$ was coated by PAM exactly.

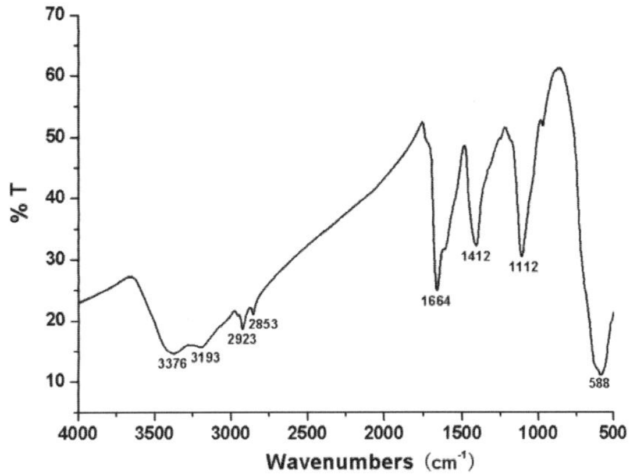


Figure 3. FT-IR spectrum of sample.

*3.3. TEM and particle size distributions results:*

Figure 4 was the TEM photo of sample respectively. The photo showed that the particles had good dispersion and ranged in diameter from 80nm to 360nm. The shapes of particles were spherical or ellipsoid. It was confirmed that $Fe_3O_4$/PAM SMNPs were synthesized in the CS-E.

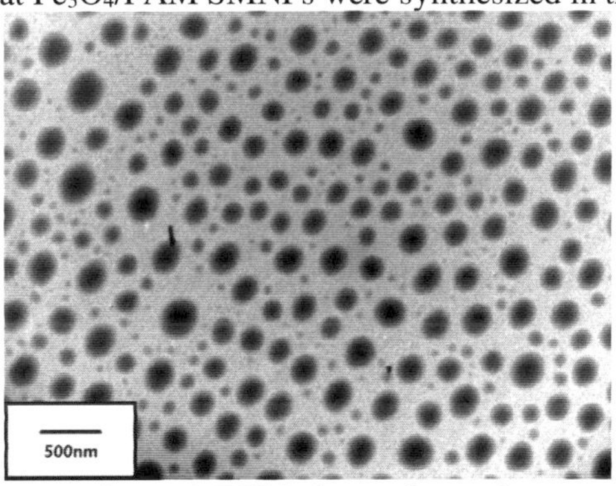


Figure 4. TEM photo of sample.

*3.4. Magnetic properties:*

The magnetic properties of $Fe_3O_4$/PAM SMNPs and $Fe_3O_4$ were studied at 25°C in figure 5. In the absence of a magnetic field, the magnetic hysteresis loops demonstrated that the $Fe_3O_4$ and $Fe_3O_4$/PAM SMNPs had no obvious remanence or coercivity. It just revealed that $Fe_3O_4$/PAM SMNPs have superparamagnetic character. The magnetic saturation values of $Fe_3O_4$ and $Fe_3O_4$/PAM SMNPs were 58.10 and 27.38emu/g respectively. The magnetic saturation value of the latter was 47.13% of the former.

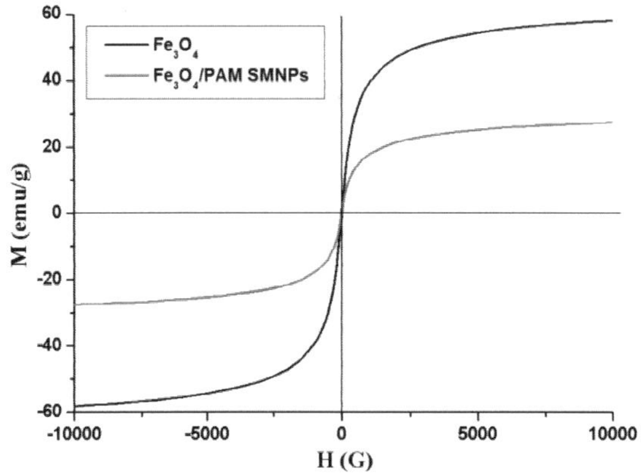


Figure 5. Magnetization graphs of $Fe_3O_4$/PAM SMNPs and $Fe_3O_4$.

## 4. Conclusions

Using the electrostatic interaction, Fe-NWDs and OH-NWDs merged quickly and efficiently and the integrated NWDs were formed. In the integrated NWDs, the coprecipitation reaction formed $Fe_3O_4$ nanoparticles and the polymerization formed $Fe_3O_4$/PAM SMNPs. The XRD graph indicated $Fe_3O_4$ was synthesized in the SMNPs. The FT-IR spectrum proved that $Fe_3O_4$ was coated by PAM exactly. The shapes of $Fe_3O_4$/AM SMNPs were spherical or ellipsoid ranged in diameter from 80nm to 360nm. The $Fe_3O_4$/PAM SMNPs had superparamagnetic property. The magnetic saturation value was 27.38emu/g.

## Acknowledgement

This work was supported by the < Daqing Normal University Scientific Research Fund > under Grant < number 14ZR13 >.

## References

[1] Oliva-Avilés A I, Avilés F and Sosa V 2011 *Carbon.* 49 2989-97.
[2] Fahrni F, Prins M W J and van IJzendoorn L J. 2009 *J. Magn. Magn. Mater.* 321 1843-50.
[3] Lu W, Shen Y, Xie A and Zhang W 2010 *J. Magn. Magn. Mater.* 322 1828-33.
[4] Feng L, Cao M, Ma X Zhu Y and Hu C 2012 *J. Hazard Mater.* 17 439-46.
[5] Yang T I, Brown R N C Kempel L.C and Kofinas P 2008 *J. Magn. Magn. Mater.* 320 2714-20.
[6] Dallas P, Georgakilas V, Dimitrios N, Komninou P, Kehagias T and Petridis D 2006 *Nanotechnology.* 17 2046-53.
[7] Huang Q Y and Wang S Z J 2019 *Disper. Sci. Technol.* 40 1379-84.
[8] Shinoda K, Yoneyama T and Tsutsumi H 1980 *J. Disper. Sci. Technol.* 1 1-12.
[9] Harusawa F, Nakajima H and Tanaka M 1982 *J. Sco. Cosmet. Chem.* 33 115-29.

IOP Conf. Series: Materials Science and Engineering **738** (2020) 012030 doi:10.1088/1757-899X/738/1/012030

# Study on Feed Processing and Application of Renewable Plant Humulus Scandens

**Guangjiao Zhou[1], Xuehua Feng[2]**

[1]Bozhou Institute of traditional Chinese medicine, Anhui Academy of traditional Chinese medicine, Bozhou, 36800, China

[2]Anhui Xinhua University, Hefei, 230000, China

The corresponding author's e-mail address is 403621620@qq.com

**Abstract.** In order to study the feed processing and development technology of the renewable plant Humulus scandens, Humulus scandens were collected, dried and comminuted at the stage of leaf unfolding, flower opening, flower blooming and yellow withering, and processed into direct raw material group, silage natural fermentation group and silage plus bacteria fermentation group. The contents of main nutrients and total flavonoids in Humulus scandens feed were determined and compared. The microbial quantity and fermentation quality of Humulus scandens in withered and yellow period in silage natural fermentation group and silage plus bacteria fermentation group were detected. The results showed that there was no significant change in calcium and phosphorus content in the three groups. The other nutrients in silage natural fermentation group and silage plus bacteria fermentation group were slightly lower than those in direct raw material group. The content of total flavonoids in silage plus bacteria fermentation group was significantly higher than that in other groups. In terms of fermentation quality and microbial quantity, the fermentation of silage natural fermentation group was not obvious, the pH value was on the high side, the fermentation of silage plus bacteria group was obvious, and the quality was also higher. In conclusion, if Humulus scandens is directly used as feedstuff, the best collection time is the blooming period of flowers. If silage and lactobacillus fermentation are used, the value of feedstuff is the highest, but the value of silage natural fermentation is the lowest, so it is not recommended to choose.

## 1. Introduction

The global livestock and fishery industry is in a period of vigorous development, the number of aquaculture is increasing, and form large-scale breeding, and the demand for forage feed is increasing. On the other hand, natural forage is seriously degraded, and the supply of forage is decreasing. This will have a serious impact on the development of the global livestock and fishery industry. There is an urgent need for new feed as a supplement. Humulus scandens is the dry overground part of Humulus scandens (Lour.) Merr. which belongs to the Humulus, a trailing herb of mulberry family. It is a perennial herb. It often grows in the ditch, wetland, ruins and other places. It has extremely strong vitality, strong renewability and high yield. Humulus scandens is distributed in many countries and regions in Asia and Africa. In addition to Xinjiang and Qinghai, all provinces and regions in China have different degrees of distribution. [1-2] As a medicinal herb, valerian is included in many provincial and municipal floras such as "Fujian Folk Herbs", "Jiangsu Wild Plants", "Anhui Medicinal Materials", etc., but it has less clinical use and occasional application by folks. Clear. However, valerian is rich in nutrients such as vegetable protein, fat, fiber and trace elements, and medicinal compounds such as flavonoids, terpenoids, volatile oils and steroids, which have the functions of detoxification, antibacterial, anti-inflammatory and anti-viral. Waste is a treasure. If weaving grass to process and develop animal feed, it can be used as a daily feed to supplement nutrition, and it can also increase the disease resistance of livestock and fish, and the prospect is objective. [3-6]

At present, most of the psyllium at home and abroad are focused on the study of its chemical properties, active ingredients and modern pharmacological effects. It has been found that flavonoids are a class of components with higher separation, And flavonoids are the main material basis for its pharmacological action such as heat-clearing and detoxification, water-saving, anti-bacterial, anti-inflammatory, anti-viral, etc., and its research as a feed processing method and application is less, has not been specifically reported, and is limited to raw materials directly as feed or feed additives. [7-10] In this study, weeds were used as test materials to prepare dry products of yarrow raw materials in different collection periods, and scientific processing technology was used to make silage natural fermentation group and lactic acid bacteria to prepare silage fermentation group, and the nutrients and totals were determined. The content of flavonoids, supplemented by the quantity of microorganisms and fermentation quality, comprehensively evaluate its quality, in order to provide a basis for accelerating the scientific feed of psyllium.

## 2. Experiment material

The test valerian samples were collected by the research team members at the north side of Hefei Railway Station at 5-10, which was the upper part of the grassland, and the weeds were removed. Dried, smashed into medium powder, spare.

## 3. Experiment method

### *3.1. Experiment design*

The sample collection period was set as follows: the leaf stage (early May), the flower opening stage (early June), the flowering period (early July and early August), and the yellowing period (mid-occupation). The aboveground parts of the whole plant of alfalfa were collected in each period, dried and pulverized, and randomly divided into 9 parts in parallel, which were respectively made into small packets, and 3 small packets were taken as the direct raw material group; and 3 small packages were mixed uniformly according to the mass ratio of the liquid to 1:2.5. Allow it to naturally ferment for 30 days, and make it into the silage natural fermentation group; the remaining 3 packets are separately added with 5% of the lactic acid bacteria dilution of 4.6×107 CFU/mL, and the concentration of the fermentation broth is 0.5 mg/mL. , fermentation for 30 days, for the silage plus bacteria fermentation group. At the end of the fermentation, the samples were taken out and dried, and the contents of nutrients and total flavonoids in each packet were determined. The microbial quantity, pH value and organic acid content of the natural fermentation and silage plus fermentation group in the flowering stage were determined. [11-12]

### *3.2. Evaluation indicators and methods*

*3.2.1. Determination of nutrients.* The crude protein (CP), crude fat (EE), crude fiber (CF) and organic matter (OM) content of each packaged dry sample were determined according to the test method recommended in "Feed Analysis and Feed Quality Testing Technology"; The routine analysis method was used to determine the contents of trace elements calcium (Ca) and phosphorus (P).[12-13]

3.2.2. *Determination of total flavonoids.* According to the results of preliminary experiments, the dry samples of valerian were ultrasonically added to 80% ethanol for 30 min according to the ratio of material to liquid, filtered, and the filtrate was continuously inspected. The rutin was used as the reference substance, the reagent was used as the blank control, the color was developed by Al (NO3)3 and NaOH, and the absorbance was measured by the colorimetric method at 510nm. The content of total flavonoids in the valerian was calculated by the standard curve method. .[13]

3.2.3. *Determination of microbial quantity, pH and organic acids.* According to the literature method, the flowering stage silage natural fermentation group and the silage plus bacteria fermentation group were separately mixed with sterile water, divided into two parts, one diluted, using lactic acid bacteria, yeast, Escherichia coli, aerobic bacteria and heat resistance. The conventional culture method of oxygen bacteria was cultured for 48 hours and counted separately; the other was filtered, and the content of acetic acid (AA), propionic acid (PA), butyric acid (BA) and lactic acid (LA) was determined by high performance liquid chromatography. The pH was measured by a pH meter.[12]

*3.2.4 Data analysis.* The test results data were statistically analyzed using Excel 2003 and SAS 9.0 software.

## 4. Test results and analysis

### *4.1. Comparison of the contents of nutrients and total flavonoids in direct raw materials at different collecting periods*

The content of nutrients and total flavonoids in the direct raw material group in different collection periods is shown in Table 1. It can be seen from Table 1 that the CP and CF contents in the different collection periods are very different, and there are significant differences ($P<0.05$). The highest CP content is the flowering period, which is 20.57%, and the lowest is only 14.35%. The content of the content in the early flowering stage and the yellowing stage was higher, there was no significant difference ($P>0.05$), but there was significant difference between the leaf expansion stage and the flowering stage. The OM content was not significantly different between the leaf expansion stage and the flowering stage, but later The content decreased gradually and there was a significant difference between them. The difference of Ca and P content was small. The content of Ca was the highest in flowering stage, reaching 3.56%. There was no significant difference with other periods. P was the smallest in the leaf-expanding stage, only 0.21%. There are significant differences with other periods; the total flavonoid content is increased first and then decreased, and there is a significant difference between them. The highest content is flowering period, which is 2.01%. In general, the EE was slightly lower in the flowering period, and the other nutrients and total flavonoids were the highest.

Table 1. Comparison of nutrient content and total flavonoids in direct raw material groups at different collection periods (%).

| Ingredient | Leaf stage | Flower opening | Flowering period | Yellow period |
|---|---|---|---|---|
| CP | $14.35\pm0.15^{d}$ | $17.03\pm0.11^{c}$ | $20.57\pm0.21^{a}$ | $18.23\pm0.37^{b}$ |
| CF | $12.37\pm0.04^{c}$ | $13.08\pm0.08^{b}$ | $15.18\pm0.17^{a}$ | $11.97\pm0.21^{d}$ |
| EE | $2.89\pm0.13^{b}$ | $3.02\pm0.24^{a}$ | $2.78\pm0.19^{c}$ | $3.03\pm0.18^{a}$ |
| OM | $89.72\pm0.25^{a}$ | $89.63\pm0.31^{a}$ | $87.37\pm0.41^{b}$ | $86.28\pm0.22^{c}$ |
| Ca | $3.34\pm0.07^{a}$ | $3.48\pm0.21^{a}$ | $3.56\pm0.52^{a}$ | $3.38\pm0.09^{a}$ |
| P | $0.21\pm0.23^{b}$ | $0.35\pm0.45^{a}$ | $0.42\pm0.06^{a}$ | $0.33\pm0.21^{a}$ |
| Total flavon | $1.03\pm0.11^{d}$ | $1.64\pm0.27^{c}$ | $2.01\pm0.08^{a}$ | $1.75\pm0.04^{b}$ |

Note: The difference between the lower-case letters of the peer data indicates that the difference is significant ($P<0.05$), and the same letter of the shoulder mark indicates that the difference is not significant ($P>0.05$). Table 2, Table 3, and Table 4 are the same.

### *4.2. Comparison of nutrient content and total flavonoids content in silage natural fermentation group in different collection periods*

IOP Conf. Series: Materials Science and Engineering **738** (2020) 012030 doi:10.1088/1757-899X/738/1/012030

The contents of nutrients and total flavonoids in the natural fermentation group of silage in different collection periods are shown in Table 2. It can be seen from Table 2 that the content of CP and CF in different collection periods is quite different, and the flowering period is significantly different from other periods (P<0.05). Both of them have increased first and then decreased, and the highest content is in flowering period. The CP was 17.31% and the CF was 16.25%, which was lower than the raw material period. The EE and OM contents were reduced to different extents relative to the raw material group, and the EE was the lowest in the leaf-expansion period, only 2.75%, and the other three phases were There were significant differences, but the difference was not significant in the other three periods (P>0.05). There was no significant difference in OM content between the leaf-expanding stage and the flowering stage, but there was a significant difference between the flowering period and the yellowing stage. The former content was higher; Ca, The content of P and total flavonoids was very close to that of alfalfa raw materials, and there was no change at all, indicating that the natural fermentation of silage had no obvious effect on the components. In general, the nutrient composition and raw material group of silage natural fermentation group decreased slightly, but the contents of Ca, P and total flavonoids did not change much. The overall period was the flowering period.

Table 2. Comparison of nutrient content and total flavonoids content in silage natural fermentation group at different collection periods (%).

| Ingredient | Leaf stage | Flower opening | Flowering period | Yellow period |
|---|---|---|---|---|
| CP | $12.61\pm0.18^{c}$ | $16.13\pm0.21^{b}$ | $17.31\pm0.19^{a}$ | $16.48\pm0.23^{b}$ |
| CF | $13.58\pm0.12^{c}$ | $14.08\pm0.03^{b}$ | $16.25\pm0.12^{a}$ | $13.47\pm0.58^{c}$ |
| EE | $2.75\pm0.16^{b}$ | $2.94\pm0.07^{a}$ | $2.88\pm0.62^{a}$ | $2.93\pm0.37^{a}$ |
| OM | $88.87\pm0.14^{a}$ | $88.71\pm0.27^{a}$ | $85.96\pm0.53^{b}$ | $85.82\pm0.07^{b}$ |
| Ca | $3.31\pm0.11^{a}$ | $3.45\pm0.18^{a}$ | $3.41\pm0.09^{a}$ | $3.40\pm0.13^{a}$ |
| P | $0.22\pm0.31^{b}$ | $0.33\pm0.37^{a}$ | $0.41\pm0.12^{a}$ | $0.33\pm0.47^{a}$ |
| Total flavon | $1.09\pm0.25^{d}$ | $1.58\pm0.06^{c}$ | $2.09\pm0.38^{a}$ | $1.81\pm0.67^{b}$ |

### *4.3. Comparison of nutrient content and total flavonoids in silage and fermentation groups in different collection periods*

Lactic acid bacteria were fermented into silage fermentation group in different collection period, and the contents of nutrients and total flavonoids were determined and compared. The results are shown in Table 3. It can be seen from Table 3 that the CP content in silage and bacteria fermentation in different collection periods is slightly lower than that in direct raw materials, which is significantly higher than that in silage. The highest value is in the flowering period, reaching 22.37%, and the lowest is the leafing stage, which is 15.57. %, flower blooming period was significantly different from other time periods (P<0.05); CF, EE, OM content decreased slightly compared with silage natural fermentation group, CF was the highest in flowering period, and significant difference with other periods, EE in flower The content of the initial opening period was the highest, but there was no significant difference between the initial flowering period and the flowering period. There were significant differences with other periods. The OM content was high-low-high-low, and the lowest was 85.31%. Significant difference, but there was no significant difference in leaf-expanding stage, initial flowering period and flowering period; there was no change in Ca and P content after fermentation; the total flavonoid content increased in different degrees at different times, and the flowering period was the highest. It reached 2.26%, which was significantly different from other periods.

Table 3. Comparison of nutrient content and total flavonoids in silage-fermented fermentation group at different collection periods (%).

| Ingredient | Leaf stage | Flower opening | Flowering period | Yellow period |
|---|---|---|---|---|
| CP | 13.57±0.23$^{c}$ | 16.98±0.15$^{b}$ | 18.47±0.24$^{a}$ | 17.22±0.14$^{b}$ |
| CF | 13.05±0.12$^{c}$ | 14.09±0.11$^{b}$ | 15.53±0.05$^{a}$ | 12.98±0.32$^{c}$ |
| EE | 2.52±0.23$^{b}$ | 2.81±0.62$^{a}$ | 2.72±0.08$^{a}$ | 2.51±0.37$^{b}$ |
| OM | 88.19±0.18$^{a}$ | 87.59±0.27$^{a}$ | 87.71±0.26$^{a}$ | 85.31±0.43$^{b}$ |
| Ca | 3.24±0.11$^{a}$ | 3.38±0.54$^{a}$ | 3.34±0.17$^{a}$ | 3.29±0.13$^{a}$ |
| P | 0.20±0.19$^{b}$ | 0.34±0.09$^{a}$ | 0.35±0.28$^{a}$ | 0.34±0.57$^{a}$ |
| Total flavon | 1.37±0.16$^{c}$ | 1.87±0.64$^{b}$ | 2.26±0.32$^{a}$ | 1.98±0.14$^{b}$ |

*4.4. Benchmarking analysis of quality and microbial quantity of alfalfa silage after fermentation*

Based on the previous test data, the mashing feed, fermentation or not, the best selection period is the flowering period. In this experiment, the fermentation quality of the natural fermentation of silage and the fermentation of silage and fermentation materials and the number of main microorganisms were compared. The results are shown in Table 4. From Table 4, the fermentation quality was improved. The PA and LA in the silage plus fermentation group increased significantly. The AA, BA and pH values decreased significantly. BA can decompose protein, its content decreased, and the protein loss will be reduced. The pH value will drop to 4.05. It meets the requirements of conventional silage pH value of 4.0-4.2. From the perspective of microbial quantity, the number of lactic acid bacteria and yeast in silage and bacteria fermentation group increased significantly, and Escherichia coli, aerobic bacteria and heat-resistant aerobic bacteria decreased significantly. On the whole, the degree of fermentation was significantly higher than that of natural fermentation, and the fermentation quality was better than natural fermentation.

Table 4. Comparison of nutrient content and total flavonoids content in silage fermented feed group at different collection periods (%).

| | Strain | Silage natural fermentation group | Silage plus fermentation group | | Ingredient | Silage natural fermentation group | Silage plus fermentation group |
|---|---|---|---|---|---|---|---|
| Number of microorganisms （lgCFU/gFM） | Lactic acid bacteria | 3.05±0.14$^{b}$ | 4.21±0.08$^{a}$ | Fermentation quality （mg/g） | PA | 1.34±0.03$^{b}$ | 3.01±0.12$^{a}$ |
| | Yeast | 2.25±0.06$^{b}$ | 3.48±0.11$^{a}$ | | LA | 12.29±0.13$^{b}$ | 15.74±0.25$^{a}$ |
| | E.coli | 0.53±0.21$^{a}$ | 0.61±0.18$^{a}$ | | AA | 0.22±0.07$^{a}$ | 0.17±0.16$^{b}$ |
| | Aerobic bacteria | 3.82±0.15$^{a}$ | 2.63±0.13$^{b}$ | | BA | 1.06±0.12$^{a}$ | 0.14±0.05$^{b}$ |
| | Heat-resistant aerobic bacteria | 0.41±0.08$^{a}$ | - | pH value | | 4.67±0.03$^{a}$ | 4.05±0.09$^{b}$ |

## 5. Results and discussion

As a renewable resource, valerian can be used as a good feed for livestock and fish. This study found that it is rich in nutrients such as crude protein, crude fiber and organic matter. It also contains a variety of amino acids in combination with the literature, and its nutritional value is high. The antibacterial, antiviral, anti-oxidant and anti-aging effects of total flavonoids can greatly increase the disease resistance of livestock and fish.

As a feed development, valerian can directly dry, pulverize and store raw materials. This method is simple and feasible, and can retain its original color, aroma and taste, but the nutritional value and disease resistance are significantly lower than the fermentation group. If this method is selected, the comprehensive nutrient content and total flavonoid content, the optimal collection period is in the flowering period; the natural fermentation of silage is compared with the unfermented, the difference of total flavonoids and trace elements Ca and P is small, not significant. The crude fiber increased slightly, but the protein, fat and organic matter decreased to varying degrees. Among them, the CP decreased most obviously, and the color was dim, the aroma disappeared, and the acidity increased. Therefore, the comprehensive evaluation of valerian is not suitable for silage natural fermentation and feed; Compared with the direct raw material group and the silage natural fermentation group, the content of total flavonoids was significantly increased, and the contents of trace elements Ca and P were basically unchanged. Other nutrients were reduced to varying degrees, but the overall content did not change much, and the protein content was higher than that of the silage natural fermentation group. The fermentation degree, fermentation quality and microbial quantity were also significantly better than the silage natural fermentation group. The herbicide feedstock not only considers the nutrient and digestibility of the material, but also increases the disease resistance of the animal and fish. Therefore, the best method is to select the flower in the flowering period and add the lactic acid bacteria to make the feed.

## References

[1] Chinese Flora Editorial Board of the Chinese Academy of Sciences. Chinese Flora,**23,**1 [M]. *Beijing: Science Press*, 1998, p220-224.
[2] Chinese Pharmacy Company. Chinese Traditional Medicine Funding [M]. *Beijing: Science Press*, 1994, p190-195.
[3] Gao Zhengquan, Meng Chunxiao. Progress in the research of alfalfa[J]. *Anhui Agricultural Sciences*, 2007, **35**(31): p9982-9984.
[4] Wang Taoyun, Ma Hongchang, Liu Jiangying, et al. Optimization of extraction process of alfalfa leaf protein[J]. *Journal of Suzhou University of Science and Technology*, **2005**(3): p59-61.
[5] Ma Fengang, Zhang Yongping. Research progress on the physiological effects and chemical constituents of Chinese herbal medicines[J].*World Health Information Digest*,2017,**17**(14):p46-52
[6] Zhou Pingping, Gong Xincheng, Zhao Xinghua. Extraction and Pharmacological Study of Active Components from Valeriana[J].*Heilongjiang Animal Husbandry and Veterinary Medicine*, **2016**(1): p188-190.
[7] Pan Xiaohua, Wang Long, Ji Wei, et al. Biological activity of valerian and its application in animal production [J]. *Guangdong Feed* 2011, **20** (12): p33-35.
[8] Liu Laiting, Zhang Huiru, Feng Jianxin, et al. Effects of valerian extract on non-specific immune function of grass carp[J].*Nutrition Feed* 2008,**44**(9):p29-31.
[9] Qiao Meihua, Ren Hongtao, Shi Ruixue, et al. Application prospect of alfalfa as a green feed additive in aquaculture [J]. *Feed wide angle*, **2017** (01): p43-44.
[10] Pang Zhizhong. Feed for development and utilization of wild alfalfa [J]. *Agricultural New Technology*, **2002** (3): p34-34.
[11] LIANG Fei, LI Yuanxiao, DING Wei, et al. Effects of fermentation methods on total flavonoids content and antioxidant activity of Valeriana officinalis[J].*Journal of Henan University of Science & Technology: Natural Science Edition*, 2018,**39**(6):p79- 83
[12] Rong Rong, Jia Yushan, Bao Jian, et al. Research on feed processing technology of willow leaves[J]. *Journal of Animal Husbandry and Feed Science*, 2019, **40**(5): p62-66.
[13] Zou Suhua, Liu Taiyu.Dynamic changes of nutrients and total flavonoids in different growth months of alfalfa[J].*China Animal Husbandry and Veterinary Medicine*,2010,**37**(5):p21-24.

# Compression property of hierarchical pyramid lattice

**Liu Tao[1], Zhu Guodong[1], Yang Na[1,2,3,4,5], Wang Jianfeng[1]**

[1]School of Automotive Engineering, Harbin Institute of Technology, Weihai 264209, P.R. China

[2]CATARC (Tianjin) Automotive Engineering Research Institute, Tianjin, Tianjin, 300300, P.R. China

[3]Automotive Engineering Research Institute, China Automotive Technology & Research Center, Tianjin 300300, P.R. China

[4]School of Mechanical Engineering, Tianjin University, Tianjin 300072, P.R. China

[5]Corresponding author's e-mail: ynhelen@126.com

**Abstract**. Hierarchical lattice structures have excellent mechanical and multi-functional properties. A novel hierarchical pyramid lattice structure was designed and the mechanical behaviours were studied using a finite element method. In order to reflect the advantages of hierarchical pyramid lattice structure, the compression simulation of the traditional pyramid lattice structure with the same relative density was also carried out. The results showed that the designed hierarchical pyramid lattice structure increased the buckling resistance, reduced the stress concentration at the joint of panel and core, and increased the energy absorption performance and specific strength. The designed hierarchical pyramid lattice structure contributes to the expansion of hierarchical lattice structure.

## 1. Introduction

Lattice structure has been widely studied due to its lightweight, high specific strength and multi-functional properties. As one of the typical lattice structures, the pyramid lattice structure has been extensive investigated by scholars for its compression [1,2], shear [3], three-point bending [4], high-speed impact [5], underwater impact [6]. These studies showed that the pyramid lattices had outstanding mechanical and multi-functional properties.

Hierarchical lattice structure has stronger energy absorption and specific strength. For the hierarchical lattice structure, scholars are interested in the hierarchical honeycomb structure, i.e., 2-D hierarchical lattice structures. Cote et al. [7] studied the compression failure modes and mechanical properties of hierarchical square honeycomb structures; Liu et al. [8] introduced flatwise compression property of hierarchical thermoplastic composite square lattice, and the results showed that the energy absorption was greatly improved; Fan et al. [9] had carried on the detailed numerical analysis to the energy absorption of hierarchical square lattice structures; Zheng et al. [10] has studied the energy absorption mechanisms of hierarchical woven lattice composite and a plastic model had been suggested; Feng et al. [11] studied compressive and shear properties of carbon fiber composite square honeycombs with optimized high-modulus hierarchical phases. However, there were few researches on the hierarchical pyramid lattice structure.

IOP Conf. Series: Materials Science and Engineering **738** (2020) 012031 doi:10.1088/1757-899X/738/1/012031

In this paper, we designed a hierarchical pyramid lattice structure based on square honeycomb lattice structure. According to the ABAQUS /Explicit, the compression behaviours were compared between the hierarchical pyramid lattices and the traditional pyramid with the same relative density.

## 2. Materials and methods

### *2.1. Hierarchical pyramid lattice structure*

Hierarchical pyramid lattice structure has two order lattice structure: square lattice structure is the first order lattice structure; pyramid lattice structure is the second order structure. Figure 1 shows the geometry: the geometry of the unit cell of square lattice is defined by its thickness of panel, $t_f$, core height, $t_c$, the wall thickness, $t_{cw}$, and inner length $L_0$ of square honeycomb; the geometry of the unit cell of pyramid lattice is defined by its length, L, thickness, T, Width, $T_W$, and angle, $\beta$ .

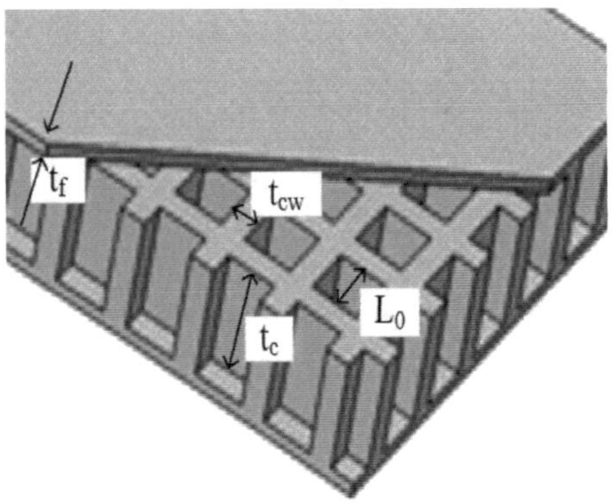


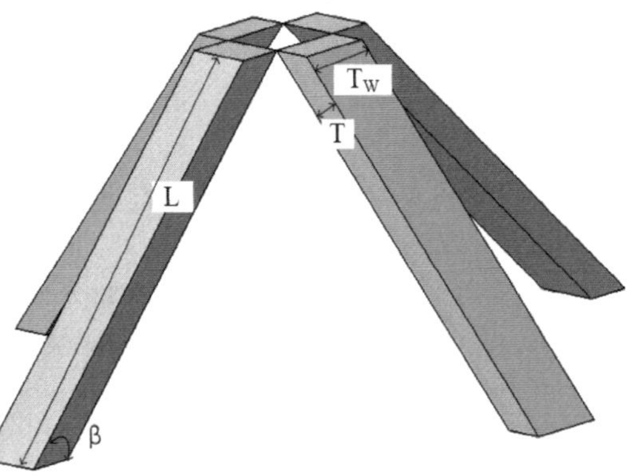


(a) (b)

Figure1. Geometry of (a) square lattice and (b) traditional pyramid lattice

The relative density of the cores of square lattice structure and pyramid lattice structure can be obtained from equations (1) and equations (2) respectively.

$$\rho_1 = 1 - \frac{L_0^2}{(L_0 + t_{cw})^2} \tag{1}$$

$$\rho_2 = \frac{2TT_W}{\sin\beta\left(T/\sin\beta + T_W/2 + L\cos\beta\right)^2} \tag{2}$$

The designed hierarchical lattice structure is shown in Figure 2.

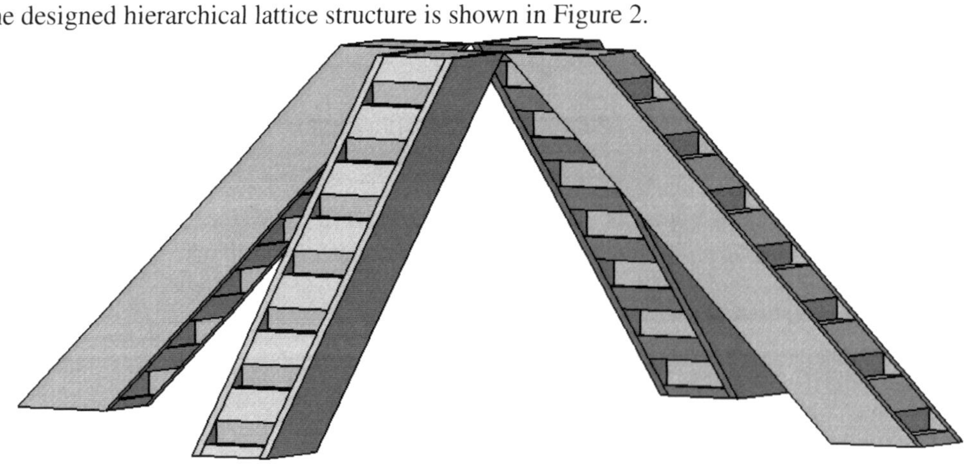

Figure 2. Hierarchical pyramid lattice structure core

According to equation (1) and equation (2), the relative density of hierarchical lattice structure can be given by equation (3).

$$\rho_3 = \frac{2T(2t_f + t_c\rho_1)}{\sin\beta(\frac{T}{\sin\beta} + \frac{2t_f + t_c}{2} + L\cos\beta)^2} \quad (3)$$

In order to analyze the compression performance of hierarchical pyramid lattice structure, we set up a traditional pyramid lattice structure with the same relative density, which was realized by changing the value of T. 4mm and 2.15mm are the thickness values of hierarchical lattice structure and traditional lattice structure, respectively. Table 1 shows the specific dimensions of two lattice structures: 1* represents hierarchical lattice structure; 2* represents traditional lattice structure.

Table 1. Lattice structure geometry

| Lattice structure | L (mm) | $\beta$ (°) | $T_W$ (mm) | T (mm) | $t_f$ (mm) | $t_c$ (mm) | $t_{cw}$ (mm) | $L_0$ (mm) | Relative density |
|---|---|---|---|---|---|---|---|---|---|
| 1* | 30 | 45 | | 4 | 0.5 | 4 | 1 | 2 | 0.04248 |
| 2* | 30 | 45 | 5 | 2.15 | | | | | 0.04227 |

*2.2. Material*

SOMOS-8000 photosensitive resin was adopted, which was produced by DuPont company (DSM) [12]. Its density is 1.16g/cm$^3$, Poisson's ratio is 0.41, and modulus of elasticity is 1400MPa, and table 2 shows the true plastic stress and strain.

Table 2. SOMOS-8000 true plastic stress and strain

| Number | Yield Stress (MPa) | Plastic Strain | Number | Yield Stress (MPa) | Plastic Strain | Number | Yield Stress (MPa) | Plastic Strain |
|---|---|---|---|---|---|---|---|---|
| 1 | 1 | 0 | 7 | 31 | 0.03 | 13 | 57.5 | 0.06 |
| 2 | 8 | 0.01 | 8 | 34 | 0.035 | 14 | 66 | 0.065 |
| 3 | 14 | 0.015 | 9 | 39.5 | 0.04 | 15 | 67 | 0.07 |
| 4 | 20 | 0.018 | 10 | 42.5 | 0.045 | 16 | 68.3 | 0.075 |
| 5 | 22.5 | 0.02 | 11 | 49.6 | 0.05 | 17 | 70 | 0.08 |
| 6 | 26 | 0.025 | 12 | 54 | 0.055 | 18 | 72 | 0.1 |

*2.3. Simulation model*

ABAQUS/Explicit was used to predict the mechanical properties of the hierarchical pyramid lattice structure under quasi-static flatwise compression. Figure 3 shows the 3*3 cores simulation model.

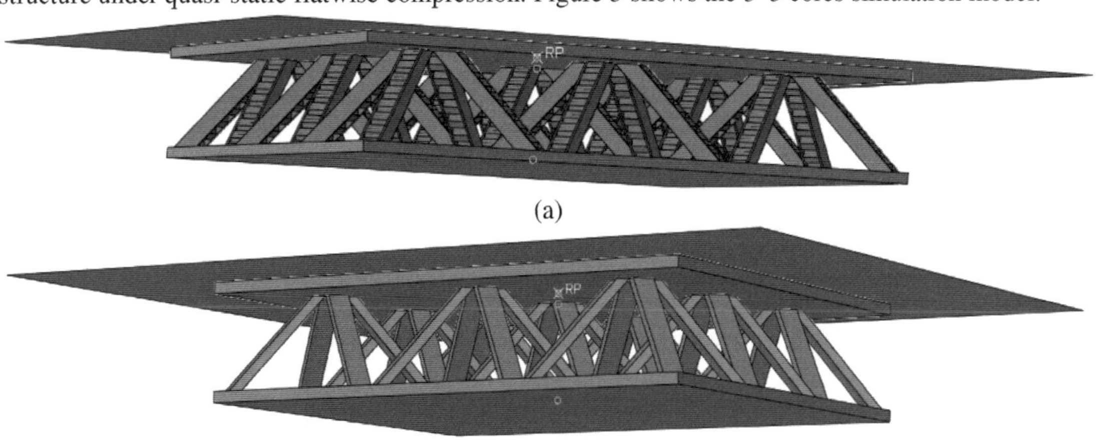


(a)

(b)

Figure 3. 3*3 cores simulation model (a) hierarchical pyramid lattice (b) traditional pyramid lattice

In the model, we designed two homogenized panels with the same thickness of 2.5mm, and these were tied with core together, respectively. Besides, the discrete rigid shell lunar was designed to simulate the pressure plate, and it was modelled with 4-node 3-D bilinear rigid quadrilateral elements(R3D4). Panels and cores were modelled with 8-node linear brick, reduced integration, hourglass control elements(C3D8R). The hierarchical pyramid lattice structure consisted of approximately 57132 C3D8R elements and 1600 R3D4 elements. At the same time, the number of C3D8R elements and R3D4 elements for traditional pyramid lattice were 46875, 1600 respectively. Besides, surface-to-surface contact (Explicit) was adopted between rigid shell planar and panel surface, and the friction coefficient was 0.36. Finally, reference point of rigid was moved down 5 mm, and another panel surface was fixed.

## 3. Results and discussion

### *3.1. Deformation process*

According to the output results, the upper and lower panels have almost no deformation due to their large thickness, and the deformation of 3*3 core is consistent with that of single core. Thus, the deformation of a single core will be used to illustrate the deformation of the whole structures. Figure 4 illustrates the deformation process of hierarchical pyramid lattice structure core, in which the rest of the simulation model is hidden. In order to explain the deformation process conveniently, we use D to express the displacement value of rigid body.

(a)

(b)

(c)

Figure 4. Deformation process of hierarchical pyramid lattice structure (a) D=0.25mm (b) D=2.75mm (c) D=5mm

When D is 0-0.25 mm, the stress distribution of the core is uniform, and there is almost no stress concentration; when D is 0.25-2.75mm, with the increase of displacement, the core of pyramid appeared obvious elastic deformation, and the stress concentration occurs in the local area of the panel of square honeycomb lattice; when D is 2.75-5mm, The value of local maximum stress, s, is kept at 72 MPa, the pyramid core appears obvious buckling, the local stress concentration area increases rapidly, and the core of square honeycomb structure occurs wrinkling.

Figure 5 shows the compression simulation results of 3*3 traditional pyramid lattice structure with the same relative density, and other parts of the simulation model are hidden.

Compared with Figure 4 (c) and Figure 5, it can be seen that the core of the traditional lattice structure has significant buckling, and the hierarchical lattice structure has markedly reduced the buckling.

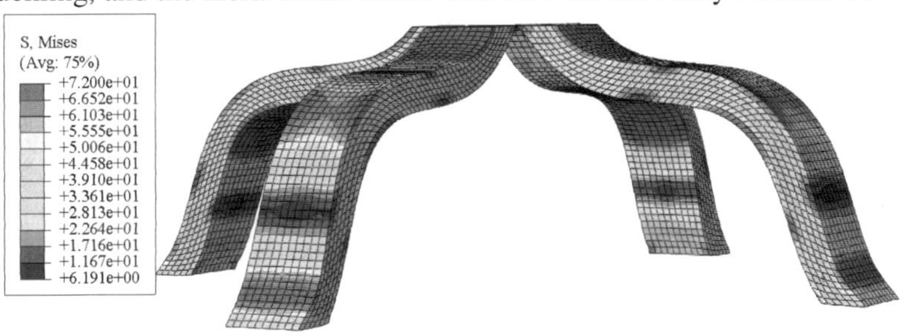


Figure 5. Deformation of traditional pyramid lattice structure (D=5mm)

### *3.2. Connection strength at the joint of panel and core*

The simulation results show that the stress concentration of traditional pyramid lattice not only appears at the joint of panel and core, but also in the middle part of core; however, the stress concentration of hierarchical lattice structure occurs in the area of square honeycomb panel rather than the connection between core and panel. This shows that the hierarchical lattice structure improves the connection strength between the panel and the core.

### *3.3. Specific strength and energy absorption properties*

Figure 6 shows the stress-strain curves of two lattice structures.

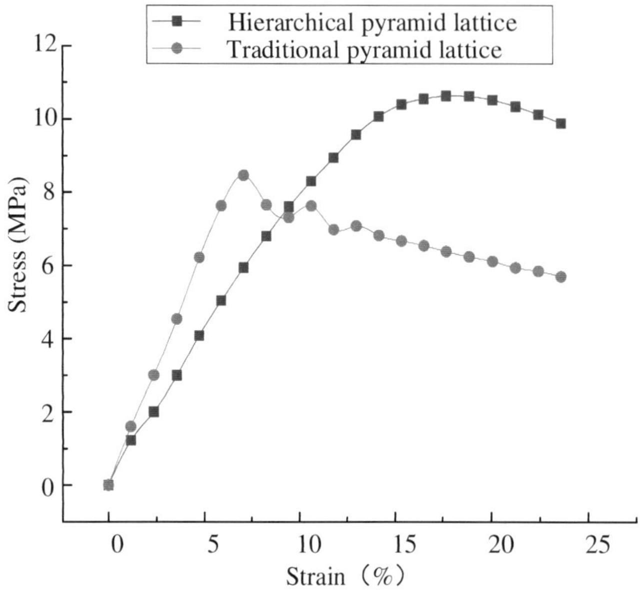


Figure 6. The stress-strain curves of two lattice structures

According to figure 6, the maximum compression strength of layered hierarchical structure is greater than that of traditional pyramid lattice, which indicates that hierarchical lattice structure has higher specific strength. According to the simple mechanical knowledge, the area surrounded by the stress-strain curve is the absorbed energy. We can see clearly that the stress-strain curve of hierarchical lattice structure includes a larger area, which indicates that it has a stronger energy absorption characteristic.

## 4. Conclusions

Based on the finite element method of ABAQUS/Explicit, the compression performance of hierarchical pyramid lattice structure was studied. According to the simulated results, the hierarchical lattice structure reduced the stress concentration at the junction of face sheets and cores, and increased the energy absorption performance and specific strength. What's more, the deformation process of hierarchical lattice structure was clarified.

## Acknowledgments

This work is supported by the National Natural Science Foundation of China (51905121), and the Fundamental Research Funds for the Central Universities (Grant No.HIT.NSRIF.2019077).

## References

[1] Queheillalt D T and Wadley H N G 2005 *Mater. Sci. Eng.* A **397** 132-7
[2] Wang B, Wu L Z, Ma L, Sun Y G and Du S Y 2010 *Mater. Des.* **31** 2659-63
[3] Li X D, Wu L Z, Ma L and Yan X Q 2019 *Mech. Adv. Mater. Struct.* **26** 866-77
[4] Smardzewski J and Wojciechowskib K Z 2019 *Compos. Struct.* **216** 340-9
[5] Yungwirth C J, Radford D D, Aronson M and Wadley H N G 2008 *Composites Part* B **39** 556-9
[6] Mori L F, Lee S, Xue Z Y, Vaziri A, Queheillalt D T, Dharmasena K P, Wadly H N G, Hutchinson J W and Espinosa H D 2007 *J Mech Mater Struct* **2** 1981-2006
[7] F Cote, B P Russell, V S Deshpande and N A Fleck 2009 *J Appl Mech* **76** 061004
[8] Liu H C, Chen L M, Du B, Peng S W, Guo Y G, Zhao Y, Chen L L, Zhou H, Li W G and Liu P 2019 *Compos Struct* **210** 118-33
[9] Fan H L, Luo Y H, Yang F and Li W W 2018 *Int J Eng Sci* **130** 12-32
[10] Zheng J J, Zhao L and Fan H L 2012 *Composites Part* B **43** 1516-1522
[11] Feng L J, Yang Z T, Yu G C, Chen X J and Wu L Z 2018 *Compos Struct* **201** 845-856

[12] Huang J J 2019 Structure Optimization of Automobile Energy Absorber Block Based on Rapid Prototyping Technology (Xi'an:Xi'an University of Technology) p 25

IOP Conf. Series: Materials Science and Engineering **738** (2020) 012032 doi:10.1088/1757-899X/738/1/012032

# Synthesis, wettability and optical properties of $xTiO_2$- (1-x) $SiO_2$ composite thin films

**Jun Wang, Lingyun Bai, Yueyuan Xiao, Yuhong Luo, Binghua Zhu, Wenjuan Zhang**

Jiangxi Key Laboratory of Surface Engineering, Jiangxi Science and Technology Normal University, Nanchang, Jiangxi, 330013, China

*corresponding author: wangjun460@jxstnu.edu.cn

**Abstract.** $xTiO_2$- (1-x) $SiO_2$ composite thin films with x=1.0, 0.9, 0.7, 0.5 were dip-coated on glass substrates by sol-gel method. The effects of composition on the microstructure, wettability and optical properties were investigated by X-ray diffraction (XRD), contact angle tester and ultraviolet visible spectrophotometer. Results showed that only the anatase diffraction peaks of $TiO_2$ were found in composite materials. With the increase of $SiO_2$ concentration, the crystallization degree of the composite material decreased slightly. The hydrophilicity was enhanced after coating on glass substrates. The water contact angle of $TiO_2$ film was the smallest in all the films. All the composite films were transparent in visible wavelength. And the average transmittance of $0.5TiO_2$-$0.5SiO_2$ was highest. The sharp decrease of transmittance was observed in the transmittance spectra, which was related to absorption of photos by electron in valence band.

## 1. Introduction

$TiO_2$ is a low cost, corrosion resistant, non-toxic and chemical stable semiconductor photocatalyst. It has been widely studied for using in solar cells, self-cleaning, biomaterials and self-sterilizing. While, the large band gap and photogenerated electron-hole recombination rate limit the application filed of $TiO_2$. Several methods have been used to reduce the band gap and improving the efficiency of photogenerated electron-hole separation by adjusting the composition and microstructure. Most popular used method is to dope materials with transition metal[1], noble metal[2-4], non-metal[5, 6] and carbon materials[7]. However, the doping modification affects the anatase structure and electron state of $TiO_2$.

Another modification method is to prepare composite material. Preparation and properties of $TiO_2$ composite has attracted a great deal of attention in recent years. ZnO[8], carbon quantum dots[9], $C_3N_4$ [10], $WO_3$[11],TiN[12] and $SiO_2$[13-15] have been used to prepare $TiO_2$ composite materials. The heterostructure formed in the interface can capture electron and extended the lifetime of the photogenerated carriers.

Although many studies have been reported about $TiO_2$-$SiO_2$ composite thin films, the correlation between composition and wettability, optical properties have not yet been reported. In this work, $TiO_2$-$SiO_2$ composite thin films with different $SiO_2$ concentration were fabricated on glass substrates by sol-gel dip-coating method. The effect of composition on the microstructure, wettability and optical properties were investigated. The results can be used as a reference for $TiO_2$ matrix composites.

## 2. Experimental

For $TiO_2$ sol, tetrabutyl titanate (TBOT) was used as titanium source, ethanol as solvent, nitric acid as acid catalyst and water as hydrolysis agent. The molar ratio of TBOT/ethanol/nitric acid/$H_2O$ was 1:20:1.5:1. In the experiment, tetrabutyl titanate was added to half of ethanol and stirred for 30 min, which was named solution A. Nitric acid and water were added to the other half of ethanol, which was named solution B. In the stirring, the solution B was gradually added to the solution A. The mixed solution was stirred for 1hour to obtain $TiO_2$ sol. In the x$TiO_2$- (1-x) $SiO_2$ (x=0.9, 0.7, 0.5) sol, partial tetraethoxy silane (TEOS) was used instead of tetrabutyl titanate (TBOT).

The films were deposited on glass substrates (25×25×2 mm). Prior to deposition, the substrates were cleaned in an ultrasonic cleaner with acetone, ethanol and distilled water for 10 min each. The films prepare process parameter by dip coating were as follows: the withdraw speed was 2000 μm/min, and the immersion time was 60 s. The wet films were dried at 100 °C for 10 min and heat treated at 300 °C for 30 min in the muffle furnace. In order to increase the thickness of the films, the coating-heating process was repeated twice. $TiO_2$-$SiO_2$ films were annealed at 500 °C for 60 min.

To study the effect of composition on the microstructure of $TiO_2$-$SiO_2$ composite materials, the powders were prepared by the corresponding sol. The sintering process of the powders was consistent with that of the films.

Structure of the fabricated $TiO_2$-$SiO_2$ composite powders were characterized by X-ray diffraction (XRD) using a Cu Kα1 (λ=1.5418 Å) X-ray radiation source in Bragg–Brentano (θ–2θ) configuration. The wettability of the composited films was evaluated by measuring the contact angle of a water droplet on the film surface. The transmittance spectra of the composite films were measured by ultraviolet visible spectrophotometer in the wavelength from 200 to 1100 nm.

## 3. Results and discussion

### *3.1. Optical properties of $TiO_2$-$SiO_2$ sol*

The stability of sol is very important for the preparation of high-quality films by sol-gel method. The experimental environment (temperature, humidity, etc.) and preparation process parameters determined the stability of the sol. In our experiment, the effect of solution concentration and PH value on the stability of sol was studied. The light yellow transparent sol was obtained and can be placed over three months at room temperature (about 25°C) and humidity of about 45%. The transmittance spectra of x$TiO_2$- (1-x) $SiO_2$ sol with x=1.0, 0.9, 0.7, 0.5 were showed in figure 1. All of the sol were high transparent in visible wavelength range.

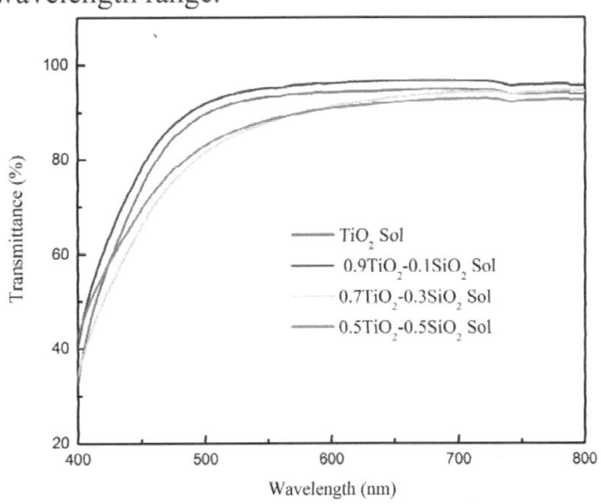


Fig. 1 Transmittance spectra of $TiO_2$-$SiO_2$ sol

### *3.2. Microstructure of $TiO_2$-$SiO_2$ composite materials*

Figure 2 gives the XRD patterns of $TiO_2$-$SiO_2$ powders. All the diffraction peaks were well matched with the tetragonal $TiO_2$ anatase phase. The diffraction peak index was marked in the figure according to the JCPDS card (No: 84-1286). No other diffraction peaks were observed in $TiO_2$-$SiO_2$ composite materials. And the angle of antase phase diffraction peak has not changed with the addition of $SiO_2$. This means that there is no bonding structure between $TiO_2$ and $SiO_2$ in the composite materials. The diffraction background for 0.5$TiO_2$-0.5 $SiO_2$ was enhanced, especially from 15 to 30°. It may be related to the formation of amorphous $SiO_2$.

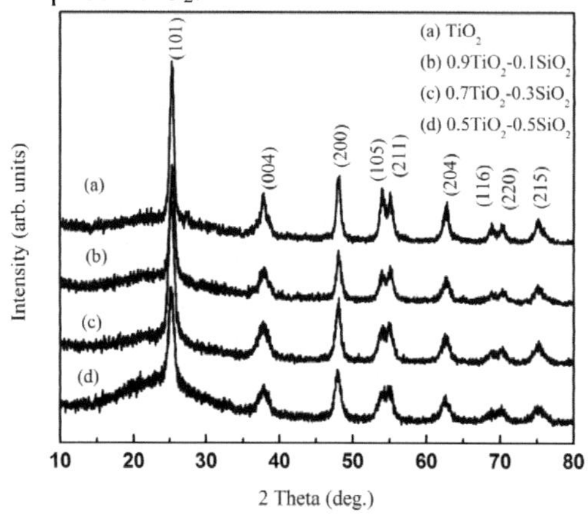


Fig. 2 XRD patterns of $TiO_2$-$SiO_2$

### 3.3. *Wettability and optical properties of $TiO_2$-$SiO_2$ composite thin films*

The wettability of material surface was characterized by water contact angle measurements. 3 mL water droplet were applied to define the wettability of $TiO_2$-$SiO_2$ composite thin films by measuring contact angles. The results were shown in figure 3. The water contact angle decreased and the hydrophilicity increased after coating on glass substrates. $TiO_2$ has the smallest water contact angle. The contact angle increased with the increase of $SiO_2$ content. This may be related to composition and the degree of crystallization of the films.

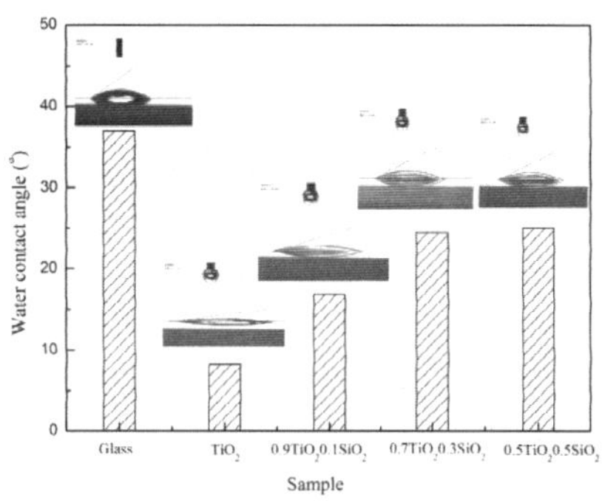


Fig. 3 Water contact angle of glass and $TiO_2$-$SiO_2$ thin film

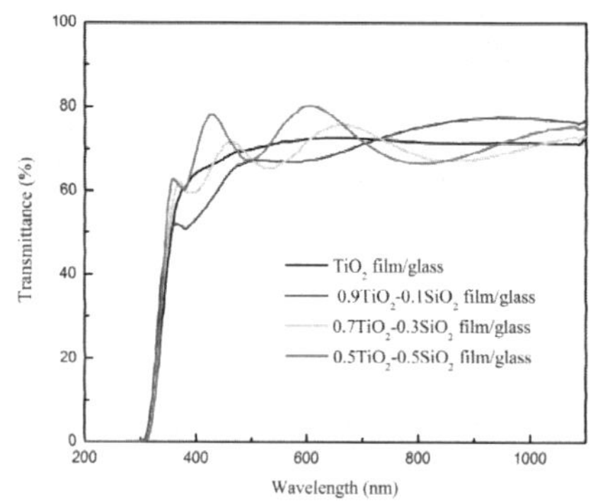


Fig. 4 Transmittance spectra of $TiO_2$-$SiO_2$ thin films on glass substrate

The transmittance was measure by normal incidence in the wavelength of 200-1100 nm. Transmittance spectra of $TiO_2$-$SiO_2$ composite thin films with different composition were shown in figure 4. It can be seen that all the samples were high transmittance in visible and near-infrared regions. 0.5$TiO_2$-0.5$SiO_2$ film has the highest average transmittance. Three interference fringes can be reveled in $TiO_2$-$SiO_2$ composite films except $TiO_2$ film. This was related the thickness of the films [8]. A strong absorption region was observed in wavelength below 360 nm. It was due to the photons with high energy (short wavelength) can be absorbed by electrons in the valence band. The electron that absorbs energy in the valence band can transition to the conduction band. The electron-hole pair was formed

IOP Conf. Series: Materials Science and Engineering **738** (2020) 012032 doi:10.1088/1757-899X/738/1/012032

and the properties of the thin films were changed.

## 4. Conclusion

$xTiO_2$-(1-x) $SiO_2$ (x=1.0, 0.9, 0.7, 0.5) composite powder and films were prepared by sol-gel dip-coating method. The composite materials were crystalized with anatase structure. As the content of $SiO_2$ increase, the crystallization degree of the composite materials decreased. The composition and crystallization degree of the films affect wettability. The contact angle of $TiO_2$ was smallest. The composite films were highly transparent in the visible region and strongly absorbed in the ultraviolet region.0.5$TiO_2$-0.5$SiO_2$ film has the highest average transmittance in the composite films.

## Acknowledgments

This work was supported by the Science and Technology Research Project of Jiangxi Education Department (Project name: Preparation and anti-fog mechanism of hollow $SiO_2/TiO_2$ nanocomposite film), the Ph.D. Start Fund Project of Jiangxi Science and Technology Normal University (No: 2018BSQD019) and innovative training program for college students of Jiangxi Science and Technology Normal University (No: 20191303030, 20191304097).

## References

[1] Wang T, Shen D, Xu T, et al. (2017) Photocatalytic degradation properties of V-doped $TiO_2$ to automobile exhaust. Science of The Total Environment, 586: 347-354.

[2] Xu M, Wang Y, Geng J, et al. (2017) Photodecomposition of $NO_x$ on Ag/$TiO_2$ composite catalysts in a gas phase reactor. Chemical Engineering Journal, 307:181-188.

[3] John Berlin I, Joy K. (2015) Optical enhancement of Au doped $ZrO_2$ thin films by sol–gel dip coating method. Physica B: Condensed Matter, 457:182-187.

[4] Lin H, Yang T, Hsi C, et al. (2014) Optical and photocatalytic properties of $Fe^{3+}$-doped $TiO_2$ thin films prepared by a sol–gel spin coating. Ceramics International, 40:10633-10640.

[5] Yoshida T, Niimi S, Yamamoto M, et al. (2015) Effective nitrogen doping into $TiO_2$ (N-$TiO_2$) for visible light response photocatalysis. Journal of Colloid and Interface Science, 447:278-281.

[6] Trevisan V, Olivo A, Pinna F, et al. (2014) C-N/$TiO_2$ photocatalysts: Effect of co-doping on the catalytic performance under visible light. Applied Catalysis B: Environmental, 160-161:152-160.

[7] Saleem H, Habib A. (2016) Study of band gap reduction of $TiO_2$ thin films with variation in GO contents and use of $TiO_2$/Graphene composite in hybrid solar cell. Journal of Alloys and Compounds, 679:177-183.

[8] Giannakopoulou T, Todorova N, Giannouri M, et al. (2014) Optical and photocatalytic properties of composite $TiO_2$/ZnO thin films. Catalysis Today, 230:174-180.

[9] Martins N C T, Ângelo J, Girão A V, et al. (2016) N-doped carbon quantum dots/$TiO_2$ composite with improved photocatalytic activity. Applied Catalysis B: Environmental, 193:67-74.

[10] Papailias I, Todorova N, Giannakopoulou T, et al. (2017) Photocatalytic activity of modified g-$C_3N_4$/$TiO_2$ nanocomposites for NOx removal. Catalysis Today, 280:37-44.

[11] Dobromir M, Apetrei R P, Rebegea S, et al. (2016) Synthesis and characterization of RF sputtered $WO_3$/$TiO_2$ bilayers. Surface and Coatings Technology, 285:197-202.

IOP Conf. Series: Materials Science and Engineering **738** (2020) 012032 doi:10.1088/1757-899X/738/1/012032

[12] Fakhouri H, Arefi-Khonsari F, Jaiswal A K, et al. (2015) Enhanced visible light photoactivity and charge separation in $TiO_2$/TiN bilayer thin films. Applied Catalysis A: General, 492:83-92.
[13] Mokhtarimehr M, Pakshir M, Eshaghi A, et al. (2013) Super-hydrophilic property of vanadium doped $TiO_2$–$SiO_2$ sol–gel derived thin film. Thin Solid Films, 532:123-126.
[14] Luna M, Gatica J M, Vidal H, et al. (2019) Au-$TiO_2$/$SiO_2$ photocatalysts with NOx depolluting activity: Influence of gold particle size and loading. Chemical Engineering Journal, 368:417-427.
[15] Xu Y J, Liao J X, Cai Q W, et al. (2013) Preparation of a highly-reflective $TiO_2$/$SiO_2$/Ag thin film with self-cleaning properties by magnetron sputtering for solar front reflectors. Solar Energy Materials and Solar Cells, 113:7-12.

IOP Conf. Series: Materials Science and Engineering **738** (2020) 012033 doi:10.1088/1757-899X/738/1/012033

# Effect of different macromonomer molecular size in polycaroxylate superplasticizer by molecular dynamics simulation

**Po-Hsiang Chuang[1,2,3], Shaomin Zhang[2], Yuliang Ke[2], Yuanqiang Guo[2]**

[1]Lets Holding Group Co., Ltd., Xiamen 361004, China

[2]KZJ New Materials Group Co., Ltd., Xiamen 361101, China

[3]College of the Environment & Ecology, Xiamen University, Xiamen 361102, China

calcium7@hotmail.com

**Abstract**. Molecular dynamics simulations were carried out to investigate the solution conformation of polycarboxylate superplasticizer (PCE) with different macromonomer molecular size. Simulation results showed that the larger size of macromonomer in the PCE make the molecules have greater intermolecular interactions. The probability of water molecules around the PCE molecules affected the water distribution in the systems. These results indicated that there were more water molecules around the main chain of PCE, while the macromonomer in PCE was larger. Knowledge of the interaction relationship between water molecules and PCE using molecular dynamics simulation opens the way for understanding the mechanism of superplasticizer in cement systems.

## 1. Introduction

In the past decades, chemical admixtures have attracted increasing attention in the cement industry. Superplasticizers are one kind of chemical admixtures that can be added to cement mixtures to improve the workability of cement and concrete. Polycarboxylate superplasticizer (PCE) is one of the superplasticizers that easy to be designed and synthesized for different functions [1]. PCEs usually composed of polymer backbone containing anionic groups and grafted side chains. The anionic groups adsorb onto the cement surface electrostatically and the side chains provide steric repulsion, resulting in an excellent cement-dispersing effect. The grafted side chains usually designed from different types and sizes of polyoxyethylene glycols (PEGs). The PEGs of different structures lead to different properties of the synthesized PCE and that affect the dispersing performance of PCE.

Generally, the PCE conformation is a key factor affecting the working properties in concrete systems. The solution conformation of PCE not only affects the adsorption ability but also affects the dispersion properties in cement. Shu et al. reported that PCE containing high backbone methyl group content exhibited a more extended polymer backbone in pore solution due to the hindered free rotation of polymer backbone and more carboxylic groups could be accessible for adsorption [2]. Ran et al. synthesized a series of PCEs with different side conformation to investigate the dispersing mechanism in terms of rheological behavior and dispersion in cement system [3]. These indicate that there is a relationship between solution conformation and performance of PCE.

Recently, molecular dynamics (MD) simulation has been widely used to study cement systems, especially in the chemical admixture. Hirata et al. used MD simulation to simulate the PCE polymers adsorbed on the surface in pore solution [4]. Chuang et al. simulated PCE molecules with different side-chain length in aqueous solution [5]. However, more insights are needed to understand and control the properties of PCE at the microscopic scale.

The present study aims to understand the effect of macromonomer molecule size in PCE systems. Toward this propose, four kinds of PCE molecules with different macromonomer molecule size were designed in aqueous solution. MD simulation was carried out to investigate the solution conformation of PCE. The interaction between molecules in the PCE system is the main factor affecting the properties of PCE. The results facilitate to clarify the molecular interaction mechanism between PCE systems that help to develop structure-property relationship of PCE molecule.

## 2. Methods

MD simulations were used to study the solution conformation of PCE molecule with different molecule size of macromonomer in aqueous solution. The PCE molecules were composed by two kinds of monomer, acrylic acid (AA) sodium and isopentenyl polyethylene glycol (IPEG), with the ratio of AA:IPEG being 4:1. The macromonomer of IPEG had different size with molecular weight of 600, 1200, 2400, and 3000 which consistent with commonly used in the chemical admixture industry. These four kinds of PCE molecules with different macromonomer molecule size were called PCE1, PCE2, PCE3, and PCE4. The molecular weight of PCE was consistent with our synthesized PCE polymer, which had been widely used in the concrete industry.

The MD simulation was used all-atom molecular dynamics simulation with COMPASS forcefield [6]. The PCE, solvent molecules and sodium ions were randomly place in the systems with the periodic boundary conditions. The sodium ions were used to balance the system charge. The concentration of PCE in the all systems was 10%, which was also consistent with the product concentration. There were 5500, 9000, 15700, and 19000 water molecules with TIP3P water model in the system respectively. All MD simulation was carried out in the canonical ensemble (NVT) with the time step equal to 1.0 fs with the system temperature set at 298 K. The total simulation time was over 1000 ps and eventually reached thermal equilibrium. An extra MD simulation run of 100 ps was carried out for analysis.

## 3. Results and Discussion

In the MD simulation, we studied the solution conformation of four different structure PCE molecules (PCE1, PCE2, PCE3 and PCE4) with same concentration at PCE in aqueous solution. Figure 1 shows the snapshots of PCE1, PCE2, PCE3 and PCE4 systems. The black, rad, and white balls represent carbon, oxygen, and hydrogen atoms, respectively. The PCE molecules have different solution conformation in the different systems. The PCE molecules with small size macromonomer aggregate themselves, on the other hand, the PCE molecules with large size macromonomer are easily cross-linked with other PCE molecules. The macromonomer with larger size in PCE molecule has more intermolecular interaction with each other molecule. That increases the steric repulsion effect in PCE system resulting in influence the dispersion properties of cement.

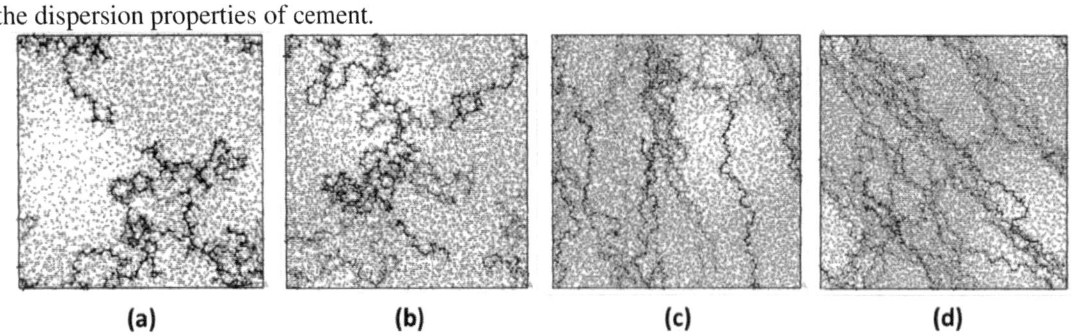


Figure 1. The snapshots of (a) PCE1 (b) PCE2 (c) PCE3 and (d) PCE4 in aqueous systems.

In order to analyze the intermolecular interaction between molecules, radial distribution functions (RDFs) of molecule in the trajectory were calculated. RDF shows the local structure and understanding the atomic distribution of substance, which obtains the effective intermolecular potentials and the probability between two objects. The intermolecular relationship between PCE molecules is shown in figure 2. Op is defined as the oxygen atom on the PCE molecule. In figure 2, the PCE molecule with larger size of macromonomer has more intermolecular interaction, while the smaller-sized macromonomer has small interaction. This result is consistent with the experiment that shows the long side chain of PCE increase the steric hindrance effect in cement systems [3].

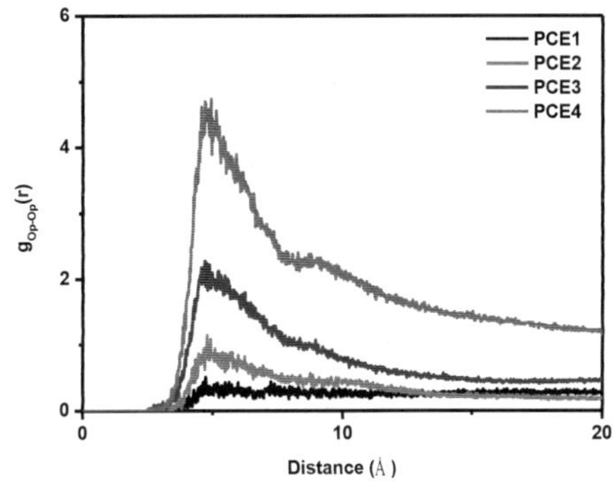


Figure 2. The radial distribution functions of $g_{Op\text{-}Op}(r)$ between PCE molecule in systems.

In addition, the distribution of water molecules around the PCE molecules affect the hydration as well as dispersion cement in system [7]. Figure 3 shows the RDF between the main chain of PCE and water molecules. Om is defined as the oxygen atom on the main chain of PCE molecule; OH is defined as the oxygen atom on the water molecule. In figure 3, all the main chain of PCE molecules have obvious hydrogen bonds with the water molecules. The probability of water molecules in four systems are PCE4 > PCE3 > PCE2 > PCE1. The hydrogen bonds make PCE molecules easily soluble and disperse in water. The distribution of water molecules around the main chain of PCE affects the water density at the cement interface due to PCE adsorption structure.

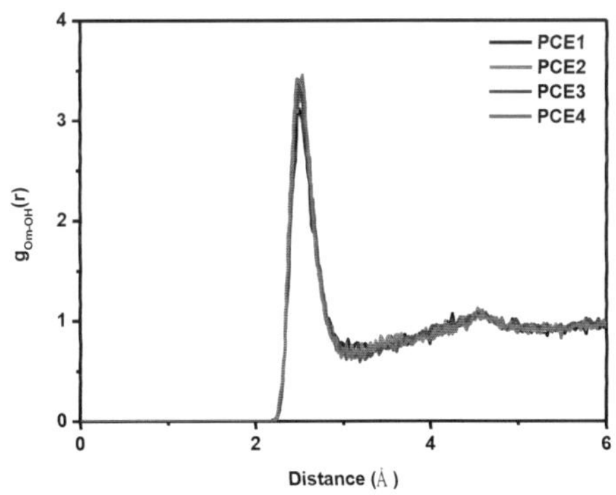


Figure 3. The radial distribution functions of $g_{Om\text{-}OH}(r)$ in systems.

To further understand the distribution of water around the main chain, the number of water molecule was calculated. The main chain of PCE was selected and calculated the space within a radius of 5 Å. The water molecules were counted in this space as shown in figure 4. The number of water molecules around the main chain of PCE increases as the molecule size of the macromonomer increases. This phenomenon indicates that the more water molecules tend to gather near the main chain of PCE with the larger molecule size in PCE macromonomer. These results show the water molecules around the main chain of PCE depend on the molecules size of macromonomer. The trapped water molecules not only affect the hydration kinetics of cement but also affect the water reducing properties of PCE. These reasons influence the working performance of PCE in the cement systems.

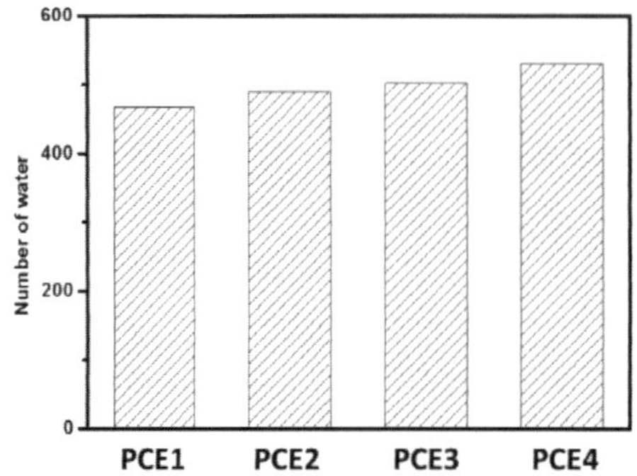


Figure 4. The number of water molecules around the main chain of PCE molecules.

## 4. Conclusion

MD simulation was carried out to investigate different macromonomer molecule size of PCE molecules with the same concentration in aqueous solution. Our results showed that the steric hindrance properties of PCE depend on the macromonomer molecule size in PCE. The larger size of macromonomer in PCE molecule had larger intermolecular interaction. On the other hand, the distribution of water molecules was also affected by the macromonomer molecule size in PCE. The more water molecules surround the main chain of PCE because the PCE had a larger macromonomer. MD simulations provided molecular insight on superplasticizers in aqueous solution. Knowledge of the distribution of molecules in the PCE systems at the microscopic scale help to further understand the mechanism of PCE.

## References

[1] Plank J, Pöllmann K, Zouaoui N, Andres P R, and Schaefer C 2008 *Cem. Concr. Res.* **28** 1210
[2] Shu X, Ran Q, Liu J, Zhao H, Zhang Q, Wang X, Yang Y, and Liu J 2016 *Constr. Build. Mater.* **116** 289
[3] Ran Q, Somasundaran P, Miao C, Liu J, Wu S, and Shen J 2009 *J. Colloid Interface Sci.* **336** 624
[4] Hirata T, Branicio P, Ye J, Zheng J, Tomike Y, Lange A, Plank J, Sullivan M 2017 *Adv. Cem. Res.* **29** 418
[5] Chuang P H, Tseng Y H, Fang Y, Gui M, Ma X, Luo J 2019 *Polymers* **11** 346
[6] Sun H 1998 *J. Phys. Chem. B* **102** 7338
[7] Hirata T, Ye J, Branicio P, Zheng J, Lange A, Plank J, Sullivan M 2017 *Sci. Rep.* **7** 16599

# Effect of different synthesis temperature on the bleeding performance of polycarboxylate superplasticizer

**Qiumin Xiong[1], Guangxing Lai[2*], Wei Dang[1], Han Xu[1], Guangzhou Li[1], Jican Guan[1], Xiujuan Guo[1], Muhe Xie[1] and Xiaobing Chen[1]**

[1]KZJ New Materials Group (Guangdong) Co., Ltd. Foshan 440600, Guangdong, China

[2]KZJ New Materials Group Co., Ltd. Xiamen 361101, Fujian, China

[*]Corresponding author's e-mail: lgx502@foxmail.com

**Abstract**. In this paper, polycarboxylate superplasticizers were synthesized by free radical solution polymerization under different synthesis temperature conditions. The bleeding performance of polycarboxylate superplasticizers under different synthesis temperature conditions was discussed by paste test and concrete test, and the structure of polycarboxylate superplasticizers was characterized by infrared spectroscopy (FTIR) and gel permeation chromatography (GPC). The results show that the bleeding and slump retention performance of polycarboxylate superplasticizer can be effectively improved by properly increasing the synthesis temperature, and the better synthesis temperature was 55 °C. With the temperature increase, the molecular weight of polycarboxylate superplasticizers increases first and then decreases. The infrared spectra of polycarboxylate superplasticizers synthesized at different temperatures were not significantly different.

## 1. Introduction

With the increasing demand for urban traffic facilities and urban buildings, and the demand for low cementitious concrete is increasing rapidly[1,2]. Therefore, it is the key issue to improve the performance of concrete and meet the use of concrete in various environmental conditions while ensuring the amount of low cementitious concrete. The bleeding rate can affect the performance of concrete in many aspects. In order to control the bleeding rate of concrete, its performance can be adjusted by adding additives[3]. Among all kinds of concrete admixtures, polycarboxylate superplasticizer (PCE) can control the bleeding performance of concrete very well, and can effectively reduce water consumption, improve slump retention performance and paste flow performance. The amount of water-reducing agent in the concrete industry is increasing year by year, so it is necessary to further improve the performance of the PCE.

The reaction temperature during the synthesis of the PCE has a great influence on its performance. With the deepening of research and the improvement of technology, the synthesis temperature of PCE gradually develops from high temperature to normal temperature, thereby reducing production energy consumption and cost[4,5]. The effects of PCE synthesized at different temperatures on the bleeding and retention properties of concrete were studied.

## 2. Experimental

### *2.1. Experimental raw materials*
The main raw materials required for the experiment were as follows: isopentenyl polyoxyethylene polyether (TPEG, molecular weight 2400, industrial grade), hydrogen peroxide ($H_2O_2$, 27.5 wt%, industrial grade), ammonium persulfate (APS), sodium hydroxide solution (30% aqueous solution, industrial grade), acrylic acid (AA, industrial grade), mercaptoacetic acid (MAA, industrial grade), sodium hypophosphite (SHP, industrial grade), TP1351 (reductant, industrial grade).

### *2.2. Test raw materials*
Cement (C, Yuexiu brand, P.O 42.5R); Sand (S, river sand with fineness modulus of 2.4 and mud content less than 0.4%); Gravel (G1, grain size of 5-10 mm; G2, grain size of 10-20 mm); Fly ash (F, Level II); Water: tap water meeting the test requirements.

### *2.3. Copolymerization*
The metered TPEG macromonomer and a certain amount of water were added into a four-necked flask. The mixture was stirred and heated in a digital thermostatic water bath at the reaction temperature (the reaction temperatures were 25 °C, 35 °C, 45 °C, 55 °C and 65 °C, respectively). After adding $H_2O_2$ into the flask, MAA aqueous solution, SHP aqueous solution and TP1351 (reductant), then AA aqueous solution was slowly added dropwise into the flask (dropping time: 3 h) and keep warm for 40 min. By adding a 30% aqueous solution of sodium hydroxide to adjust the pH to 6.0-7.0, and then PCE with solid content of 49% was obtained. According to the initial reaction temperature 25 °C, 35 °C, 45 °C, 55 °C and 65 °C, the polycarboxylate superplasticizers were named PCE-1, PCE-2, PCE-3, PCE-4 and PCE-5, respectively.

### *2.4. Performance test method*

*2.4.1. Fluidity of cement paste.* The fluidity of cement paste is conducted in accordance with GB/T8077-2012 "Test method for homogeneity of concrete admixtures".

*2.4.2. Concrete test.* The concrete test is conducted in accordance with GB/T50080-2016 "Standard Test Methods for Performance of Common Concrete Mixtures" and GB/T50081-2002" Standard Test Methods for mechanical properties of Common Concrete".

*2.4.3. Gel permeation chromatographic (GPC) measurement.* American Waters 1515 Isocratic HPLP pump/Waters 2414 differential detector was used for testing and analysis.

*2.4.4. Fourier transformer infrared spectra (FTIR) measurement.* FTIR were obtained from the pressed disc of PCE and KBr. The spectra in the range of 4000-400 $cm^{-1}$ were recorded on Perkin Elmer Spectrum 100 FTIR spectrophotometer.

## 3. Experimental results and discussion

### *3.1. Paste fluidity test*
Temperature is an important factor affecting the polymerization during the synthesis, which not only affects the molecular weight of the product, but also affects the dispersion of PCE. The influence of synthesis temperature on the dispersion of PCE was shown in Figure 1.

It can be seen from Figure 1 that with the increase of PCE synthesis temperature, the fluidity of the paste first increases and then decreases. When the synthesis temperature was 55 °C, the fluidity of the paste reaches the maximum. The reason for the analysis is that when the synthesis temperature was near room temperature, the unsaturated bond of the macromonomer does not break, and the copolymerization reaction was difficult to occur, resulting in too low conversion rate and low dispersibility of the PCE.

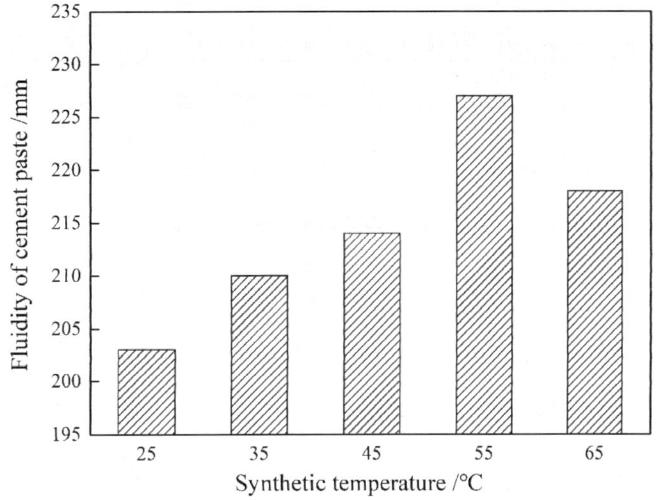


Figure 1. Fluidity of cement after adding PCEs synthesized at different synthesis temperatures

When the synthesis temperature is enough high, a large number of free radicals produced by TPEG will easily lead to the concentration of local free radicals is too high, resulting in clots, and increase the number of branched chains in the graft, which is unfavorable to the polymerization of macromolecular polymers; High temperature will increase the number of side reactions and the reaction process is difficult to control. The side products produced by these side reactions reduce the dispersibility of cement particles. In addition, the acrylic acid (AA) monomer is easy to self-polymerize with the increase of reaction temperature, which makes it difficult for large monomers to polymerize, therefore, high synthesis temperature results in low fluidity and poor dispersion of cement paste.

### *3.2. Concrete test*

The concrete test was carried out under the same conditions without any slump retaining and retarding components. The concrete test mix proportion (kg/m$^3$) was m (cement): m (fly ash): m (river sand): m (big stone): m (water) = 250: 100: 800: 1050: 165. The concrete test results were shown in Table 1.

Table 1. Concrete test results

| Samples | Synthesis temperature /°C | Dosage /% | Slump /mm | | Dispersion /mm | | 1 h Bleeding rate /% |
|---|---|---|---|---|---|---|---|
| | | | initial | 1 h | initial | 1 h | |
| PCE-1 | 25 | 1.6 | 165 | 115 | 520 | 380 | 14.4 |
| PCE-2 | 35 | 1.5 | 210 | 180 | 510 | 420 | 13.7 |
| PCE-3 | 45 | 1.5 | 215 | 190 | 525 | 425 | 12.5 |
| PCE-4 | 55 | 1.4 | 210 | 160 | 540 | 460 | 8.5 |
| PCE-5 | 65 | 1.4 | 205 | 160 | 520 | 420 | 12.4 |

As shown in Table 1, when the synthesis temperature was low, the polymerization time of PCEs need more time. Therefore, properly increasing the synthesis temperature can effectively improve the reaction rate and shorten the production cycle. The concrete performance of PCEs synthesized at different temperatures was tested. It can be seen from the Table 1 that when the reaction temperature was 55 °C, the synthesized polycarboxylate superplasticizer had a higher water reduction rate, better slump retention effect and lower bleeding rate.

### *3.3. GPC test*

In order to compare the molecular weight and distribution of PCEs synthesized at different temperatures, GPC tests were performed on PCE-1~PCE-5. The results were shown in Table 2.

Table 2. Gel permeation chromatographic data of PCEs

| Samples | Synthesis temperature /°C | $M_n$ | $M_w$ | $M_p$ | $M_w/M_n$ | Conversion rate /% |
|---|---|---|---|---|---|---|
| PCE-1 | 25 | 14312 | 25237 | 21159 | 1.76 | 69.61 |
| PCE-2 | 35 | 15240 | 27280 | 22617 | 1.79 | 80.10 |
| PCE-3 | 45 | 15192 | 27720 | 22293 | 1.82 | 80.14 |
| PCE-4 | 55 | 15467 | 28614 | 22882 | 1.85 | 88.85 |
| PCE-5 | 65 | 14919 | 25521 | 22051 | 1.71 | 77.26 |

It can be seen from Table 2 that the weight average molecular weight ($M_w$) of the PCE synthesized at different temperatures was about 27000. As the temperature increases, the number average molecular weight ($M_n$) of the PCE increases first and then decreased, but they were all around 15000. When the reaction temperature was 55 °C, the conversion rate of PCE-4 was as high as 88.35% and the material polydispersity index ($M_w/M_n$) was 1.85, the result showed that the PCE synthesized at 55 °C had good dispersibility and high conversion rate.

### *3.4. FTIR test*

FT-IR was used to character the compositional information of PCE-1 and PCE-4 and the results were shown in Figure 2. It can be seen from the figure that the positions of the infrared absorption peaks of the PCE-1 and PCE-4 were similar. The change of absorption peak at 3489 $cm^{-1}$, 2828 $cm^{-1}$ and 1731 $cm^{-1}$ ascribed to the stretching vibration of hydroxyl -OH, C-H stretching vibration of methyl and methylene and stretching vibration of C=O in ester bond, respectively. The absorption peak at 1456 $cm^{-1}$ and 1349 $cm^{-1}$ were generated by bending vibration of methylene $-CH_2$ and methyl $-CH_3$. The position of infrared absorption peak of the PCE synthesized under room temperature and heating conditions basically coincide, but with the increase of temperature, the chemical reaction in the synthesis process was accelerated. Therefore, the peak intensity of the PCE synthesized under heating conditions was slightly higher than that of the PCE synthesized at normal temperature.

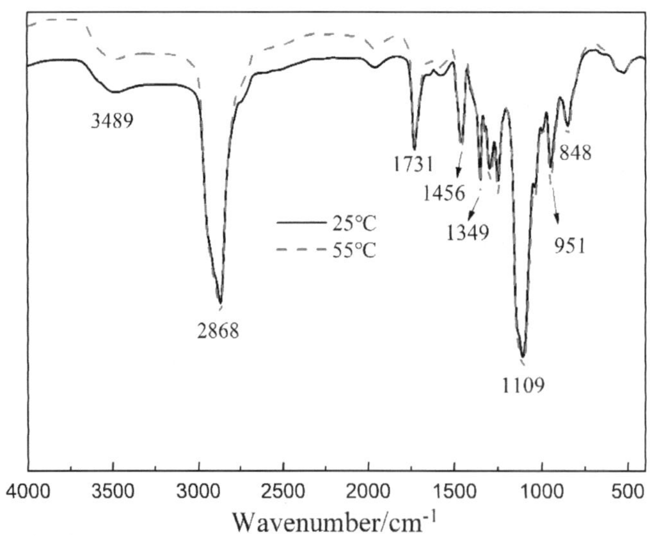


Figure 2. Infrared spectra of the PCEs

IOP Conf. Series: Materials Science and Engineering **738** (2020) 012034 doi:10.1088/1757-899X/738/1/012034

## 4. Conclusions

(1) Proper heating can increase the chemical reaction rate of PCE, promote the polymerization of free radicals, improve the dispersion performance, and then effectively improve the water bleeding, fluidity and slump retention of concrete.

(2) The PCE-4 synthesized at 55 °C has low bleeding rate, high water reduction and good slump retention effect.

(3) With the temperature increase, the molecular weight of polycarboxylate superplasticizers first increase and then decrease.

(4) The infrared spectra showed that the peak position of polycarboxylate superplasticizers synthesized at room temperature was basically the same as that synthesized under heating condition, but the peak intensity of the PCE synthesized under heating conditions was slightly higher than that synthesized at room temperature.

## References

[1] Chao Guo. Problems and solutions of lagging bleeding of concrete [J]. Commodity concrete, 2010, (9): 49-51.

[2] Lin Li. Application of polycarboxylate superplasticizer in self compacting concrete with low glue content [J]. Commodity concrete, 2012, (6): 40-43.

[3] Xinhua Yan. Influence and control of concrete cement water-from the hazard of cement water in deep pile foundation [J]. Commodity concrete, 2012, (3): 53-54.

[4] Zhiqin He, Guanghua Zhang, Song Qin, et al. Synthesis and properties of polycarboxylate superplasticizer at room temperature [J]. Chemical Industry and Engineering Progress, 2015, 34(11): 4014-4018.

[5] Tian Yingfeng, Zhu Xiaping, Xiang Niannian, et al. Study on low temperature synthesis of modified polyether polycarboxylate superplasticizer [J]. New building materials, 2013, (12): 88-92.

IOP Conf. Series: Materials Science and Engineering **738** (2020) 012035 doi:10.1088/1757-899X/738/1/012035

# Research of Properties and Preparation of Waterborne Polyurethane Adhesives for Shoes

**LI XI-LAN, SONG JIA-LONG, GUO XIU-FEN, SUN GUAG-YAN**

Zhuhai College of Jilin University, Zhuhai 519041

Corresponding author's e-mail: 381855458@qq.com

**Abstract:** In this paper, a waterborne polyurethane adhesive for shoes has been prepared by a composite method, and a series of tests such as peel strength, heat resistance, yellowing resistance, and hydrolysis resistance were performed on the sample. The results show that the performance of the water-based polyurethane adhesive prepared by this method has reached the level of similar products in the market. It has great promotion significance in shoe factories.

## 1. Introduction

Adhesives for shoes have undergone a development process from low-end to high-end, from non-environmental protection to environmental protection. The first generation is ordinary chloroprene adhesive, the second generation is graft-modified chloroprene and solvent-based polyurethane adhesive, and the third generation is methacrylic-modified "triphenyl-free" chloroprene adhesive, graft-modified Neoprene adhesive and polyurethane (PU) adhesive [1]. Although the third generation has solved the problem of benzene pollution, it still cannot solve the problem of organic matter volatility (VOC). In the 1990s, the fourth generation of environmentally-friendly adhesives appeared, the main representative of which is water-based adhesives, and the outstanding performance and outstanding performance of water-based polyurethane adhesives are particularly noticeable [2-5].

Water-based polyurethane is a binary colloid system with water as the dispersion medium. The adhesive products made from it are highly adhesive, non-toxic, odorless, do not generate static electricity, and have good flame retardancy. It is suitable for metals, fabrics, leather, The bonding of rubber, EVA and other shoe materials can basically replace solvent-based polyurethane adhesives.

At present, except for a small number of Taiwan-funded enterprises in China-and only limited to high-end travel shoes such as NIKE, ADIDAS, REEBUK, and special requirements from European and American importing countries-most of them still use solvent-based polyurethane adhesives except water-based polyurethane adhesives.

This situation is caused by various reasons. First, in some aspects, the performance of water-based polyurethane is worse than that of solvent-based polyurethane adhesives [6]; second, domestic environmental protection legislation is incomplete or enforcement is not good enough; third, although there has been some domestic industrialization of water-based polyurethane in recent years The degree of progress has been improved, but the gap is still relatively large compared with foreign countries. Fourth, the matching adhesive additives are basically in the initial stage.

According to the author's research, the water-based polyurethane adhesive products currently used in shoe factories are mainly from Taiwan Dadong, South Korea Dongcheng, Taiwan Nanbao and other companies.

In this article, the author has prepared a waterborne polyurethane adhesive for shoes by a compound method. The test results show that it can basically meet the on-site bonding requirements of shoe factories.

## 2. Experimental part

### *2.1. raw material*
Dispercoll U 54, Bayer Material Technology Trading (Shanghai) Co., Ltd.; MYT 6996A, Bayer Material Technology Trading (Shanghai) Co., Ltd.; L75N, Borchers; BYK 024, Germany's Becker.; BYK 375, Germany's Becker.

### *2.2. Preparation of water-based polyurethane adhesive*
Add the appropriate amount of U 54 emulsion and 6996A emulsion to a suitable stainless steel container. Turn on the stirring; add the leveling agent BYK 375 and antifoaming agent BYK 024 in this order. After stirring for 5 minutes, L75N was slowly added dropwise. After adjusting to the target viscosity, stirring was continued for 30 minutes and then stopped. Then use a 60-mesh filter to filter the prepared glue solution, and the filtered glue solution is the target sample.

### *2.3. Performance test*

#### *2.3.1. Appearance*
Visually check its glue state.

#### *2.3.2. Viscosity*
The target sample was placed in a constant temperature biochemical incubator at 25 ° C for 24H. After stabilization, the viscosity was measured with a rotor viscometer (BROOKFILDE, LVDVE230).

#### *2.3.3. Solid content*
Weigh 2 to 5 g (W2) of the sample in a clean and dry watch glass, and place it in a 120 °C constant temperature blast drying oven (Shanghai Jinghong, DHG-9246A) for about 2 hours. Cool down to room temperature and weigh. (W1), then the solid content = (W1 / W2) × 100%.

Note: In the three tests of 1.3.1-1.3.3, the target sample is directly taken for testing, and from 1.3.4-1.3.8, 3 to 5% of the curing agent is added to the target sample and stirred Test under mixing.

#### *2.3.4. Determination of hydrolysis resistance*
The test piece uses canvas with a length × width = 60 mm × 25 mm specification. Both test pieces are coated with glue and placed in a drying box at 55 ~ 60 °C. After drying for 3 ~ 4 minutes, the two bonding surfaces are bonded to each other. Then press it under the pressure of 3-4 KG for 10s, and leave it at room temperature for a period of time, store it in a hydrolysis-resistant machine at 70 ° C × 95% humidity, and then take it out every 7 days to test its peel strength.

#### *2.3.5. Determination of peel strength*
The test piece uses a PVC sheet with a length × width × thickness = 120 mm × 25 mm × 3 mm. The surface of the test piece is cleaned with methyl ethyl ketone. Both test pieces are glued and placed in a drying box at 55 ~ 60 °C for 3 ~ After 4 min, the two bonding surfaces were bonded to each other, and the pressure was applied with a press (Wenling Tingwei Vacuum Equipment, TW-1A type) at a pressure of 3 to 4 KG for 10 s, and it was left at room temperature for 5 min. High-speed rail inspection instrument, AI-7000S). During the test, clamp one end of it in the upper and lower clamps symmetrically, and the clamping part cannot slide. Start the tensile tester. The upper and lower clamps were separated at a rate of (100 ± 10) mm / min, and the mean value of the peeling force was read on a digital display to obtain the initial peeling strength. The peel length of the specimen is (50 ± 10) mm. The unpeeled part was left at room temperature for 24 h. The peeling test was performed in the same procedure, and the average peeling force was read to obtain the later peeling strength, and the form of failure was recorded.

*2.3.6. Determination of heat resistance*

The test piece uses a PVC sheet with a length × width × thickness of 60 mm × 25 mm × 3 mm. The surface is cleaned with methyl ethyl ketone. Both test pieces are glued and placed in a drying box at 55 ~ 60 °C and dried for 3 ~ After 4 minutes, the adhesive surfaces of the test pieces were put on each other, and then pressed with a press at a pressure of 3 to 4 KG for 10 seconds. The test pieces were placed for 5 minutes after gluing, then placed in a constant-temperature drying oven at a temperature of (70 ± 1) °C, and then taken out after holding for 10 minutes, and the length of the cracked surface was measured to determine the heat resistance value.

*2.3.7. Determination of viscosity retention time*

Take a rubber strip with a length × width = 60 mm × 25 mm specification, roughen it, and then brush and dry it. Immediately after drying, take it out of the oven and place it in a 25 ° C constant temperature room, and then start counting from this time. At an interval of 1min, apply 5mm from the end of the rubber to each other. When applying the next time, you must cut off the last bonding part with scissors., Until it can not be posted, this time is recorded as the sticky maintenance time.

*2.3.8. pot life determination*

Pot life is the normal usage time after the glue and curing agent are prepared. This is because the water-based polyurethane begins to react and crystallize after the hardener is added. If the crystal is dried and activated after the crystallization is complete, the formed adhesive film will no longer be sticky. This phenomenon is called "dead glue". Dead glue is irreversible. Therefore, the prepared glue must be used up in the pot life of the glue.

*2.3.9. Determination of yellowness*

Take a white EVA cube with length × width × thickness = 10 mm × 10 mm × 10 mm. After cleaning the surface, apply the glue. When applying the glue, make it as thin and uniform as possible to avoid the build-up and lack of glue. Then place it at 55~ 60. Dry in a drying box at °C. After drying, place it in a yellowing machine (Chinese high-speed rail detection instrument, GT-7035UB) and irradiate it with 2H, take it out and compare it with a standard color chart, and read the level of yellowing.

## 3. Results and discussion

The U 54 water-based polyurethane emulsion produced by Bayer has 50% high solid content, no yellowing, low activation temperature, high initial peel strength, and stable performance. It has been in the domestic and foreign markets since its introduction to the market. leading position. However, the disadvantage of this product is its poor heat resistance, which greatly limits the promotion and application of the emulsion in the shoe adhesive industry. Bayer is also aware of this, and in 2008 developed another water-based polyurethane emulsion MYT 6996A. The emulsion has a large molecular weight and is particularly outstanding in heat resistance, but it has the disadvantage of poor initial viscosity, especially compared with U54.

Therefore, U54 and 6996A are highly complementary. Through the method of orthogonal experiments, the author found an optimal ratio of the two. At this ratio, supplemented with auxiliary additives, a shoe adhesive product with excellent performance and comparable to market rivals was formulated.

The prepared adhesive sample has a milky appearance, a viscosity of about 5000 CPS / 25 ° C, and a thixotropic coefficient of 1.8 to 2.0. It is very suitable for brushing the surface of rubber, MD EVA, leather, canvas and other shoe materials.

The solid content of the gel sample reaches 50%, and the moisture can be dried by drying at 55-60 ° C for 2 to 5 minutes. The glued surface has high bonding strength, and the year-on-year performance has reached the level of solvent-based polyurethane adhesive. For rubber and PVC materials, the initial peel strength can be more than 6.0 N / mm, and the later peel strength can be more than 10.0 N / mm or the material is damaged.

For shoe adhesives, in addition to being easy to apply and having good initial tack, other properties are also indispensable, such as heat resistance, yellowing resistance, and hydrolysis resistance. In the heat resistance test of this rubber sample, the heat resistance value can reach 25 mm, and under the environmental condition of 70 ° C × 10min, it does not pop open at all. In the yellowing test, the yellowing resistance value can reach 4.8, which is basically not yellowing. This can be used for the bonding of white or light-colored shoe materials, and there is no need to worry about the yellowing of the manufactured shoes to affect the appearance. In the hydrolysis resistance test, after five weeks of hydrolysis resistance test, it was found that the bonding surface can still maintain more than 70% strength in the first 3 weeks.

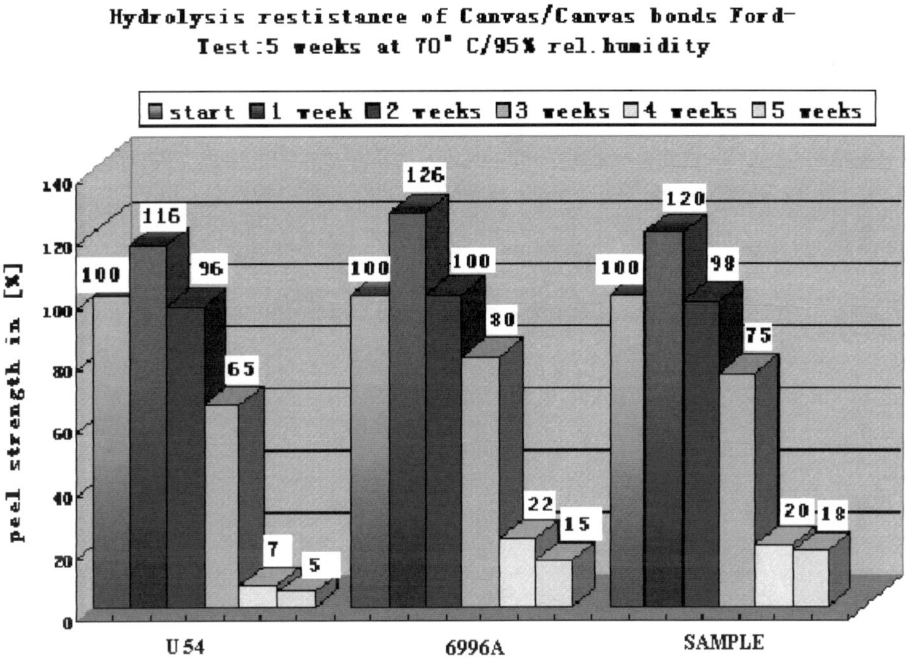


Canvas / Canvas - Bonds: adhesive coating on both sides by brush
bonding: 10 sec. at 4 kgf/cm2 and 60°C adhesive temperature; pre-storage: 2 days and 7 days 1 week at 23°C/50% rel. humidity
storage under hydrolysis conditions: 70°C/95% rel. humidity up to 5 weeks; reconditioning: 1 week at 23°C/50% rel. humitity

Fig. 1 Hydrolysis resistance of U 54, 6696A and the prepared sample

## 4. Conclusion

Bayer water-based polyurethane emulsions U54 and 6996A were used as the main raw materials, followed by adhesive samples prepared by the corresponding functional assistants. The samples were tested for peel strength, heat resistance, and hydrolysis resistance. The test results show that the glue sample has high peel strength, good heat resistance, excellent hydrolysis resistance, etc., and its comprehensive performance has reached the level of waterborne polyurethane adhesives in the market for similar shoes.

## References

[1] WENG HAN-YUAN. Development of water-based polyurethane adhesives[J] Chemical propellants and polymer materials,2007, 5(4): 1-5.
[2] Noble Karl - Ludwig Waterborne polyurethane [J], Progress in Organic Coatings, 1997, 32 (1 - 4) : 131 - 136.
[3] LIU JIAN-XIONG, ZHANG BAO- LING, BAI GUO-BAO, et al. Domestic waterborne

IOP Conf. Series: Materials Science and Engineering **738** (2020) 012035 doi:10.1088/1757-899X/738/1/012035

polyurethane adhesive products and application status[J]. China Adhesive, 2004, 14 (12): 35~39.

[4] ZHANG SHOU-WEN, WANG WEN –JUN, LI HONG –XU, et al. High-performance water-based polyurethane adhesive[J]. China Adhesive, 2001, 10 (4): 40~41.

[5] DUAN YOU-LU, ZHOU HE-XIAN. Aqueous polyurethane dispersion used as adhesive and coating and its equipment [P]. CN 1362436A, 2002 - 08 – 07.

[6] DU HUI, ZHAO YU-HUA, YIN NING, et al. Preparation. and properties of waterborne polyurethane adhesives modified by polystyrene [J]. Journal of Applied Polymer Science, 2008, 110: 3 156-3 161.

IOP Conf. Series: Materials Science and Engineering **738** (2020) 012036 doi:10.1088/1757-899X/738/1/012036

# Research on Short-term Power Load Forecasting Method Based on Temperature Accumulation Effect Embedded in Time Series

**Duan Jinchang[2], Li Zhiyong[1], Yao Haicheng[1], Li Bao[1], Zhou qiang[2]**

[1]China Southern Grid Power Dispatching Control Center, Guangzhou, Guangdong, 510663, China

[2]NARI-TECH Nanjing Control Systems Ltd., Nanjing, Jiangsu, 211106, China

**Abstract.** Short-term power load forecasting is an important link in power grid dispatching, which has an important impact on unit combination, economic dispatching, optimal power flow, etc. As meteorological factors have great influence on load, the influence of meteorological factors should be considered reasonably in short-term load forecasting. The accuracy of short-term electric charge prediction results can help senior personnel of the power system to make accurate and feasible power operation methods. In order to ensure the safe and stable operation of power grid in various special periods and to ensure the economic benefits of related power enterprises to the greatest extent, it is imperative to establish a highly accurate prediction model. The accuracy of power load forecasting will directly affect the position of each power enterprise in the market. Based on the function of time series embedding, this paper analyzes the rule of cumulative effect on load. The correlation between temperature and load is greatly improved after considering the cumulative effect to deal with temperature correction.

## 1. Introduction

Short-term load forecasting is an important link in power grid dispatching, which has an important impact on unit combination, economic dispatching, optimal power flow, etc. Improving the accuracy of load forecasting is conducive to improving the utilization rate of power equipment and reducing the risk of power grid operation. Short-term load forecasting of power system usually refers to forecasting the power load index for one month, one week or one day in the future. It plays an important role in real-time control of power system and ensuring economic, safe and reliable operation of the system [2]. Short-term power load has strong nonlinear characteristics, and its changes are affected by many factors, and many factors cannot give the influence weight quantitatively [3]. The accuracy of short-term electric charge prediction results can help senior personnel of the power system to make accurate and feasible power operation methods. In order to ensure the safe and stable operation of power grid in various special periods, and to ensure the economic benefits of related power enterprises to the greatest extent [4]. There are many factors that affect the accuracy of load forecasting, including the change rule of load itself, the level of economic development in the region where it is located, the structure of electricity consumption, the level of electricity price, and climate change [5]. Different power load structures and climatic conditions in different regions lead to different load changes [6]. Therefore, the change in short-term load shows a certain degree of unpredictability, which is exactly the problem to be solved in the current short-term load forecast.

With the continuous development of the economy and the continuous improvement of people's living standards, high-power and high-energy-consuming appliances are becoming more and more popular, and the proportion of cooling and heating loads in the total electricity load is increasing. [7]. Among the many factors that affect load changes, meteorological factors are the main factors. Among the meteorological factors such as temperature, humidity and precipitation, the influence of temperature is the main factor. The change of power load is a non-linear function related to many factors. Because the traditional algorithms of load forecasting, such as the traditional time series method, cannot consider the non-linear influence of weather on the load, it is predicted in the face of sudden weather changes or sudden events. The error is large [8]. The short-term power load forecast is not only closely related to the average temperature of the day, but also has a great relationship with the temperature of the previous days [9]. The accuracy of power load forecasting will directly affect the position of each power company in the market, so it is imperative to establish a highly accurate forecasting model. In order to further improve the accuracy of load prediction, this paper combines the effect of time series embedding to analyze the influence of cumulative effects on load. The cumulative effect has different strengths in different situations. After considering the cumulative effect on the temperature, the correlation between temperature and load is greatly improved.

## 2. Typical Phenomenon of Cumulative Effect Affecting Load.

If a place is in high temperature for a long time, the load in this area will be at a higher level. In this case, even if the temperature drops, the degree of load reduction is not obvious, and may even rise instead of falling. There are many factors that affect short-term load. Due to the fact that relevant information is often not available before prediction and the degree of influence of some factors on load is very complex, it is unrealistic to consider all factors. Due to the uncertainty and locality of the change trend of humidity, wind force and rainfall, the impact on the total load in the area with large coverage area is uncertain on the whole [11]. Most of the current short-term load forecasting methods consider external factors such as meteorology and major political events, and the important influence of meteorological factors on short-term load forecasting has reached a consensus. Human senses have a process of adapting to temperature changes, which is the root cause of cumulative effect. The higher the proportion of air conditioning load, the more obvious the cumulative effect of air temperature, which is manifested as cumulative effect under the condition of continuous high temperature and cumulative effect under the condition of continuous low temperature. As the daily maximum load is not only affected by the daily maximum temperature, but also by other factors, such as holidays and week types. In this way, in the analysis process, the data on holidays and weekends must be eliminated first.

According to the close relationship between daily maximum load and daily maximum temperature, a quadratic regression model between daily maximum load and daily maximum temperature can be initially established:

$$L = at^2 + bt + c \quad (1)$$

Where: L is the daily maximum load, MW; t is the daily maximum temperature, ° C; a, b, and c are coefficients to be determined.

It can be seen from Fig. 1 that the daily maximum load and the daily maximum temperature change trends in summer are very similar, which indicates that temperature is one of the important factors affecting the load change. Some peaks or valleys of the two curves do not overlap in time series. The specific manifestation is that the load change lags behind the temperature change, which is the so-called cumulative temperature effect. The cumulative temperature effect refers to the phenomenon that the human senses have an adaptive process to temperature changes, and the daily load change to be predicted lags behind the temperature changes. In order to improve the fitting accuracy of the regression model, it is necessary to consider the cumulative effect of temperature.

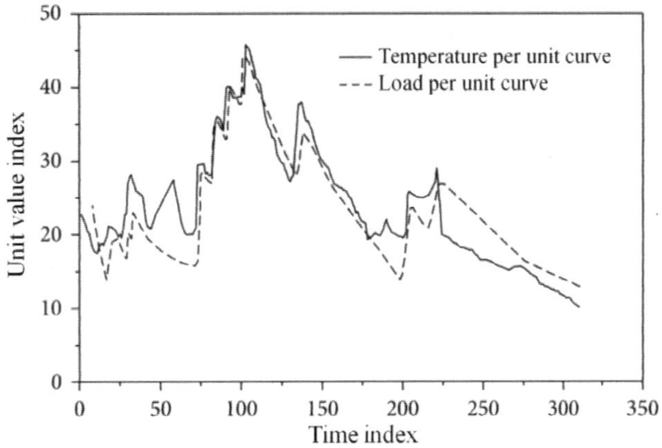

Fig. 1 Curve of daily maximum temperature and daily maximum load of the power system

For provincial power grids in relatively large areas, the regional power grid has small capacity, the overall meteorological conditions and change trends are relatively consistent, the load composition is relatively simple, and the influence factors of power grid load by temperature are more obvious. Under the continuous high temperature or cool weather, when the temperature changes suddenly, the degree of load change is not obvious due to the influence of the temperature for several days before. Among the meteorological factors, the main factors that affect the load are temperature and climate [12]. The influence of temperature on load is not only related to random climatic factors such as sunshine, shade and rain, but also related to factors such as temperature level, rise, threshold value, specific time of change, duration of action, etc. Generally speaking, the occurrence of cumulative effect of temperature needs to meet the condition that the daily temperature to be predicted is in the temperature range where human body feels more sensitive.

## 3. Considering the Cumulative Effect of Time Series Embedding

### *3.1. Factors Affecting Cumulative Effect Intensity*

The cumulative effect intensity is different under different conditions. Using temperature correction value to reflect cumulative effect can not only reflect the impact of cumulative effect on load, but also make full use of existing load forecasting methods. Research shows that when the average temperature is higher than 20°C, the temperature and load show a positive correlation. In other words, for central China, the influence of temperature and temperature accumulation on load can only be reflected in high temperature period. For example, Table 1 shows the correlation between daily summer load and temperature in a certain area.

Table 1 Correlation between daily load and temperature in a certain area in summer

| Temperature category | Maximum load | Minimum load |
|---|---|---|
| Mean temperature of air | 0.775 | 0.763 |
| Maximum air temperature | 0.791 | 0.584 |
| Minimum air temperature | 0.736 | 0.536 |

There is a good correlation between the maximum load and the highest temperature in this area. Similarly, similar conclusions can be drawn from the analysis of the average temperature, the highest temperature and the lowest temperature. In order to explore the influence of temperature accumulation effect on daily load, the highest load and temperature data for several consecutive days in summer in this region were selected from historical data and Table 2 was formed.

Table 2 Summer daily load and temperature data in a certain area

| Date | Maximum load /MW | Minimum load /MW | Maximum temperature /(°C) | Average temperature /(°C) |
|---|---|---|---|---|
| 8-1 | 631.5 | 402.8 | 34 | 25 |
| 8-2 | 692.8 | 496.5 | 36 | 28 |
| 8-3 | 721.7 | 507.9 | 38 | 31 |
| 8-4 | 711.6 | 522.8 | 38 | 29 |
| 8-5 | 748.9 | 547.3 | 37 | 29 |
| 8-6 | 732.3 | 556.2 | 38 | 30 |
| 8-7 | 685.7 | 537.7 | 38 | 32 |
| 8-8 | 722.8 | 541.5 | 37 | 32 |
| 8-9 | 736.7 | 505.2 | 37 | 31 |
| 8-10 | 547.1 | 388.6 | 28 | 25 |

In the short-term dispatching, due to the randomness and uncontrollable nature of power, it will cause the increase of the system's rotational reserve capacity and the change of the conventional unit start-up and stop strategy after the power is connected to the network, which may lead to the increase of the operation cost of the power system. The algorithm evolution curve is shown in Fig. 2.

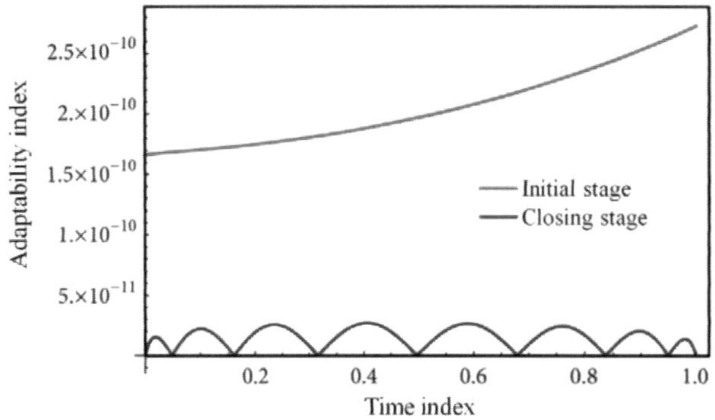


Fig. 2 Algorithm evolution curve

The intensity of temperature accumulation effect is not only affected by the temperature of the day to be predicted, but also closely related to the difference between the temperature of the day to be predicted and the temperature of the previous N days. The greater the temperature difference, the greater the cumulative effect of the temperature in the previous N days on the day to be predicted. Time series decomposition model is widely used in short-term load forecasting of power system. Through a large number of data comparison, time series decomposition model improves the accuracy of load forecasting model to the greatest extent by dividing periodic components and non-periodic components. Time series method takes time as independent variable and the predicted target as dependent variable to establish an appropriate mathematical model. When using this method to predict, it is generally necessary to use a lot of historical data, so that the corresponding mathematical model can accurately reflect the change trend of things. In actual load forecasting, it is often encountered that the error increases due to incomplete training samples or obvious load trend. The time series decomposition model can reflect the multiple factors affecting the load value and reduce the actual error value of load forecasting.

### *3.2. Temperature Correction Method Considering Cumulative Effect*

The historical data of power load is an ordered set sampled and recorded at certain time intervals, so it is a time series. It is difficult to change samples in real time for real-time training, because the number of samples used to train neural networks is usually large due to the inclusion of various patterns, and it takes a long time to train.

The cumulative effect can be considered according to the following formula to modify the daily temperature to be predicted:

$$T' = (1-k)T_0 + kT_1 - \sum_{i=0}^{p} k^{i+1}(T_i - T_{i+1}) \quad (2)$$

Where: T' is the maximum temperature correction value of the day to be predicted after considering the cumulative effect; T0 is the highest temperature on the day to be predicted; Ti is the true value of the temperature I days before the day to be predicted; K is the cumulative effect coefficient; P= min(n,3), n is the number of consecutive days when daily maximum temperature is higher than 28°C.

The power load time series forecasting technology is to try to establish a mathematical model of time series according to the historical data of load. On the one hand, the model is used to describe the regularity of the time series change process of power load, on the other hand, the mathematical expression of load forecasting is established on the basis of the model to forecast the future load. Most load forecasting problems are nonlinear programming problems, and the solution is to solve the maximum and minimum values of multivariate nonlinear functions in mathematical functions.

## 4. Summary

Meteorological factors are important factors that affect the accuracy of summer load forecasting, especially the temperature. An in-depth analysis of the relationship between load and temperature is of great significance to improve the accuracy of load forecasting. Short-term load forecasting under high temperature in summer is still a difficulty in the whole load forecasting work. The long-lasting meteorological characteristics determine that it is not enough to consider only the meteorological factors on that day. Theoretical and practical applications show that no single power load forecasting method can be applied to all occasions. Considering that the power system is a multi factor system, combined with the analysis of relevant factors, we can try to mine the historical data and find out the leading factors that affect the prediction accuracy. In this paper, the rule of cumulative effect on load change in summer high temperature season is studied. It is found that the intensity of cumulative effect is different under different conditions. With the improvement of technology and service of meteorological department, it is possible to obtain the meteorological information in time. If we can use these more accurate meteorological information, the short-term and ultra short-term load forecasting work will be further developed and improved.

## References

[1] Meng Huanan, Hu Xiaobo. Short-term power load forecasting based on similar model functional time series [J]. Shaanxi Electric Power, 2014, 42(5):56-60.
[2] Wang Wei, Yang Huihua, Liu Zhenbing, et al. Short-term power load forecasting based on extreme learning machine [J]. Computer simulation, 2014, 31(4):137-141.
[3] Ning Xiaoguang, Zhu Yongli. Power load forecasting based on incremental optimization extreme learning machine [J]. Computer simulation, 2016, 33(2):163-166.
[4] Guo Chong, wang zheng. Power load forecasting algorithm based on correlation dimension index analysis [J]. Power and Energy, 2016(2):202-206.
[5] Guo Wei, Ba Xiuling, Ma Wenyuan, et al. Research on Load Forecasting of Power System Based on Neural Network [J]. Automation and Instruments, 2017(10):192-194.
[6] Lu Shan. Applying Accumulated Temperature Effect to Predict Xi 'an Electric Power Meteorological Load in Summer [J]. Meteorological Science and Technology, 2017, 45(6):1090-1094.
[7] Fu Z, Heuer N, Sayas F J. A non-symmetric coupling of Boundary Elements with the Hybridizable Discontinuous Galerkin method[J]. Computers & Mathematics with Applications, 2017: S0898122117305278.
[8] Cockburn B, Fu Z, Hungria A, et al. Stormer-Numerov HDG Methods for Acoustic Waves[J]. Journal of Scientific Computing, 2017.

[9] Li Qianyi, Wu Yuan. Forecast of Winter Gas Daily Load Based on Cumulative Effect of Temperature-Taking Xi 'an City as an Example [J]. Oil and Gas Storage and Transportation, 2016, 35(9):1014-1017.
[10] Fu Guiqin, You Fengchun, Cao Xin, et al. Application and Test of Accumulated Temperature Effect in Daily Peak and Valley Load of Electric Power [J]. journal of applied meteorological science, 2015, 26(4):492-499.
[11] Fu Z, Heuer N, Sayas F J. Coupling of HDG with a double-layer potential BEM[J]. Mathematics, 2012.
[12] Huang Liangliang, Wang Yong, Yang Heng, et al. Short-term power load forecasting based on Fuzzy fuzzy processing of meteorological factors [J]. computer applications and software, 2014(2):171-173.
[13] Xia Bo, Yang Chao, Zheng Kewen. Design and Implementation of Power Load Forecasting System Based on Big Data Platform [J]. Automation Instruments, 2018(6):81-84.
[14] He Yaoyao, Wen Caixi, Xu Qifa, et al. Probability density prediction method of medium-term power load considering temperature factor [J]. Power Grid Technology, 2015, 39(1):176-181.
[15] Cao Jing-Jin, Qin Lijun. Application of Improved Bayesian Classifier in Power System Load Forecasting [J]. Science and Technology Innovation and Productivity, 2014(5):108-110.
[16] Li Yanan. A fuzzy neural network prediction method for time series [J]. Science and technology horizon, 2017(8):148-149.

# AUTHOR INDEX